Benchmark Papers
in Microbiology

Series Editor: Wayne W. Umbreit
Rutgers—The State University

PUBLISHED VOLUMES

MICROBIAL PERMEABILITY / John P. Reeves
CHEMICAL STERILIZATION / Paul M. Borick
MICROBIAL GENETICS / Morad Abou-Sabé
MICROBIAL PHOTOSYNTHESIS / June Lascelles
MICROBIAL METABOLISM / H. W. Doelle
ANIMAL CELL CULTURE AND VIROLOGY / Robert J. Kuchler
PHAGE / S. P. Champe
MICROBIAL GROWTH / P. S. S. Dawson
MICROBIAL INTERACTION WITH THE PHYSICAL ENVIRONMENT / D. W. Thayer
MOLECULAR BIOLOGY AND PROTEIN SYNTHESIS / Robert A. Niederman
MARINE MICROBIOLOGY / Carol D. Litchfield
INDUSTRIAL MICROBIOLOGY / Richard W. Thoma

Benchmark Papers
in Microbiology / 12

A BENCHMARK® Books Series

INDUSTRIAL MICROBIOLOGY

Edited by
RICHARD W. THOMA
E. R. Squibb & Sons, Inc.

Dowden, Hutchinson & Ross, Inc.
STROUDSBURG, PENNSYLVANIA

Copyright © 1977 by **Dowden, Hutchinson & Ross, Inc.**
Benchmark Papers in Microbiology, Volume 12
Library of Congress Catalog Card Number: 76-28259
ISBN: 0-87933-251-4

All rights reserved. No part of this book covered by the copyrights hereon may be reproduced or transmitted in any form or by any means—graphic, electronic, or mechanical, including photocopying, recording, taping or information storage and retrieval systems—without written permission of the publisher.

79 78 77   1 2 3 4 5
Manufactured in the United States of America.

LIBRARY OF CONGRESS CATALOGING IN PUBLICATION DATA
Main entry under title:
Industrial microbiology.
  (Benchmark papers in microbiology ; 12)
  Includes indexes.
  1. Industrial microbiology—Addresses, essays, lectures.
I. Thoma, Richard W.
QR53.T48   660'.62'08   76-28259
ISBN 0-87933-251-4

Exclusive Distributor: **Halsted Press**
A Division of John Wiley & Sons, Inc.
ISBN: 0-470-98938-6

SERIES EDITOR'S FOREWORD

We have had some hesitation in calling this collection simply *Industrial Microbiology*, since the subject is in fact much broader than the area covered here. Yet this group of papers covers the part of industrial microbiology that has been more dramatic in the recent past and that most people think of as truly "industrial" microbiology—i.e., the preparation of specific, perhaps somewhat exotic, substances by way of microorganisms. While the production of both industrial and beverage alcohol has been in progress for years, the aseptic "fermentation" at neutral pH with high aeration has provided the challenge to engineers and microbiologists. The development of the methods, the approach, and the apparatus for conducting such processes on a 50,000-liter scale is thus at the core of modern aspects of microbiology; it is this area that is this volume's primary concern.

The papers in this volume have been selected by a man with a very considerable experience in the field—a man who lived through these developments, was in contact with the principal investigations and advances, and certainly contributed significantly to the latter. Dr. Thoma serves as an excellent guide—sound, capable, and perceptive—and he has brought together papers that give the flavor and the essence of this field. This selection was difficult because, as he mentions, each could be replaced by one or more of equal merit, yet the selection as made provides a real basis for an understanding of modern industrial microbiology.

WAYNE W. UMBREIT

PREFACE

Industrial microbiology is a vast and important field of knowledge—surely too large to be encompassed in a single volume. Industrial microbiologists may find themselves engaged not only in the production of valuable products, but in problems of waste disposal, ecology, health hazards, packaging, and even in problems concerning the evaluation of immediate and long-term effects of the microbial products involved. They may make vaccines, or single-cell protein, or industrial butanol, or antibiotics. They may deal with engineers, with physicians, with production people, with sanitarians, and more frequently than they possibly would like, with administrators, accountants, efficiency experts, and salespeople. But which of these aspects are the core of industrial microbiology? True, some of the industries that depend on microbiology and/or on the control of microorganisms are of ancient origin—the alcoholic beverage industry, for example, and certainly food production and food preservation. And clearly, these *are* industries, and microbiologists working in them are "industrial" microbiologists.

Yet in the past half century an important development has taken place: the use of microorganisms to prepare specific, sometimes rather exotic, substances. This development has been especially significant because the preparation is done on a large scale with relatively rich media, under aseptic conditions at a neutral pH. It is one thing to prepare a gram of penicillin in the laboratory but quite a different thing to prepare a ton of it in a factory. It is this area—the application of microbiology to the large-scale production of somewhat unstable substances under aseptic conditions—that has constituted the challenge to microbiologists and engineers. It is this area that seems to be the "core" of modern industrial microbiology. Without the developments described here, antibiotics would be only laboratory curiosities, not familiar substances. Their use would be restricted to the most desperate cases on an experimental basis, and hence there would be a mad scramble for them. Our world would indeed be quite different. But industrial microbiologists have met this challenge; the substances are available. New methods, new approaches, and new concepts have been devised. It is this area of industrial microbiology that is covered in this volume.

RICHARD W. THOMA

CONTENTS

Series Editor's Foreword   v
Preface   vii
Contents by Author   xiii

Introduction   1

PART I: OVERVIEWS

Editor's Comments on Papers 1 and 2   8

1   JOHNSON, M. J.: Fermentation—Yesterday and Tomorrow   9
Chem. Technol., **1**, 338–341 (June 1971)

2   PERLMAN, D.: Prospects for the Fermentation Industries, 1974–1983   13
Chem. Technol., **4**(4), 210–216 (1974)

PART II: CULTURES: ACQUISITION AND CONSERVATION

Editor's Comments on Paper 3   22

3   HESSELTINE, C. W., and W. C. HAYNES: Sources and Management of Micro-organisms for the Development of a Fermentation Industry   23
Progr. Ind. Microbiol., **12**, 3–6, 24–46 (1973)

PART III: STRAIN DEVELOPMENT

Editor's Comments on Papers 4, 5, and 6   50

4   BACKUS, M. P., and J. F. STAUFFER: The Production and Selection of a Family of Strains in *Penicillium chrysogenum*   52
Mycologia, **47**(4), 429–463 (1955)

5   THOMA, R. W.: Use of Mutagens in the Improvement of Production Strains of Microorganisms   87
Folia Microbiol., **16**, 197–204 (1971)

6   PONTECORVO, G., and G. SERMONTI: Parasexual Recombination in *Penicillium chrysogenum*   95
J. Gen. Microbiol., **11**(1), 94–104 (1954)

Contents

PART IV: MEDIUM DEVELOPMENT

Editor's Comments on Papers 7A Through 9   108

7A   MOYER, A. J., and R. D. COGHILL: Penicillin. VIII. Production of Penicillin in Surface Cultures   111
J. Bact., **51**(1), 57–78 (1946)

7B   MOYER, A. J., and R. D. COGHILL: Penicillin. IX. The Laboratory Scale Production of Penicillin in Submerged Cultures by *Penicillium notatum* Westling (NRRL 832)   133
J. Bact., **51**(1), 79, 80, 93 (1946)

8   RAKE, G., and R. DONOVICK: Studies on the Nutritional Requirements of *Streptomyces griseus* for the Formation of Streptomycin   136
J. Bact., **52**(2), 223–226 (1946)

9   DAVEY, V. F., and M. J. JOHNSON: Penicillin Production in Corn Steep Media with Continuous Carbohydrate Addition   140
Appl. Microbiol., **1**, 208–211 (1953)

PART V: APPARATUS AND EQUIPMENT

Editor's Comments on Papers 10 Through 13   146

10   SUMINO, Y., S. AKIYAMA, and H. FUKUDA: Performance of the Shaking Flask (I) Power Consumption   148
J. Ferment. Technol., **50**(3), 203–208 (1972)

11   KROLL, C. L., S. FORMANEK, A. S. COVERT, L. A. CUTTER, J. M. WEST, and W. E. BROWN: Equipment for Small Scale Fermentations   154
Ind. Eng. Chem., **48**(12), 2190–2193 (1956)

12   WEST, J. M., G. P. STICKLE, K. D. WALTER, and W. E. BROWN: An Improved pH Electrode Assembly for Pilot Plant and Plant Fermentors   158
J. Biochem. Microbiol. Technol. Eng., **3**(2), 125–137 (1961)

13   JOHNSON, M. J., J. BORKOWSKI, and C. ENGBLOM: Steam Sterilizable Probes for Dissolved Oxygen Measurement   171
Biotechnol. Bioeng., **6**(4), 457–468 (1964)

PART VI: STERILIZATION

Editor's Comments on Papers 14 and 15   184

14   DEINDOERFER, F. H., and A. E. HUMPHREY: Analytical Method for Calculating Heat Sterilization Times   186
Appl. Microbiol., **7**(4), 256–264 (1959)

15   HUMPHREY, A. E., and E. L. GADEN, JR.: Air Sterilization by Fibrous Media   195
Ind. Eng. Chem., **47**(5), 924–930 (1955)

PART VII: OXYGEN TRANSFER: AERATION AND AGITATION

Editor's Comments on Papers 16 and 17   204

16 BARTHOLOMEW, W. H., E. O. KAROW, M. R. SFAT, and R. H. WILHELM: Oxygen Transfer and Agitation in Submerged Fermentations   206
Ind. Eng. Chem., **42**(9), 1801–1809 (1950)

17 MAXON, W. D.: Aeration-Agitation Studies on the Novobiocin Fermentation   215
J. Biochem. Microbiol. Technol. Eng., **1**(3), 311–324 (1959)

PART VIII: METABOLIC REGULATION

Editor's Comments on Paper 18   230

18 DEMAIN, A. L.: Cellular and Environmental Factors Affecting the Synthesis and Excretion of Metabolites   231
J. Appl. Chem. Biotechnol., **22**, 345–362 (1972)

PART IX: KINETICS AND MODELING

Editor's Comments on Papers 19 and 20   250

19 GADEN, E. L., JR.: Fermentation Process Kinetics   252
J. Biochem. Microbiol. Technol. Eng., **1**(4), 413–429 (1959)

20 CALAM, C. T., S. H. ELLIS, and M. J. McCANN: Mathematical Models of Fermentations and a Simulation of the Griseofulvin Fermentation   269
J. Appl. Chem. Biotechnol., **21**, 181–189 (July 1971)

PART X: AUTOMATION

Editor's Comments on Paper 21   280

21 NYIRI, L. K.: A Philosophy of Data Acquisition, Analysis, and Computer Control of Fermentation Processes   281
Develop. Ind. Microbiol., **13**, 136–145 (1972)

Author Citation Index   291
Subject Index   297
About the Editor   317

CONTENTS BY AUTHOR

Akiyama, S., 148
Backus, M. P., 52
Bartholomew, W. H., 206
Borkowski, J., 171
Brown, W. E., 154, 158
Calam, C. T., 269
Coghill, R. D., 111, 133
Covert, A. S., 154
Cutter, L. A., 154
Davey, V. F., 140
Deindoerfer, F. H., 186
Demain, A. L., 231
Donovick, R., 136
Ellis, S. H., 269
Engblom, C., 171
Formanek, S., 154
Fukuda, H., 148
Gaden, E. L., Jr., 195, 252
Haynes, W. C., 23
Hesseltine, C. W., 23

Humphrey, A. E., 186, 195
Johnson, M. J., 9, 140, 171
Karow, E. O., 206
Kroll, C. L., 154
McCann, M. J., 269
Maxon, W. D., 215
Moyer, A. J., 111, 133
Nyiri, L. K., 281
Perlman, D., 13
Pontecorvo, G., 95
Rake, G., 136
Sermonti, G., 95
Sfat, M. R., 206
Stauffer, J. F., 52
Stickle, G. P., 158
Sumino, Y., 148
Thoma, R. W., 87
Walter, K. D., 158
West, J. M., 154, 158
Wilhelm, R. H., 206

INDUSTRIAL MICROBIOLOGY

INTRODUCTION

A field of technology—that is, any special area of application of one or more scientific disciplines—evolves sometimes smoothly, sometimes in spurts, in response to human needs. Progress in a field is usually made by exploiting existing theory or by adapting technology from related fields. Sometimes if the driving force is great enough, a field of technology is developed by trial and error, and it may in fact advance beyond the base of its theoretical support. Thus rapidly advancing technology may lead scientific research and provide the incentive for it, but probably just as often the gap is filled by belated interpretation or reinterpretation of evolving technology in the light of newer theoretical advances. Fermentation technology, especially the technology that fills the need for the production of antibiotics and other fine chemicals, has developed in this pattern.

We have not selected these Benchmark papers to prove this thesis, although the first two papers in the volume deal with the evolution of fermentation technology from the perspective of the present day. If we had been concerned primarily with the history of the art, and if we had had unlimited space, we would have included not only many original papers in which concepts were exposed for the first time, but also reviews published in several different eras by scientists with different points of view. For example, the recent essay by Perlman (1975) would have been included; but for balance we should have included all or part of two excellent reviews from the pre-antibiotic era (Frey 1930 and Frey, Kirby, and Schultz 1936). We might also have included in

Introduction

whole or part, for variety, as well as for their historical value, these two reviews: "Antibiotics" (Chapter XXXVI in Prescott and Dunn 1949), and *A Scientific Symposium* [Pasteur Fermentation Centennial, 1857-1957 (1958)]. We hope that those who feel the need for historical enlightenment will read not just these and other reviews and general surveys, but more importantly, some of the original papers cited in these reviews.

Now, we would like to leave what some may believe is an apology for not looking more intently at the past and to apologize indeed for what may seem to be careless omissions from a proper portrait of the present. We have explained in the prefaces to this volume that we are focusing our attention on fermentation processes for production of specific substances, implicitly useful and implicitly produced more economically by fermentation than by any other means. We have selected Benchmark papers that for the most part are relevant to the development of the methods, approach, and apparatus for conduction of large-scale, highly aerated processes for production of antibiotics. We have not specifically referred to processes for producing fine chemicals by selective modification of closely related substances by exposure to growing cells, resting cells, washed cells, stabilized cells, or cell fractions. Such processes are usually called "bioconversion processes" to distinguish them from processes in which the desired product is formed by more complex metabolic activities from materials that bear little or no structural relationship to the product. Much has been published in the last 25 years on bioconversion, or transformation of steroids and other drugs or drug candidates. The primary literature as well as the review literature is rich in detail with respect to the types of reactions recognized, the variety of organisms involved, the spectrum of substrates exposed, and the products identified (Beukers, Marx, and Zuidweg 1972). However, only a few papers have dealt with the kinetics of bioconversion processes from the bioengineering viewpoint (Mateles and Fuld 1961, Maxon, Chen, and Hanson 1966). Likewise, only a few publications have addressed the problem of metabolic regulation in complex bioconversion systems (Lee *et al.* 1969, Lee *et al.* 1971). The decision not to include any paper dealing specifically with bioconversion processes was somewhat arbitrary, but it was based on the opinion that we could make more valuable selections from the literature dealing with biosynthetic processes or with more fundamental and generally applicable techniques for large-scale cultivation of microorganisms.

Introduction

Although the recovery of products from fermentation broth is an essential part of any fermentation process, the art and science of isolation and purification of biosynthetic or bioconversion products, together with the unit operations of extraction, selective absorption, precipitation, evaporation, crystallization, and derivitization are not dealt with in this Benchmark volume. These activities are usually not considered to be within the microbial physiologist's sphere of competence or interest. Moreover, in larger fermentation process–based industries, the traditional job of recovering fermentation products has been a special one delegated to chemists and engineers who too often know little and care less about microorganisms, and hope to be presented with only a concentrated, clean aqueous solution of a single product. The parochial attitude of both types of specialists will undoubtedly change—and must change—as fermentation and recovery operations become more highly automated and as fermentation manufacturing plants are managed as integrated systems (Young and Koplove 1972, Okabe and Aiba 1975).

We have explained why recovery operations are not considered to be within the scope of this volume, and we have explained why we have excluded, with some regret, papers dealing specifically with enzymes and bioconversion processes. We must explain further that certain papers concerned with the principles of kinetics and dynamics of microbial growth, although they are definitely relevant and might well have been included in our volume, are not offered here because they appear in *Microbial Growth* (Dawson 1974), a companion volume in this series of Benchmark Papers in Microbiology. If we had not had the benefit of claiming Dawson's volume as complementary to ours, we would have included, space permitting, several papers he used, most notably the paper on the continuous culture of bacteria, by Herbert, Elsworth and Telling (1956). If we had been able to expand our volume, we might also have included several papers not used by Dawson, which illustrate problems and principles of growth and product formation in continuous culture (Pirt and Callow 1958, 1959). A more thorough treatment of continuous fermentation from the industrial viewpoint would have necessitated the inclusion of a few papers on continuous fermentation processing (Reusser 1961, Unger *et al.* 1942, and Bilford *et al.* 1942).

In concluding our introductory remarks we would like to offer two final comments on the nature of the literature from which we have made our selections for this Benchmark volume.

Introduction

First, because much of the effort expended in advancing fermentation technology is made in industrial laboratories, we would like to suggest that many findings that might be made the subject of scientific papers are disclosed first in the patent literature or not disclosed at all, until it is no longer of importance to exclude others from use of the special information, or until the findings have become common knowledge by independent invention or informal disclosure. Second, because we believe that industrial microbiology in general and fermentation technology in particular are in a dynamic state, we would like to predict that, as difficult as our task has been, the compilers of a Benchmark volume 10 years hence will have a much more difficult time than we have had; they will have to exclude many more papers of real merit because there will be far more papers of high quality from which to choose.

REFERENCES

Beukers, R., A. F. Marx, and M. H. Z. Zuidweg. 1972. Microbial conversion as a tool in the preparation of drugs. In E. J. Ariens, ed. *Drug Design*, Vol. III. Academic Press, New York.

Bilford, H. R., R. E. Scalf, W. H. Stark, and P. J. Kolachov. 1942. Alcoholic fermentation of molasses: Rapid continuous fermentation process. *Ind. Eng. Chem.*, **34**:1406–1410.

Dawson, P. S. S., ed. 1974. *Microbial Growth*. Benchmark Papers in Microbiology Series, Vol. 8. Dowden, Hutchinson & Ross, Stroudsburg, Pa.

Frey, C. N. 1930. History and development of the modern yeast industry. *Ind. Eng. Chem.*, **22**:1154–1162.

———, G. W. Kirby, and A. Schultz. 1936. Yeast: Physiology, manufacture, and use. *Ind. Eng. Chem.*, **28**:879–884.

Herbert, D., R. Elsworth, and R. C. Telling. 1956. The continuous culture of bacteria: a theoretical and experimental study. *J. Gen. Microbiol.*, **14**:601–622.

Lee, B. K., D. Y. Ryu, R. W. Thoma, and W. E. Brown. 1969. Induction and repression of steroid hydroxylases and dehydrogenases in mixed culture fermentations. *J. Gen. Microbiol.*, **55**:145–153.

———, W. E. Brown, D. Y. Ryu, and R. W. Thoma. 1971. Sequential 11α-hydroxylation and 1-dehydrogenation of 16α-hydroxycortexolone. *Biotechnol. Bioeng.*, **13**:503–515.

Mateles, R. I., and G. J. Fuld. 1961. Continuous hydroxylation of progesterone by *Aspergillus ochraceus*. *Antonie van Leeuwenhoek J. Microbiol. Serol.*, **27**:33–50.

Maxon, W. D., J. W. Chen, and F. R. Hanson. 1966. Simulation of a steroid bioconversion with a mathematical model. *Ind. Eng. Chem. Process Design and Develop.*, **5**:285–289.

Okabe, M., and S. Aiba. 1975. Optimization of a fermentation plant: example of antibiotic production. *J. Ferm. Technol.*, **53**:730–743.

Pasteur Fermentation Centennial, 1857–1957. 1958. *A Scientific Symposium*. Chas. Pfizer and Co., New York.

Perlman, D. 1975. Influence of penicillin fermentation technology on processes for production of other antibiotics. *Process Biochem.*, **10**:23–27.

Pirt, S. J., and D. S. Callow. 1958. Exocellular product formation by microorganisms in continuous culture. I. Production of 2:3-butanediol by *Aerobacter aerogenes* in a single stage process. *J. Appl. Bact.*, **21**:188–205.

———. 1959. Continuous-flow culture of the filamentous mould *Penicillium chrysogenum* and the control of its morphology. *Nature*, **184**:207–310.

Prescott, S. C., and C. G. Dunn, eds. 1949. *Industrial Microbiology*. McGraw-Hill, New York.

Reusser, F. 1961. Theoretical design of continuous antibiotic fermentation units. *Appl. Microbiol.*, **9**:361–366.

Unger, E. D., W. H. Stark, R. E. Scalf, and P. J. Kolachov. 1942. Continuous aerobic process for distiller's yeast. *Ind. Eng. Chem.*, **34**:1402–1405.

Young, T. B., and H. M. Koplove. 1972. A systems approach to design and control of antibiotic fermentations. Pages 163–166 in G. Terui, ed. *Proc. 4th Int. Fermentation Symp.* Society for Fermentation Technology, Tokyo.

Part I

OVERVIEWS

Editor's Comments
on Papers 1 and 2

1   **JOHNSON**
    Fermentation—Yesterday and Tomorrow

2   **PERLMAN**
    Prospects for the Fermentation Industries, 1974–1983

These two papers give, as well as can be given in so few pages, an excellent overview of the status of fermentation technology and the industries based on fermentation processes.

Johnson's essay is philosophical, perhaps whimsical, but his ideas deserve hours of reflection. Perlman's paper is also thought-provoking, but it is valuable chiefly for its remarkable content of useful data concerning the fermentation process–based industries, especially the pharmaceutical industry.

Johnson and Perlman are products of the University of Wisconsin. Although Johnson was Perlman's thesis advisor and major professor, they were in those days rivals in a sense, and students together under the great William H. Peterson. Both Perlman and Johnson are deeply concerned with the humanitarian aspects of their work; each has lived by the principle that knowledge shared is knowledge gained.

Copyright © 1971 by the American Chemical Society
Reprinted from *Chem. Tech.*, **1**, 338–341 (June 1971)

FERMENTATION—YESTERDAY AND TOMORROW

M. J. Johnson

Although I was originally going to discuss "Fermentation—Today and Tomorrow," I changed the title to "Fermentation—Yesterday and Tomorrow." The reason for the change is easily understood. The only way we have of predicting the future is to extrapolate from the past. It is a pretty poor method, but it's the only one we have. If we have a curve with a point on it for each year, we can, with some assurance, predict the course of the curve for the next few years. If we have only one point, that is, data for the present only, we are in difficulty. It is hard to extrapolate a point.

Let us, therefore, begin with a look at the past of fermentation technology. Let us divide it into periods. The first period we will consider runs from the beginning of time up to about 1860—i.e., the pre-Pasteur period. Food fermentations, such as kraut, pickles, soy sauce, sourdough bread and the like, will not be considered. Before 1860, the alcoholic fermentation was known, the aerobic *Aspergillus oryzae* fermentation was known, and baker's yeast was manufactured (anaerobically) and sold. The vinegar generator was in general use. Techniques included pasteurization, inoculation, and what might be called semiasepsis. It was recognized that air was necessary for some fermentations. It might be argued that since the nature of the causative agents of fermentation were not known, there could have been no real knowledge of fermentation. This is a good argument, but we must bear in mind that much of what we do is still done without knowledge of why it works.

Be that as it may, before 1860 the anaerobic fermentor was known as were two types (vinegar generator and *Aspergillus oryzae*) of aerobic fermentors. Besides this knowledge of equipment, knowledge of techniques included inoculation, pasteurization, aeration, and cleanliness (semiasepsis).

Between 1860 and 1900, one additional fermentation (lactic acid) was reduced to practice, and one additional technique was developed; aeration in liquid fermentors. People had noticed that yeast grew better if the fermentor was aerated. Trickling filters for sewage disposal were invented, but these were an adaptation of the known vinegar generator. This was not a very impressive showing for 40 years of post-Pasteur work.

During the next 20 years, from 1900 to 1920, there was much more activity due, partially at least, to World War I. Fermentative production of glycerol, acetone, and butanol, and of bacterial and fungal enzymes, was worked out. The Imhoff tank for anaerobic sewage digestion was introduced, and the activated sludge process for aerobic sewage disposal was invented. Aerobic production of yeast with continuous sugar addition was also invented during this period, which might be thought of as the birth period of the fermentation industry.

During the next 20 years, from 1920 to 1940, fewer new fermentations were introduced, and fewer new techniques were invented. The chief new fermentations were sorbose production and gluconic acid production. Improvements were, of course, made in older fermentations. Only two important techniques were invented: the use of an agitated and aerated fermentor and the sterilization of air with fibrous filters. It looked as though the fermentation industry had reached a plateau, that it would grow no faster than the economy.

Then, from 1940 to 1950, the industry exploded. One usually attributes this explosion to the discovery of penicillin, which, by encouraging the discovery of other antibiotics, brought about the familiar revolution in therapy. The situation is not quite that simple. Was the fermentative production of riboflavin and vitamin B_{12}, which was developed during the same period, a result of the discovery of penicillin? I doubt it. I think that a fairer view of the course of events would be the following: The development between 1930 and 1940 of (a) the aerated, agitated vessel as an aerobic fermentor, (b) the sterilization of air by passage through fibrous filters, and (c) the invention of the shake flask as an aerobic laboratory tool, made it possible to exploit observations that previously had not been exploited. The observation that certain organisms, in growing aerobically, make large amounts of riboflavin can be exploited easily if shake flasks are available for laboratory work and if good fermentors are available for production.

During the decade from 1950 to 1960, the new fermentations that were developed included amino acid fermentations, steroid conversions, some new antibiotics, and gibberelins. Since 1960, there has been no important development in the "new fermentation" category; with the possible exception of nucleotide production for food flavor use. If we try to think of new techniques invented since 1950, we conclude that the chief advances have been in measurement and control; i.e., pH measurement and control, foam control, automatic chemical analysis, oxygen and CO_2 analyses, and the like. Since 1950, the rate of invention of new techniques, as well as of new fermentations, has slowed down. If we extrapolate present tendencies, we must conclude that the fermentation industry will continue to exist, but will go into a condition of stasis similar to its condition during most of the last 5000 years. It is, of course, a question of great interest to many of us whether this is actually going to happen or whether the recent renaissance in fermentation can be prolonged for a while by suitable geriatric treatment. My own belief is that we need look forward to neither of these dismal alternatives. I should like to explain why I think so, and what I think we can do about it.

Let us ask ourselves whether we are about to experience another renaissance similar to the one we underwent between 1940 and 1950. The answer is no, because what happened between 1940 and 1950 was not a renaissance of an earlier period of activity. It was rather a period of exploitation of some of the basic advances made during an earlier period. As we have pointed out, the techniques that were exploited were the use of aerated and agitated vessels for growth of aerobes, the sterilization of air, the use of previously developed techniques for large-scale asepsis, and the use of the shake flask for laboratory fermentations.

Our question about the future of fermentation technology should therefore be changed to read something like this: Are there techniques, invented recently or not so recently, that are not now being exploited, but that deserve exploitation? Also, we should probably not ask what new fermentation products are on the horizon, but what new techniques are about to be invented.

With this point of view, one can consider known techniques in order to guess which of them might be better exploited. When I did this, I was reminded of two techniques. The first technique that I want to talk about is the technique of growing a microorganism, not on a given medium, but under a given set of environmental conditions. This technique was originated by the yeast-makers just after World War I. I do not think they thought it to be revolutionary, and I suspect that I am going to have trouble convincing you that it was revolutionary. What the makers of baker's yeast first found out was this: If yeast was grown in high sugar concentrations, it used its anaerobic metabolic system even in the presence of excess oxygen and made alcohol. I should explain that what I meant by a high sugar concentration is a concentration roughly high enough to be analytically detectable.

The yeast-makers, by feeding a sugar solution at a rate less than the rate at which the yeast could use it, achieved a steady-state low sugar concentration, and the sugar turned into yeast instead of into alcohol, and the company made money. Of course they did not add a simple sugar solution, but a more or less complete medium. They soon realized that their new technique had lots of advantages. They fed at an exponentially increasing rate as the yeast population increased, but the rate of growth was under control. By changing the concentration of ammonia, phosphate, and the like in the feed, they were able to change the steady-state pH, and the nitrogen and phosphorus content of the yeast being produced. They were able to adjust the environment of the yeast either for the production of maximum yeast yield, for maximum CO_2-producing activity in the yeast, or for maximization of some other desirable property such as ability to withstand drying. One often can, through most of the production period, maintain an environment that is conducive to good cell yield, and only toward the end change the environmental conditions to obtain some desired special property. Dr. Fry has been particularly active in growing yeast that will do what he wants it to do. He is a real expert in what we might call continuous environmental control during growth of a microorganism.

It is remarkable how little this ability for continuous environmental control has been exploited. We haven't exploited it because we have been carefully trained to think wrongly. When I was pretty young, I took some courses in microbiology. I learned that you grow a microorganism on a medium. All good media had names: Czapek-Dox medium, Burk's medium, Eosin–methylene blue medium, or maybe 214 medium, and the like. I was taught to think that the best way to grow an organism was to mix up a recipe, preferably one taken out of a book, inoculate it with the organism, and then sit down and wait.

I knew that the organism continually changed the medium in which it grew. But I was taught to have great faith in the medium, which would magically provide at all times an optimum environment for the organism. I was not the only one who was fooled. Everyone else, with the exception of Fry and the rest of the yeast-makers, was also fooled. Let me give you a couple of examples. I will take them from the penicillin fermentation, because I am familiar with its history.

In the very early days of penicillin production, people relied on the phenylacetic acid that was naturally present in corn steep liquor to provide an R-group for the penicillin molecule. Later, it was found that with the newer high-yielding cultures, yields could be increased by adding phenylacetic acid to the fermentation. But it was found to be toxic, so only a little could be added to the medium. Phenylacetic acid was oxidized by the mold, and disappeared. The obvious thing to do, one would think, would be to add phenylacetic acid continuously or at intervals during the fermentation. This, however, was not done. It violated the unwritten rule that you weren't supposed to touch a medium once it had been inoculated. If you did fool with it after inoculation, you did so in private, without admitting that you were too stupid to make a good medium on the first try. What was done was that a large number of derivatives of phenylacetic acid: phenylacetamide, phenylethylamine, phenylacetyl glycine, and a number of others, were used in the medium, in the hope that they could be added at the beginning, and would slowly be converted by the mold to phenylacetic acid, at just the right rate. It was not until it had been demonstrated that much better penicillin yields could be obtained much more easily by continuous or repeated addition of phenylacetic acid during the fermentation, that people abandoned the idea of a precursor that could be added only before inoculation.

A similar example from the penicillin fermentation is the case of lactose. Moyer discovered that adding lactose to the medium greatly increased penicillin yields. The reason it did so was that partial energy-source starvation is necessary during the penicillin-forming phase of the fermentation, and lactose is used only slowly by the mold. Neither Moyer nor anyone else knew or cared about this. The rules were that you tried out various media, in a blind unthinking manner, until by sheer luck you hit upon a medium that would do that almost miraculous thing: provide, at all times during the fermentation, precisely the right set of conditions for optimal end results. We did a lot of work on penicillin fermentation, but even after we had found out why lactose worked, it never occurred to us to achieve partial starvation in the simple way, the way the yeast people had been doing for 30 years. It was not until our colleagues in the Department of Botany had given us some high-yielding mutants whose ability to handle lactose was a little unreliable that it slowly dawned on us that the best way to get good control of the degree of partial starvation was simply to control the rate of addition of energy source, rather than to try to use an energy source that had exactly the right degree of availability.

This matter of prejudice against adding things during the fermentation is still with us, and is still real. People still use odd media to get results that could better be obtained with continuous pH control. People are still prejudiced against changing the temperature, or even the aeration rate during a fermentation. People still try to improve a fermentation by changing the medium rather than the environmental conditions; i.e., they refuse to repeal the rule that states that the only legal method of changing the conditions in a 24-hr old fermentation is to add at zero time some constituent that will begin to have an effect only at 24 hr. I could quote many examples in which repeal of the rule has resulted in great improvements. I therefore believe strongly that here is an old technique that we are not adequately exploiting. If we want fermentation to prosper in the future, we should make use of some of the well-known but unfashionable techniques, as well as invent new ones.

I should like to give another example of a technique that is not exploited because it is not fashionable. Pasteur and his followers showed us how to purify cultures of microorganisms, and how to handle these cultures in ways that would keep them from becoming contaminated with other cultures. It is obvious that from that period on, anyone working with microorganisms worked with pure cultures. This was certainly a good thing, and made many of the advances in fermentation possible. However, I suspect that this like many other good procedures, can be overdone. When I was a brand new graduate student, a number of people in the laboratory where I worked were studying anaerobic cellulose fermentation. With crude cultures, which were enrichments from horse manure, they got rapid conversion of cellulose to organic acids, and rapid growth of their organisms. Pure cultures were very hard to obtain, and when they were obtained, were slow-growing, indolent, and unimpressive. I was young and inexperienced enough to wonder why they insisted on using pure cultures when the crude cultures were so much "better." It wasn't until I had been exposed to a couple of advanced courses in microbiology that I saw the light and realized that it was very sinful to work with mixed cultures, even if they did work better.

There are many well-known examples of the vigor of mixed cultures. The example that comes to mind is the classic work of Boruff and Buswell on anaerobic methane production from various organic materials. The cultures they used were very active, and would convert almost anything quantitatively into CO_2 and methane. But they were crude cultures, which was of course bad, and the cultures had an aura of sewage and sewage disposal, which was even worse. There are many papers in the literature on anaerobic methane formation by pure cultures. Pure cultures are not as vigorous as crude cultures, and do not attack nearly as wide a range of substrates. But they are much more scientific. Pure cultures are certainly better for studies of biochemical pathways, but in many cases they are not particularly good for preparative purposes. In our laboratory we found that mixtures of yeasts would outperform pure cultures of yeasts when grown on normal alkanes. We are at present working on cultures that grow well on methane as a sole carbon source. Again we find that our active cultures are mixtures, and that pure cultures capable of rapid growth are difficult to find.

I believe that the use of mixed cultures in fermentation has not been given the attention it deserves. One can speak of two different types of mixed cultures. They may be called (a) stable crude cultures and (b) mixed pure cultures. A mixed pure culture is a mixture of pure cultures, propagated under conditions of strict asepsis, the ratio of the numbers of each of the component cultures present being determined by the environmental conditions. A stable crude culture is a culture growing under nonaseptic conditions, the composition of which is entirely determined by the environmental conditions. It may be essentially a pure culture (as in vinegar manufacture) or have a large number of components (as in activated sludge). In actual practice, the situation is generally somewhere in the intermediate area bounded by a pure culture, a stable crude culture, an unstable crude culture, and a pure mixed culture.

The stable crude culture was invented a long time ago. Wine, vinegar, beer, and the like have been made by it for a long time. More modern examples are production of yeast cells from sulfite waste liquor and other materials. People who attempt to develop nonaseptic fermentations usually approach the problem of achievement of a stable crude culture in a very naive and unscientific way. Sometimes they just fervently hope for no contamination. Sometimes they adopt the view that reasonable cleanliness should be, for their fermentation, a suitable substitute for asepsis. Very seldom do they face up to the necessity of so arranging environmental conditions that a contaminant has little chance of causing trouble. There are many possible fermentations that will be sufficiently low-cost only if asepsis can be made unnecessary.

In the matter of pure mixed cultures, we have a virtually untouched field. There have been a few examples, such as beta-carotene production, of the use of pure mixed cultures, but the amount of information available is just about enough to point up the possibilities in the field.

What I have been trying to say is that there will be no great boom in fermentation unless we make one. In the past, advancements have been possible only by the exploitation of new techniques. We need not only new techniques, but we need further exploitation of old ones. Among the old techniques that deserve better use are what I have called the environmental rather than the "fixed medium" approach. Another old technique that deserves more consideration is the use of mixed cultures, which have already been shown to be effective in a large variety of situations.

Professor Johnson was one of the founders of the ACS Division of Microbial Chemistry and Technology, and has served as its Chairman, Councilor, and committeeman. He has held similar positions in the International Organization. Professor Johnson did both his undergraduate and graduate work at Wisconsin where, with the exception of a year at the Deutsche Technische Hochschule, Prague, he has been ever since. His 70 graduate students have worked on the production of antibiotics, enzymes, fermentation mechanisms, mammalian cell culture, and, quite recently, on the metabolism of methane by a mixed culture under nonsterile conditions. Perhaps as far-reaching as Dr. Johnson's technical accomplishments is his reputation as wit and poet:

> Once upon a time I blamed on fate
> The fact that I am bland and placid;
> But now I put the onus on
> Deoxyribonucleic acid.

Dr. Johnson has also been quoted as saying "Xerox is a mechanism for asexual reproduction of sterile material."

Prospects for the fermentation industries, 1974–1983

David Perlman

While the U.S. fermentation industries date back to 1881, when a commercial process for producing lactic acid was started, it wasn't until 1960 that the major expansion in these industries occurred. Now a decade later it is moot to ask how long the rapid growth period will last and what the future holds.

Prior to World War II a few companies were using fermentation to supply industrial chemicals, pharmaceuticals, and food and feed ingredients. Among these products were ethanol, acetone, n-butanol, citric and gluconic acids, crude enzymes with protease, invertase or amylase activity, and bakers' yeast. Similar technology has been used in the 30 years since the war as the basis for current activities in the fermentation industries whose major efforts in 1973 included production of some 60 antibiotics, citric and glutamic acids, yeast, and a variety of enzymes for industrial, food, household, and pharmaceutical uses.

Now, despite this record of expansion and financial success, there is concern in some quarters that the boom is over; that in the immediate future we must expect less return for invested effort than was achieved 10 or 15 years ago. This pessimistic attitude seems to me to be without valid foundation. This essay has as its purpose the evaluation of the status of the fermentation industries, in order to make some predictions of what we might expect in the next 10 years. First we will examine the structure of the industry as a whole and then we will look at some individual products and manufacturers. Then we can prognosticate.

Evolutionary patterns

In earlier analyses (1–3) we concluded that the U.S. fermentation industries seem to go through a "boom and bust" cycle that is independent of economic cycles. Further study suggests that the industries' recent cyclic behavior is more related to political and related forces than to availability of scientific and methodological advances or the availability of funds for expansion and/or demonstration of new needs for fermentation products.

Generally, once a product has been shown to be of commercial value, new and/or improved processes are developed to meet market demands. This is as true for fermentation process as it is for others. In a few cases, process improvement did not effect selling price. Notable here is citric acid, where process know-how for the *Aspergillus niger* method was kept secret. In most instances, however, continued successful commercialization resulted in significant price reduction, which has sometimes led to new uses. For example, over a 25-year period the price of penicillin G dropped about 100-fold, and the product's use expanded from its initial role in human medicine, to a substantial application in veterinary products and, until recently, wide acceptance as an animal and poultry feed supplement. Worldwide production of penicillin G is now measured in thousands of tons per month. Nevertheless the current inability to meet demand, together with currency devaluations, has resulted in a wholesale price increase of about 50%.

Other fermentation products being produced in similar volume include citric and glutamic acids, tetracyclines, and penicillin V. Most are in short supply and additional fermentation capacity is being constructed, while other facilities that were used for

---

David Perlman, a native of Madison, Wis., received his three degrees at the University of Wisconsin. He joined the staff of the Squibb Institute for Medical Research in 1947, and during nearly twenty years there he did research on antibiotics, vitamin B_{12}, microbial transformations of steroids, and growth of mammalian cells in tissue culture. In 1967, Dr. Perlman returned to the University of Wisconsin as Professor of Pharmaceutical Biochemistry; he is also Dean of the School of Pharmacy. Prof. Perlman has been an ACS member since 1941, and is Chairman-elect of the Wisconsin Section. This is Dr. Perlman's second CHEMTECH paper (cf, Sept., 540, 1971).

other processes will be used for these products. It is thus possible that the supply will soon be able to meet the demand.

In general, there are remarkably few fermentation processes that have suffered obsolescence because of either competition from chemical processes or loss of markets because of changing patterns of product use. Competition from chemical process industries, until recently, hurt fermentation production of ethanol, acetone, and n-butanol, all of which could, until now, be made more cheaply from petrochemicals. However, in 1974, with restrictions on petrochemical feedstocks, and with prospects of a plentiful supply of starchy grains, these products may once again be largely made by fermentation. The same situation has been noted for lactic acid and riboflavin where, until 1974, the chemical process efficiency was higher than the fermentation route, which kept the latter's products in a secondary position.

Recent political pressures and other "external" forces have been influencing fermentation industries. Thus novobiocin and microbial protease (intended for washing powder use) have been withdrawn from the general market, and it is possible that the use of penicillin, tetracyclines, and erythromycin as animal feed supplements may be restricted soon by governmental regulation. This latter anticipation has given impetus to a search for antibiotics not related to those used to treat human infections and that, when added in small amounts to feeds, will promote animal and poultry growth. Among those already approved by the FDA for this purpose are the relatively old virginiamycin (staphylomycin) and bambermycin. Others including thiopeptin, macarbomycin, and mikamycin are used for these purposes in Japan. Antibiotics now being evaluated as alternatives include moseimycin, diumycin, and parvulin.

In a few spectacular situations, demonstration of the utility of a product has been followed by development of a fermentation process for its manufacture: The value of cortisone in treatment of rheumatoid arthritis and other inflammatory disease set off many searches for methods to produce the drug. In 1952 the processes with microbial hydroxylation of the steroid nucleus at C-11 achieved manufacturing status. It is still the most economic. More recently, there has been new demand for L-dihydroxyphenylalanine (L-DOPA) in high-dosage treatment of Parkinson's disease.

The economic success of these microbial hydroxylation steps and the importance of the microbial oxidation of sorbitol to sorbose in the ascorbic acid process has encouraged many to consider combined microbial/chemical syntheses. This cooperation, which involves microbiologist, chemist, biochemist and engineer, has proved feasible in locations where teamwork is encouraged. Thus some years ago starting materials for complicated steroid syntheses were made available by enzymic hydrolyses of cholesterol and plant sterols (4). More recently the preparation of a series of prostaglandins by syntheses involving microbiological and chemical steps was described by Sih et al. (5).

Processes first operated in the fermentation industries from 1905 to 1940 were mainly concerned with conversion of carbohydrates. Among the products of

Upjohn's new fermentation plant in Arecibo, Puerto Rico, opened early in 1974. Expansion is already underway for production of lincomycin (and conversion to clindamycin)

Table 1a. Antibiotics manufactured by fermentation in 1973

Product	Some producers
Adriamycin	38
Alazomycin F	90
Aminosidin	38
Amphomycin	61
Amphotericin B	99, 100
Antimycin A	13, 56
Bacitracin	11, 12, 26, 45, 70, 74, 75, 79, 56
Bambermycin	37
Bicyclomycin	41
Blastocidin S	52, 67
Bleomycin	66
Cactinomycin	36
Candicidin B	74
Capreomycin	32, 59
Carzinophillin	56
Cellocidin	67
Cephalosporins	20, 39, 41, 44, 59, 80, 104
Chromomycin A_3	104
Colistin	14, 54, 57, 56
Cycloheximide	52, 105, 109
Cycloserine	4, 26, 38
Dactinomycin	64
Daunorubicin	38, 85
Destomycin	63
Enduracidin	104
Erythromycins	1, 32, 59, 78, 80, 88, 95, 109
Fumagillin	1, 22
Fusidic acid	58
Gentamicins	93
Gramicidin A	74, 111
Gramicidin S	3, 63, 68
Griseofulvin	44, 51, 66, 104

(continued)

Table 1a. Antibiotics manufactured by fermentation in 1973

Product	Some producers
Hygromycin B	18, 59, 104
Josamycin	91, 113
Kanamycins	14, 20, 38, 63, 68, 78, 85
Kasugamycin	14, 52, 66, 91
Leucensomycin	38
Lincomycins	109
Lividomycin	55
Macarbomycin	63
Maridomycin	104
Midecamycin	63
Mikamycins	14
Mithramycin	75
Mitomycin G	56
Monensin	59
Myxin	48
Neomycins	11, 18, 66, 70, 74, 75, 78, 85, 88, 99, 100, 104, 109
Nigericin	18
Novobiocin	109
Nystatin	22, 58, 78, 99, 100
Oleandomycin	18, 75, 76, 78
Paromomycin	73
Penicillin G	3, 4, 6, 10, 14, 15, 16, 17, 18, 34, 36, 37, 38, 42, 44, 47, 50, 59, 63, 64, 66, 70, 75, 76, 85, 99, 100, 106, 112
Penicillin O	109
Penicillin V	1, 3, 4, 14, 17, 18, 20, 32, 37, 38, 42, 44, 63, 66, 70, 75, 76, 85, 99, 100, 106, 112
Penicillins (semi-synthetic)	3, 10, 14, 15, 16, 20, 36, 38, 42, 63, 75, 80, 83, 84, 99, 100, 104, 106, 112
Pentamycin	68
Pimaricin	42
Polymyxins	32, 70, 75
Polyoxins	52
Pristinamycins	85
Ribostamycin	63
Rifamycins	39, 46
Sarkomycin	14, 63
Siccamin	90
Siomycin	95
Spectinomycin	1, 109
Spiramycins	56, 85
Staphylomycins	84
Streptomycins	10, 34, 37, 38, 42, 44, 47, 52, 56, 58, 63, 64, 70, 75, 84, 85, 100, 112
Tetracycline	3, 7, 10, 14, 20, 21, 24, 27, 31, 37, 38, 46, 50, 63, 70, 75, 76, 80, 82, 83, 85, 88, 97, 100, 104
Chlortetracycline	7, 31, 38, 47, 50, 60, 78, 80, 82, 84, 85, 97, 104
Demethylchlorotetracycline	7, 38, 80, 85, 104
Oxytetracycline	18, 27, 42, 51, 75, 76, 78, 80, 82
Thiopeptin	41
Thiostrepton	99
Trichomycin	41
Tylosin	32
Tyrothricin	17, 74, 111
Tyrocidine	111
Validamycin	104
Vancomycin	59
Variotin	66
Viomycin	39, 76

commercial interest were citric acid, gluconic acid, lactic acid, acetone, butanol, and ethanol. The only "complicated" products of this period were a few crude enzyme preparations.

Production of the "less complicated" chemicals increased gradually over the 1920–1940 period. n-Butanol, initially a by-product of the acetone process, became the more important product after its use as a component of quick-drying lacquers was appreciated. Eventually the solvent market became so large that chemical production was developed. This alternate process gradually increased its share of the market so that now, 50+ years after the start of the commercial fermentation process, fermentation is no longer an important source of either butanol or acetone. Lactic acid has a similar process history; until the 1950's it was produced primarily by fermentation. Ethanol somewhat earlier met and lost out to competition from chemical processes. Now, in 1974, with prices of petroleum increasing, and surplus starchy grains in prospect, it is likely that fermentation processes for these products will be able to compete economically once again.

Coincident with the decline in importance of fermentation processes for ethanol, acetone, butanol, and lactic acid there was an increase in food uses of sugars and carbohydrates, and a resulting reduction in the quantity of these materials available for fermentations. The usefulness of hydrocarbons and other relatively cheap carbon-containing materials was thus studied and the production of citric acid by a process based on *Candida* yeasts (6) and glutamic acid appeared feasible. A somewhat similar economic squeeze was recently seen in certain antibiotic fermentations that use soybean meal in the fermentation media: It may have been competitive with soybean meal costing $20 per ton but when the price topped $300 per ton in July 1973, this process may have become too costly to meet competition.

Research management is now focusing on microbial processes for producing "complicated" molecules that cannot easily be prepared chemically. Antibiotics, enzymes, vitamin B_{12}, and selected amino acids fall in this class of complicated substances. With certain of the amino acids research is being directed toward substitution for carbohydrates of methanol, ethanol, and acetic acid, as well as cheaper hydrocarbons, as energy sources for the metabolism of the microorganisms. The success of such a substitution in glutamic acid fermentation is bound to encourage use of hydrocarbons as energy sources in other fermentations once the raw material prices stabilize.

New equipment and better control of the variables has resulted in improved fermentation economics. A number of these developments have been included in Solomons' publications (7, 8) and mentioned by Dawson and Phillips (9). This type of equipment should make possible rapid evaluation of the effect of fermentation variables on productivity.

Fermentation products—1973/74

Products now manufactured by fermentation are listed in Table 1 (a–e). Some of the products including acetone, n-butanol, ethanol, and lactic acid are now prepared mainly by chemical synthesis, and mention of fermentation processes is more for accuracy of reporting than industrial significance. Others such as citric acid and amylases have been produced in increasing quantities since the 1920's and manufacturing capacity appears to be expanding continually.

The majority of the items listed have been produced on a manufacturing scale for less than 20 years. Although accurate production figures are not available, there seems to be no doubt that production of most of these items has increased annually; sometimes this has been coincident with a drop in the number of manufacturers.

With the advent of semi-synthetic antibiotics, e.g., chemical derivatives of 6-aminopenicillanic acid, there has been a remarkable continuation of interest in the "old timers." There are now 24 penicillins being distributed on a large scale. Those available in the U.S. include Penicillin G, Penicillin O, Penicillin V, Amoxicillin, Ampicillin, Carbenicillin, Cloxacillin, Dicloxacillin, Hetacillin, Nafcillin, Methicillin, Oxacillin, and Phenethicillin. Additional penicillins used in foreign countries include Azidocillin, Ancillin, Cyclacillin, Epicillin, Flucloxacillin, Phenbenicillin, Pivampicillin, Propicillin, Quinacillin, Pyrazocillin, and Sulfobenzylpenicillin. The total weight of penicillin used keeps increasing, and at the moment the supply cannot meet the demand. This phenomenon, the ability of the market to absorb a new product without marked decrease in demand for an older, related product, has also been noted with the tetracyclines (where we have 7) and the cephalosporins (where 7 are in use and 3 more are to be introduced in the U.S. in 1974). Information on structure-activity relationships in the penicillins (10), tetracyclines (11) and cephalosporins (12) suggests that more useful compounds can be expected to result from better definition of the objectives of these chemical studies.

Newer uses for enzymes include alternatives to calf stomach rennet (13), alternatives to barley malt, and agents for converting glucose to fructose, which is sweeter. The popularity of alkali-stable protease as a component of washing powder mixtures apparently will not again reach the heights of 1971. These uses stimulated planning for increased manufacturing capacity for enzymes in general. In each of these instances it seems apparent that the need for an enzyme to carry out a transformation or function was determined prior to the search for a microbial enzyme to fill the need. The value of enzymes as reagents in organic chemistry is just beginning to be appreciated and expansion of this use at least on a laboratory scale can be expected now that insolubilized enzymes are becoming more widely used (14).

Enzymes useful in medicine are in many senses specialty products, and do not represent a significant portion of the total volume of the enzyme market. Among those that are produced from microorganisms are penicillinase (Riker), streptokinase and streptodornase (Lederle), and asparaginase (Merck). Other enzymes have been considered for therapeutic use (15), and may eventually become commercial products.

In the animal health field there is an increased use of antibiotics including erythromycin and tylosin to control disease of poultry and other animals. The latest reports on annual sales of these products in the U.S. run as high as $125 million. Monensin, an antibiotic with coccidiostat properties, has displaced

Table 1b. Enzymes, organic acids, and solvents manufactured by fermentation in 1973

Product	Some producers
Enzymes	
Amylases	18, 25, 42, 45, 65, 70, 79, 81, 86, 87, 101, 111
Amyloglucosidase (Glucamylase)	25, 29, 44, 45, 65, 70, 111
Catalase	65, 94, 111, 56
Cellulase	3, 42, 65, 70, 86, 87, 111
Glucanase	111
Glucose isomerase	25, 70
Glucose oxidase	65, 94, 111, 56
Hemi-cellulase	65, 70, 87, 111
Invertase	42, 70, 94, 101, 108, 111
Lactase	42, 65, 56
Lipase	30, 65, 87, 111
Pectinase	23, 65, 70, 86, 87, 94, 111
Pentosanase	87, 111
Proteases	3, 18, 22, 42, 45, 52, 65, 70, 79, 86, 87, 104, 111
Rennet substitute	30, 65, 70, 75, 111
Organic acids and solvents	
Citric acid	65, 75, 76, 85
Gluconic acid	21, 41, 45, 75, 79, 94
Itaconic acid	75, 85
2-ketogluconic acid	64
α-ketoglutaric acid	2, 56
Erythorbic acid	41
Lactic acid	19, 25, 85, 104
Pyruvic acid	2
Urocanic acid	2
2,3-Butanediol	42
Ethanol	42, 45, 81, 104

Table 1c. Amino acids manufactured by fermentation in 1973

Product	Some producers
L-alanine	2, 56
L-arginine	2, 56
L-aspartic acid	2, 56, 21
L-citrulline	2, 56
L-glutamic acid	2, 26, 56, 72, 91, 102, 104
L-glutamine	2, 56
L-histidine	2, 56
L-isoleucine	2
L-leucine	2
L-lysine	2, 56, 85
L-ornithine	2, 56
L-phenylalanine	56
L-proline	2, 56
L-serine	2, 56
L-threonine	2, 56
L-tryptophan	56
L-tyrosine	56
L-valine	2, 56

Table 1d. Vitamins, yeast, growth factors, nucleotides and nucleosides manufactured by fermentation in 1973

Product	Some producers
Gibberellic acids	1, 51, 56, 59, 64, 104
5'-ribonucleotides and nucleosides	2, 56, 104
Riboflavin	45, 64, 75
Vitamin B_{12}	38, 44, 45, 64, 78, 85, 88, 113
Yeast	5, 8, 9, 28, 33, 35, 40, 42, 43, 53, 56, 62, 69, 71, 90, 96, 98, 101, 103, 107, 108, 110, 114
Zearalenol	26

Table 1e. Miscellaneous products manufactured by fermentation in 1973

Product	Some producers
Acyloin	64
Desferrioxamine	39
Dihydroxyacetone	111
Dextran	3, 18, 77
Sorbitol oxidation	48, 49, 64, 75, 104
Steroid oxidation	39, 45, 75, 76, 92, 93, 99, 100, 109, 112
Xanthan	64, 85
Insecticides (bacterial)	1, 57, 89

Table 2. Some fermentation products commercialized since 1968

Antibiotics

Use	Product
anti-tumor compounds	Adriamycin, Bleomycin, and Daunorubicin
animal feed supplements— growth promotants	Bambermycin, Destomycin, Macarbomycin, and Thiopeptin
treatment for human TB	Capreomycin
semi-synthetic, for treatment of bacterial infections caused by gram-positive and by gram-negative organisms	Cephalosporins: Cefazolin, Cephaloglycine, Cephalexin, Cephapirin, Cephradine
for use against gram-positive bacterial infections	Josamycin and Maridomycin
for use against gram-negative bacterial infections	Lividomycins and Bicyclomycins
treatment of coccidial infections in poultry	Monensin and Nigericin
semi-synthetic Penicillins for treatment of bacterial infections	Amoxicillin, Cyclacillin, Epicillin, Flucloxacillin, Pivampicillin
semi-synthetic Rifamycin for treatment of human TB	Rifampicin
treatment of human gonorrhea	Spectinomycin
semi-synthetic Tetracyclines for treatment of infections caused by gram-positive and by gram-negative bacteria	Doxycycline, Minocycline

Other products
Rennet substitute: microbial enzyme preparation
Zearalenol: anabolic agent for animals

a number of chemically produced drugs. Zearalenol and perhaps other agents will be economically useful as anabolic agents in fattening farm animals and poultry now that diethylbesterol is under fire as feed supplement. The use of antimycin A as a selective fish poison (16) has now reached the stage of commercial exploitation and is an example of possible uses for microbial metabolites that may not be immediately contemplated when the biological properties are first examined.

Prospects for the next decade (1974–1983)

Forecasting for the fermentation industries is not an assignment to be taken without a sense of responsibility. In setting earlier forecasts (17–19) against historical developments, it will be noted that most often the projections have been complicated by the appearance of unexpected new products that opened new markets: penicillin and other antibiotics; vitamin B_{12}; monensin; gibberellic acids; antimycin A. None could have been foreseen in 1930, 1950, or even more recently. These examples should strengthen our confidence in a sound future for fermentation. The list of items that have reached commercialization in the past 5 years (Table 2) is impressive, and we can expect similar performance in the future.

A number of new fermentation products are currently being evaluated; some may prove to be as successful as suggested by their earliest evaluation. Microbial insecticides may replace DDT and malathion in view of the more careful controls on the use of these chemicals. The antibiotic needs of the future are becoming better defined, and it seems likely that fermentations will be involved for some time to come. With a growing understanding of the limitations of antibiotics we can set the requirements of new compounds rather carefully:

• A penicillin that is non-allergenic
• An antibiotic effective when taken orally against systemic Gram-negative infections
• An antifungal agent effective systemically and less toxic than the polyene amphotericin B
• A growth-promoting antibiotic for animal feed supplements that is not cross-resistant with antibiotics used in therapy of human infections (20). All of these are currently being investigated, and candidate compounds may be available shortly.

Finally, the potential of animal cell culture as a fermentation process should be mentioned: It is likely that as more experience is gained with the technology it will be possible to use these cells for antiviral vaccines (21) or for producing hormones and other useful medicinals (22, 23). Part of the problem at present is selection of the cells for study and then their production in large quantities (23, 24); both of these problems are nearing solution. Years ago "therapeutic bacteriophage" was considered useful in cholera treatment, and modern technology may now make it possible to produce large amounts of phage in systems similar to those used for virus vaccines (25).

The list of fermentation products that "never made it" is much larger than desirable. Some are produced in such low yields in fermentations that they are poor commercial risks: fumaric acid from sucrose; propionic acid from glucose; iso-propanol from glucose; acetoin from glucose; ustalagic acids;

Aerial view of Eli Lilly Fermentation Plant in Clinton, Indiana, for manufacturing of cephalosporins and monensin

and a host of compounds from hydrocarbon fermentations (26). Others are formed in good yield, but have not yet been found broadly useful. For example, the food process industries annually use about 120 million pounds of acidulants. At present 65 million lbs of citric acid, 30 million lbs of phosphoric acid, 10 million lbs of fumaric acid, and 12 million lbs of a mix of tartaric, adipic, and malic acids are the acids used (27). Other fermentation acids for which acidulant uses might be developed are araboascorbic and alloisocitric, both of which are produced in high yields from sugar by fungal fermentations. A fermentation process for direct production of erythorbic acid may also swell the supply of acidulants or antioxidants for the food process industries, even though the present process via 2-ketogluconic acid has been a "reasonable" process.

The potential protein shortage has spurred research into production of cells of microorganisms to meet some of the demand. Growth of these organisms on hydrocarbons has been attractive technologically until recently when the price of raw materials rose sharply. Even with this raw material cost increase, and the concern about nutritional value and toxicity, the economics may not be "impossible." Fermentation processes making use of hydrocarbons for conversion by yeast to the more valuable citric acid have reached commercialization (28), and other hydrocarbon fermentations may soon be economically successful.

Technology

As mentioned earlier, fermentation technology has usually been forthcoming once problems were characterized. Fermentors of 100,000 to 250,000 gallon capacity were developed for the acetone-butanol fermentation when it was shown that there was a market for the fermentation products (29). Pilot plant and manufacturing equipment for aerated fermentations (7, 8) and "fancy" hardware such as pH controllers, aeration controllers, and computerized controllers (9, 30) have come into the picture to make fermentation processes more reproducible.

The idea of continuous fermentations has challenged bio-engineers, and although successful for lactic acid, ethanol, and wine, there is some reluctance to use continuous fermentations for other products where experience with the batch-type operation has been so satisfactory (31). The increased costs of construction and fabrication of fermentation facilities has led to new fermentor designs that are likely to be more efficient than the more conventional stirred vats.

Genetics

Perhaps equal in importance to the development of new equipment is the application of genetics to the selection of microorganisms with greater capacity for carrying out the desired metabolic process. So many successes have been reported, that many fermentation research groups have accepted geneticists as valued members (32). Some managers have questioned the economics involved, and a single example might be cited to show the advantage of culture selection programs: If a manufacturing plant produces annually 1 million lbs of a product selling for $1 per lb, then increase in efficiency of the fermentation by 10% will be worth approximately $100,000 per year. Under such circumstances an investment of $35,000 for a man-year in genetics research may be a very fine investment.

The economics of the fermentation industries will change in the next few years as they have in the past. Some costs such as raw materials including carbohydrate and protein that have followed a cyclic pattern in the past will not do so in the future because they will be valued as food and will not be in oversupply as in the past. Continuous fermentation techniques may be economic under certain circumstances, especially when there is a short residence-time in the fermentor and the manufacturer wishes to devote as few vessels to the process as possible (31). With increased research costs there will be more interest in patent licensing and purchase of patented processes (33) even though present governmental policy is tending against such arrangements.

Research for the future

Research in fermentations during the 1915-1940 period was concentrated in colleges and universities, notably Iowa State, Cornell, Wisconsin, and MIT. With the support of Congress, the U.S. Department of Agriculture started regional research laboratories, and fermentation research was promoted at several of these, notably the Northern Regional Research Laboratory in Peoria, Illinois (34).

Research interest in the U.S. has shifted in the past 20 years toward molecular biology, especially in schools. There is also evident de-emphasis at the U.S. Department of Agriculture on fermentation research programs. Both trends will reduce the number of "new findings" and force industries either to carry out in-house research programs, or go abroad for new processes.

The priority given to fermentation research in Japan (35) nearly 20 years ago has paid off. Many new products listed in Table 2 have been manufactured on an industrial scale, and others are likely to materialize in the not-too-distant future. Unless present trends in the U.S. are reversed, we can expect that in the future it will be desirable to send our young students to Japan to learn the techniques that will assure the continuation of the fermentation industries in the United States.

References

(1) Perlman, D., *Chem. Week*, **101** (No. 25), 44 (1967).
(2) Perlman, D., *Wallerstein Lab. Commun.*, **33**, 165 (1970).
(3) Perlman, D., *ASM News*, **39**, 648 (1973).
(4) Nagasawa, M., et al., *Agr. Biol. Chem.*, **34**, 838 (1970).
(5) Sih, C. J., et al., *J. Amer. Chem. Soc.*, **95**, 1676 (1973).
(6) Tabuchi, T., et al., *J. Agr. Chem. Soc. Jap.*, **43**, 154 (1969).
(7) Solomons, G. L., *Advan. Appl. Microbiol.*, **14**, 234 (1971).
(8) Solomons, G. L., *Materials and Methods in Fermentations*, Academic Press, Inc., London and New York, 1969.
(9) Dawson, P. S. S., and Phillips, K. L., *Advan. Appl. Microbiol.*, **17**, in press (1974).
(10) Price, K. E., *Advan. Appl. Microbiol.*, **11**, 17 (1969).
(11) Blackwood, R. K., and English, A. R., *Advan. Appl. Microbiol.*, **13**, 237 (1970).
(12) Sassiver, M. L., and Lewis, A., *Advan. Appl. Microbiol.*, **13**, 163 (1970).
(13) Sardinas, J. L., *Advan. Appl. Microbiol.*, **15**, 39 (1972).
(14) (a) Smiley, K. L., and Strandberg, G. W., *Advan. Appl. Microbiol.*, **15**, 13 (1972); (b) Vieth, W., and Venkatasubramanian, K., *CHEM-TECH*, Jan., 47, 1974, and references therein.
(15) Sizer, I. W., *Advan. Appl. Microbiol.*, **15**, 1 (1972).
(16) Lennon, R. E., and Vezina, C., *Advan. Appl. Microbiol.*, **16**, 56 (1973).
(17) May, O. E., and Herrick, H. T., *Ind. Eng. Chem.*, **22**, 1172 (1930).
(18) Johnson, M. J., In *Pasteur Fermentation Centennial, 1857-1957*, Charles Pfizer and Co., New York, 1958, p 33.
(19) Langlykke, A. F., In *Fermentation Advances* D. Perlman, Ed., Academic Press Inc., New York, 1969, p 883.
(20) Jukes, T. H., *Advan. Appl. Microbiol.*, **16**, 1 (1973).
(21) Telling, R. C., *Proc. Biochem.*, **4** (No. 6), 49 (1969).
(22) Tashjian, A. H., *Biotechnol. Bioeng.*, **11**, 109 (1969).
(23) Johnson, I. S., and Boder, G., *Advan. Appl. Microbiol.*, **15**, 215 (1972).
(24) Telling, R. C., and Radlett, P. J., *Advan. Appl. Microbiol.*, **13**, 91 (1970).
(25) Sargeant, K., *Advan. Appl. Microbiol.*, **13**, 121 (1970).
(26) Abbott, B. J., and Gledhill, W. E., *Advan. Appl. Microbiol.*, **14**, 249 (1971).
(27) *Oil, Paint, Drug Rep.*, p 5, Apr. 11, 1970.
(28) *Chem. Mkt. Rep.*, p 3, Aug. 13, 1973.
(29) Gabriel, C. L., and Crawford, F. M., *Ind. Eng. Chem.*, **22**, 1163 (1930).
(30) Nyiri, L. K., *Microwaves*, publ. by New Brunswick Sci. Co. (April 1973).
(31) Righelato, R. C., and Elsworth, R., *Advan. Appl. Microbiol.*, **13**, 399 (1970).
(32) Demain, A. L., *Advan. Appl. Microbiol.*, **16**, 177 (1973).
(33) Whittenberg, J. V., *Advan. Appl. Microbiol.*, **13**, 383 (1970).
(34) Ward, G. E., *Advan. Appl. Microbiol.*, **13**, 363 (1970).
(35) Sakaguchi, K., In *Fermentation Technology Today*, G. Terui, Ed. Society of Fermentation Technology, Osaka, Japan, 1972, p 7.

Selected Further Reading

General Information

(1) *Chemical Oxidations with Microorganisms*, by G. S. Fonken and R. A. Johnson, Marcel Dekker, Inc., New York, N.Y., 1972.
(2) *Microbial Technology*, H. J. Peppler, Ed., Reinhold Publishing Corp., New York, N.Y., 1967.
(3) *Fermentation Advances*, D. Perlman, Ed., Academic Press, New York, N.Y., 1969.
(4) *Fermentation Technology Today*, G. Terui, Ed., Society of Fermentation Technology, Osaka, Japan, 1972.
(5) *The Microbial Production of Amino Acids*, K. Yamada, S. Kinoshita, T. Tsunoda, and K. Aida, Eds., Kodansha Ltd., Tokyo, and John Wiley & Sons, New York, N.Y., 1972.

Author's address: School of Pharmacy, The University of Wisconsin, Madison, Wisconsin 53706.

World Index to Fermentation Companies

1. Abbott Laboratories, North Chicago, Illinois
2. Ajinomoto Company, Tokyo, Japan
3. Aktiebolaget Astra, Sodertalje, Sweden
4. Aktiebolaget KABI, Stockholm, Sweden
5. Aktiebolaget S. J. A., Stockholm, Sweden
6. Alembic Chemical Works Co., Ltd., Baroda, India
7. American Cyanamid, Wayne, N.J.
8. Anchor Yeast (Pty.) Ltd., Johannesburg, South Africa
9. Anheuser-Busch, Inc., St. Louis, Mo.
10. Antibioticos, Madrid, Spain
11. Apothekernes Laboratori für Specialpraeparater A/S, Oslo, Norway
12. Asahi Chemical Industry, Osaka, Japan
13. Ayerst Laboratories, New York, N.Y.
14. Banyu Pharmaceutical Company, Tokyo, Japan
15. Beecham, Inc., Clifton, N.J.
16. Beecham Research Laboratories, Brentford, England
17. Biochemie G.m.b.H., Kundl, Austria
18. Biogal, Debrecen, Hungary
19. Bowmans Chemicals, London, England
20. Bristol Laboratories, Syracuse, N.Y.
21. Istituto Carlo Erba, Milano, Italy
22. Chinoin, Budapest, Hungary
23. Ciba-Geigy, Basle, Switzerland
24. Cipan, Lisbon, Portugal
25. Clinton Corn Processing Co., Clinton, Iowa
26. Commercial Solvents Corporation, Terre Haute, Indiana
27. Compagnie Européenne de Fermentation, Villeneuve La Garconne, France
28. Compania Argentian de Levaduras S.A., Buenos Aires, Argentina
29. CPC International Inc., Argo, Illinois
30. Dairyland Food Industries, Inc., Waukesha, Wis.
31. Diaspa, Milano, Italy
32. Dista Products Ltd., Liverpool, England
33. Distillers Company (Yeast) Ltd., Grant Burgh, Epsom, Surrey, England
34. Dumex A/S, Kobenhavn, Denmark
35. Establissements Fould-Springer, Maisons-Alfort(Seine), France
36. Farbenfabriken Bayer AG, Leverkusen, Germany
37. Farbwerke Hoechst AG, Frankfurt (Main), Germany
38. Farmitalia S.p.A., Milano, Italy
39. Fervet, S.p.A., Naples, Italy
40. Finnish State Alcohol Monopoly(Alka), Helsinki, Finland
41. Fujisawa Pharmaceutical Company, Osaka, Japan
42. Gist-Brocades, N. V., Delft, Netherlands
43. Gist & Spiritusfabrieken, Ghent, Belgium
44. Glaxo Laboratories, Ltd., Greenford, England
45. Grain Processing Company, Muscatine, Iowa
46. Gruppo Lepetit, Milano, Italy
47. Hindustan Antibiotics, Ltd., Pimpri, India
48. Hoffmann-LaRoche, Inc., Nutley, N.J.
49. F. Hoffmann-LaRoche & Co. Ltd., Basle, Switzerland
50. Icar, S.p.A., Roma, Italy
51. Imperial Chemical Industries, Ltd., Manchester, England
52. Kaken Chemical Company, Tokyo, Japan
53. Kanegafuchi Chemical Industry Company, Osaka, Japan
54. Kayaku Antibiotics Research Company, Tokyo, Japan
55. Kowa Company, Nagoya, Japan
56. Kyowa Hakko Kogyo Company, Tokyo, Japan
57. Laboratoires Roger Bellon, Neuilly, France
58. Leo Pharmaceutical Products, Ballerup, Denmark
59. Eli Lilly and Company, Indianapolis, Indiana
60. Lohmann & Company, AG., Cuxhaven, Germany
61. H. Lundbeck & Company, Valby, Denmark
62. Mauri Brothers and Thomsen, Ltd., Sydney, Australia
63. Meiji Seika Kaisha Ltd., Tokyo, Japan
64. Merck & Company, Inc., Rahway, N.J.
65. Miles Laboratories, Inc., Elkhart, Indiana
66. Nihon Kayaku Company, Tokyo, Japan
67. Nihon Nohyaku Company, Tokyo, Japan
68. Nikken Chemicals Company, Ltd., Tokyo, Japan
69. Norddeutsche Hefe Industrie G.m.b.H., Hamburg-Wandsbeck, Germany
70. Novo Indutsri A/S, Kobenhavn, Denmark
71. Oriental Yeast Company, Ltd., Tokyo, Japan
72. Orsan, S. A., Paris, France
73. Parke, Davis and Company, Detroit, Michigan
74. S. B. Penick and Company, Orange, New Jersey
75. Chas. Pfizer and Company, New York, New York
76. Pfizer International, New York, New York
77. Pharmacosmos, Valby, Denmark
78. Pierrel S.p.A., Milano, Italy
79. Premier Malt Products, Milwaukee, Wisconsin
80. Proter S.p.A., Opera, Italy
81. Publicker Industries, Inc., Philadelphia, Pennsylvania
82. Rachelle Laboratories, Inc., Long Beach, California
83. Quimasa S. A.—Quimica Industrial Santo Amaro, Santo Amaro—Sao Paulo, Brazil
84. Recherche Industrie Therapeutique, Genval, Belgium
85. Rhone Poulenc, Paris, France
86. Rohm G.m.b.H. Chemische Fabrik, Darmstadt, Germany
87. Rohm & Haas, Philadelphia, Pennsylvania
88. Roussel UCLAF, Romainville, France
89. Sandoz-Wander, Inc., Homestead, Florida
90. Sankyo Company Ltd., Tokyo, Japan
91. Sanraku Ocean, Tokyo, Japan
92. Schering AG, Berlin, Germany
93. Schering Corporation, Inc., Bloomfield, New Jersey
94. Searle Biochemics, Arlington Heights, Illinois
95. Shionogi & Company, Ltd., Osaka, Japan
96. Societe Industrielle de Saffre, Marcq-en-Baroeul, France
97. Societa Prodotti Antibiotici, Milano, Italy
98. Societe Francaise des Petroles BP, Lavere, France
99. E. R. Squibb & Sons, Princeton, New Jersey
100. Squibb International, New York, New York
101. Standard Brands, Inc., Stamford, Connecticut
102. Stauffer Chemical Company, Stamford, Connecticut
103. Syndicat des Producteurs de Levure Aliment de France, Paris, France
104. Takeda Chemical Industries Ltd., Osaka, Japan
105. Tanabe Seiyaku Company, Ltd., Osaka, Japan
106. Toyo Jyozo, Shizuoka-ken, Japan
107. Union Yeast Products, Ltd., Johannesburg, South Africa
108. Universal Foods Corporation, Milwaukee, Wisconsin
109. Upjohn Company, Kalamazoo, Michigan
110. Vereinte Maunter Markhofsche Presshefe Fabriken, Vienna, Austria
111. Wallerstein Laboratories, Deerfield, Illinois
112. Wyeth Laboratories, Philadelphia, Pennsylvania
113. Yamanouchi Pharmaceutical Company, Tokyo, Japan
114. Zellstoffabrik Waldhof, Mannheim-Waldhof, Germany

Part II
CULTURES: ACQUISITION AND CONSERVATION

Editor's Comments
on Paper 3

3   HESSELTINE and HAYNES
Excerpt from *Sources and Management of Micro-organisms for the Development of a Fermentation Industry*

A culture used in a fermentation process is often genetically mixed or inherently unstable. Thus it should be preserved by a procedure that imposes minimal stress during storage, in any of the steps taken to prepare it for storage, or in any step taken to recover it from storage. Although the ability to generate typical productive fermentations is the prime requisite, retention of viability is the best indication that no undue stress has been imposed.

Hesseltine and Haynes have developed an appreciation for the general as well as the special problems of managing a collection of industrially useful microorganisms through years of association with a collection recognized as a public repository by the U.S. Patent Office, and through long involvement in the fermentation research activities of a government laboratory with close ties to industry.

Paper 3 is included only in part; pages 7–23, which contain most of a table of names and addresses of culture collections, have been eliminated to conserve space.

SOURCES AND MANAGEMENT OF MICRO-ORGANISMS FOR THE DEVELOPMENT OF A FERMENTATION INDUSTRY

C. W. Hesseltine and W. C. Haynes

1. INTRODUCTION

The micro-organism used in a fermentation is the key to the success or failure of the process. It is the catalyst that makes the fermentation work. A microbial culture must have certain general attributes if the process it generates is to be operable, regardless of the nature of the product and the simplicity or complexity of the engineering process:

1. The strain must be genetically stable. A culture that constantly and spontaneously produces one or more different forms is undesirable.

2. The strain must readily produce many vegetative cells, spores, or other reproductive units. Since Basidiomycetes produce only mycelium they are rarely, if ever, used in industrial fermentation.

3. The strain should grow vigorously and rapidly after inoculation into seed tanks or other containers used to prepare large amounts of inoculum before an industrial fermentation.

4. The strain should be a pure culture, not only free of other microscopically visible micro-organisms, but also free of phages.

5. The strain should produce the required product within a short period of time, preferably in 3 days or less.

6. The strain should produce the desired product to the exclusion of all toxic substances. The desired product should be easily separated from all others.

7. The strain should be able to protect itself against contamination, if possible. Self-protection might take the form of lowering the pH, growing at high temperature, or rapidly elaborating a desirable microbial inhibitor.

8. The strain should be readily maintained for reasonably long periods of time.

9. The strain should be amenable to change by certain mutagens or group of mutagenetic agents. A mutation program may be conducted with the object of developing strains that give enhanced yields of the product.

10. The strain must give a predictable amount of desired product in a given fermentation time.

Micro-organisms that meet these conditions may be either isolated from nature or obtained from a culture collection. To isolate, purify, screen, and test a culture from nature requires trained microbiologists. Since, in developing nations, such trained people are often in shorter supply even than money, it seems to us that culture collections would be the best source of micro-organisms for setting up a fermentation industry. But plenty of time and money would still not guarantee success. To obtain the proper culture, sometimes one must isolate the micro-organism from a special, ecological niche that may not even exist in a particular country. For example, *Blakeslea trispora,* which produces large amounts of β-carotene, cannot be isolated in temperate regions of the United States, but rather one must seek wild strains in the tropics growing on flowers of certain higher plants. For such cultures, collections are almost always the only logical source.

Another source of cultures in the food industry, which should not be overlooked, is the micro-organisms selected through the centuries for preparing native fermented food products. The principal micro-organisms can be obtained with little difficulty. Since the micro-organisms have been used in a particular food fermentation for centuries, there has been a constant purposeful selection of the best strains. The yeast strains used in the municipal Bantu beer breweries of South Africa were acquired in this fashion. One of us (Dr. C. W. Hesseltine) was told the original strains were isolated from the better native brews. After a number of strains were tested, the best were chosen and are now the ones used in an industry producing 150 million imperial gallons of the beverage yearly.

In this paper, we have tried to be realistic in our approach to the problem of obtaining the proper micro-organisms for use in industrial fermentations. Our views are based upon first hand knowledge of the operation of a large industrial culture collection supported entirely by government funds; experience of several years operating a culture collection in a large industrial fermentation company; an understanding of the problems faced by fermentologists in developing countries; contact with microbiologists working in our fermentation laboratory from developing countries; and an acquaintance with some of the primitive food fermentations of the world.

2. SOURCES OF MICRO-ORGANISMS FOR INDUSTRY

The ultimate sources of culture of micro-organisms for industry are soil; water; fresh, fermenting, and rotting vegetables; living plants and animals; sewage; fresh and spoiled food; frass and insect droppings; and the like.

The immediate sources of cultures, however, are permanent culture collections. Almost all large industrial firms dealing in fermentations have their own collections of micro-organisms secured from a continuous program of isolation. New isolates and variant substrains derived from concurrent mutation studies swell the numbers of strains so that many of the proprietary collections are quite large. However, most of their micro-organisms never get into general circulation, being intended solely for exploitation by the parent company.

A few cultures from proprietary industrial collections are in general and private collections in the United States. In 1949, the U.S. Patent Office took the position that a culture is an essential part of a patent process and that the culture must be disclosed. Hence, it must be deposited in a recognized culture collection and be available to the public at the time the patent issues. Two U.S. collections — the American Type Culture Collection at Rockville, Md., and the ARS Culture Collection in Peoria, Ill. — are recognized as a result of this practice as official depositories for cultures from industrial concerns both domestic and foreign. As might be expected, the depositing companies do not advertise the fact that particular strains are placed in outside culture collections, and the named depositories agree not to reveal possession of patent cultures or to distribute them without authorization by the depositor, if this is his wish, until the U.S. patent issues.

The holdings of the companies are supplemented also by accessions from public and private culture collections whose culture distributions are not so rigidly controlled.

Private collections do not have as a principal purpose of existence the distribution of cultures. They usually are specialist collections, by which is meant their scope is confined to one or a few taxa of special interest to the scientists who operate or control them. Generally, private collections are associated with a university or research institute. Although their curators decline to distribute cultures far and wide to anyone who asks, they nevertheless often send cultures to other investigators with like interests, or to research institutes and to industrial men who might continue research they no longer can pursue or who might continue development of an industrial process. Private collections generally do not charge fees for their cultures. Like proprietary collections, they usually do not publish or distribute lists of their cultures.

Public collections have as one of their principal reasons for existence the accumulation of a diverse collection of salable micro-organisms. They send cultures to any bonafide investigator anywhere in the world who is willing to pay their price. As might be expected, they publish catalogs listing the micro-organisms that are for sale. They also often provide other services such as identification of micro-organisms and preservation of cultures by lyophilization or liquid nitrogen refrigeration. Their diversity may be as wide as that of the American Type Culture Collection, which maintains actinomycetes, algae, bacteria, cell lines, molds, protozoa, viruses, and yeasts.

Among the specialized culture collections, some concentrate on industrially useful micro-organisms. Such micro-organisms are bacteria, yeasts, molds, actinomycetes, algae, and protozoa that are used in the food, pharmaceutical, and fermentation industries and in research and development laboratories to convert selected substrates to products of enhanced nutritional, medicinal, or industrial value or to reduce the biochemical oxygen demand (BOD) in

sewage and industrial effluents. Such collections are of principal interest to the United Nations Industrial Development Organization (UNIDO) and its adherent groups and members.

We concluded that a list of such collections giving addresses, names of curators, and types of micro-organisms contained would be useful (Table 1). We are indebted to Dr. S. M. Martin of the Division of Biosciences, National Research Council, Ottawa 7, Ontario, Canada, for most of the names of collections and information about them. Under the aegis of the World Federation of Culture Collections (WFCC, formerly the Section on Culture Collections) of the International Association of Microbiological Societies, Dr. Martin is preparing and will soon publish a World Directory of Culture Collections and List of Species in which most of the collections in the world will be named and described.

Names and addresses of additional collections may be found in some of the larger culture collection catalogs listed at the end of this paper.[2, 3, 4]

Fees for cultures vary from one collection to another. In the United States the American Type Culture Collection charges $30 per strain for all cultures to profit-making institutions. The cost is reduced to $20 for non-profit institutions (except for some special teaching strains which are $10). The Centraalbureau voor Schimmelcultures in The Netherlands charges 40 guilders for cultures that are to be used for industrial purposes. There is a reduction in cost if 10 strains or more are purchased in 1 year. This collection, like some others, does not guarantee the production of chemical substances by its cultures.

As a general rule, collections which advertise their cultures in printed catalogs charge a fee for their strains. Some collections, such as the one with which we are associated, do not issue a catalog, do not charge a fee but do exert considerable restraint on the number of strains sent at any one time to any individual or institution.

[*Editor's Note:* Table 1 has been omitted because of lack of space.]

3. CHARACTERISTICS OF A GOOD CULTURE COLLECTION

Although much has been spoken and written about culture collections, to our knowledge no one has ever laid down the characteristics of a good applied or industrial culture collection. Many of the following points apply equally well to other types of collections:

1. The collection must be part of, or closely related to, a fermentation research laboratory or to a fermentation plant, or both. For example at the Northern Regional Research Laboratory the ARS Culture Collection is part of one of four research units in the Fermentation Laboratory.

Interactions between fermentologists and culture collection staff work to the mutual benefit of both. The fermentologists, being aware of general trends in fermentation research, are able to anticipate future areas of interest and to give guidance as to what micro-organisms the culture collection should accession to meet future needs. The microbiologists, with their knowledge of the relationships among genera and the physiological requirements of various micro-organisms, can make valuable suggestions regarding screening programs. In their detailed studies on individual strains, they may make observations leading to new fermentation products or higher yields of known products. Their ready recognition of contamination or degeneration of the culture being used in the development of a process acts as a type of quality control.

2. A culture collection must be well funded, and this funding must be at a relatively uniform level each year. In many other research operations, a program may be increased or decreased readily with changes in the amount of budgeted money. Personnel can be shifted easily from one project to another. On the other hand, a culture collection is a continuing operation, which must be sustained without great fluctuations in budget or people from year to year. Many of the culture collection projects become long-term studies of a genus or family, and years are required to assemble wild cultures and known type materials in order to do a first-rate job.

3. A culture collection must have adequate facilities and equipment, including transfer rooms, refrigerator space, incubators, microscopic and photographic equipment, autoclaves, and lyophilizers. Usually these facilities should be separate from those of other research groups.

4. Library facilities are necessary so that personnel may have access to the taxonomic and fermentative literature being published, not only in the region or country of location, but also in the world.

5. The collection should have an active and continuous program of isolating new strains of micro-organisms from nature. This goal will lead to the discovery of new products and reactions. New material will also add to the understanding of the classification of special groups of micro-organisms. New material makes it possible to discover species and genera new to science.

6. The collection must have an adequate staff to support the curators. By this requirement we mean technical help to prepare media, sterilize glassware, and perform routine techniques; secretarial help to keep the voluminous records and to handle correspondence; and shops to construct special apparatus. At the Northern Laboratory, our glassblower devised an automatic machine to make lyophil tubes. Reliable sources of supplies are also necessary. We have had experience in setting up a lyophil apparatus in a developing nation. Although the lyophil equipment was readily made to our specifications, not a culture could be processed for 2 years because there was no dry ice.

Optimally, each curator should have a careful, intelligent, and dedicated assistant with some microbiological training. In our experience, technicians need not be specialized because they always must be trained in the special techniques required for the collection. These assistants should handle periodic transfers, lyophilization and associated records, inoculation of cultures for study by the curator, seeding of flask cultures for preliminary surveys for new products, and making and recording routine observations on all cultures.

7. The curator(s) must do research, as well as maintain the collection. Each must have an active research program either in taxonomy or genetics

with preference to the former. Thus a curator will have an intimate knowledge of the strains he is maintaining and will develop a reputation as an expert in his field. Consequently, important material will be sent to him for safekeeping, for identification, and for other purposes. Other microbiologists will know from whom they may get expert advice, cultures, and information. This point is an important one that we have stressed before.[7,8,9]

8. College-trained members of a culture collection staff must be aware of the field of applied microbiology, appreciate the work being done in fermentation research and development, and understand the operation of fermentation plants. They must comprehend the problems of geneticists, fermentologists, engineers, biochemists, and organic chemists. Probably the most difficult job from an administrator's standpoint is indoctrinating the curators of a culture collection. They must understand fully the point of view of other scientists and must realize that they are part of a team. They must be made aware of the needs of other research people. Anyone in a culture collection who does not appreciate other areas of work should never be in the position of decision making.

In turn, the members of a culture collection should be informed of developments in associated fermentation research areas and of problems in a fermentation plant. Currently, all reports and papers from our Fermentation Laboratory are circulated to all the other senior scientists. Also, before papers and reports are given at scientific meetings the authors present them before other members of the Fermentation Laboratory for review and criticism. By this means errors are detected, speaking time is adjusted, and lastly, the staff is kept informed of progress in related areas.

9. Although the training of curators should be in taxonomy, the overall background of the staff should have balance. If the collection has more than one senior man, then the broader the interests of the group the better. It does no good to have three specialists on bacteria and yet have no mycologists, or vice versa.

In the ARS Culture Collection, we currently have a zymologist, two bacteriologists, two mycologists, a plant pathologist, and a biochemist. Although a geneticist is not part of the Collection, a microbial geneticist works closely with the Collection members.

10. At least in larger collections, young people with new ideas and knowledge of new techniques should be brought into the group periodically. This means of rejuvenation may be supplemented with postdoctoral fellows and exchange of personnel from other institutions. They should not necessarily be people from other collections. In turn, the resident staff needs periodically to travel or study in other laboratories.

11. Members of a culture collection should be like playmakers in a basketball game. They should spot profitable new ideas of fermentation research. They should be the originators of new processes and products. On the other hand, once a profitable research area has been discovered, they should become advisors to other groups who are responsible for developmental research. They should not do developmental research beyond this point.

12. Culture collection people should not only be engaged in research, but they should be actively reporting their research in the form of papers published in scientific journals, giving lectures, and occasionally taking out initial patents.

4. LOCATION OF CULTURE COLLECTIONS

The questions of how many kinds of culture collections (general, medical, reference, or agricultural) should be sponsored and where they should be located are under study by the WFCC. Our concern, which overlaps theirs, is the narrower one of where collections of industrial micro-organisms should be situated so that they will be accessible to, and do the most good for, people in developing nations. Should each nation have its own collection? Will the existing ones suffice? We think the answer to both questions is, 'No'.

It seems unrealistic to support the idea of national collections of any sort because scientists trained in culture collection science in developing countries are scarce. It seems to us that the source of cultures of micro-organisms should be limited to a few well-equipped collections located at various places around the world. They need to be adequately staffed and financially supported on a long-term basis. A culture collection in each country would be wholly unrealistic and unworkable. Money spread over so many places would be utterly wasted. On the other hand, public collections that distribute industrial micro-organisms are rare and often are located too far from the emerging nations to furnish the expert assistance that is needed in handling the cultures. Also, the cost of the cultures is prohibitive because hard currency is difficult to come by in many of the developing areas. These are our conclusions based upon practical experience and on frank discussions with a number of knowledgeable people from various countries.

We think it would be well if regional collections could be set up in strategic places where governmental stability would allow proper development of a collection and where political considerations would not restrict the free flow of cultures and information to fermentologists in the service areas. Certainly the existing ones must suffice for now, and in certain regions they can provide satisfactory service. Thus, culture collections in existence in the United States, U.S.S.R., United Kingdom, Netherlands, Japan, Canada, and South Africa serve many areas. The excellent Dutch collection would, and does, fill the needs of Central Europe. The U.S.S.R. All-Union collection can adequately meet the requirements of the U.S.S.R., Poland, Romania, and Bulgaria. Well-staffed and financed collections need to be established in: South America (perhaps in Brazil or Argentina), India, Central Africa, and the Middle East and North Africa. We do not mean to imply that none exist in these regions but rather to suggest that they need to be enlarged in size, better equipped, and more adequately staffed. We believe that if these proposed collections were established, the fermentation industry in these areas would be adequately backstopped with sources of cultures, culture information, and technical expertise.

Existing collections, which have gained stature over the years, are associated with either a research institute or with a university that is famous for its fermentation studies. For example, the University of Tokyo collection is housed in the Department of Agricultural Chemistry and Applied Microbiology. The famous Dutch collection of yeasts is housed with the Institute of Microbiology at Delft. Government or organizations establishing new collections should bear this kind of location in mind. A culture collection of fermentation micro-organisms not placed in close proximity with an active institution of fermentation and applied microbiological research would be like planting a seed on a rock.

The question can then be justly asked, 'What should one do with a small plant producing a given product, say a fermented food destined for human consumption?' In this instance, the developmental work should be done in some central research laboratory. To ensure reliability, inoculum should be prepared and supplied in a dry, stable form which the workers in the plant can use to seed the fermentation to the degree that the process will go to completion in spite of contaminants. For example, in the Bantu beer process, even the larger plants do not keep cultures or prepare inoculum, instead it is supplied to them in 1-pound packages which a technician uses to inoculate a given quantity of media in full confidence that he can depend on obtaining a certain type of product at a certain time. Little or no formal microbiology needs to be known by the plant operator. The original starter culture can be kept in a central culture collection and supplied to a company who makes, packages, and distributes the inoculum.

5. PROCEDURES FOR ISOLATION AND SELECTION OF MICRO-ORGANISMS FROM NATURE

Innumerable techniques for isolation of micro-organisms are described in the literature. No attempt will be made here to give specific details because they vary from group to group and sometimes even a special technique is required for a single species. Information can usually be found in textbooks on microbiology or taxonomic monographs. Currently, the Mycological Society of America has a project involving the preparation of a manual on methods for the isolation and study of all the groups of fungi. An expert on each fungus family or genus has prepared a section on how to isolate and to study his group. The material was written more than 2 years ago and now is being evaluated by graduate students to discover which parts are workable and which must be revised. The entire text will be edited to give a uniform style. Eventually, the whole will be published as a source book of information.

Techniques for isolating micro-organisms, which can grow free of other living things, can be classified into several general categories. Micro-organisms are here interpreted to include bacteria, fungi, algae, and protozoa. One of the oldest techniques is culture enrichment. It involves the transfer of soil, sewage, or some other material with a large and diverse population of micro-organisms into a selective medium followed by incubation of the culture under such conditions of temperature, aeration, and pH, among others, that the growth of desired forms is favored. A small amount from the initial culture is transferred to a second that is set up in the same manner as the first, and this procedure is continued until the desired flora predominates and can be isolated in pure culture.

If a strain of *Clostridium* that would produce acetone on corn meal is wanted, a sterile corn mash would be inoculated with soil, sewage, and other material containing large bacterial populations. The corn mash would be kept under anaerobic conditions at a desirable fermentation temperature, for example, 37°C. A large number of flasks would be started. Certain ones that gave the appearance of vigorous anaerobic growth and that had a solvent odor would be used to inoculate new flasks. These, in turn, would be incubated until a culture of exceptional solvent-producing ability emerged. Finally, the selected *Clostridium* would be plated out on a corn meal medium under anaerobic conditions, and many strains would be isolated as pure cultures. Each would then be tested in the proposed industrial fermentation for which corn meal is the basic ingredient. Assays of the solvent yields would be determined, and the best strain selected for evaluation in the scale up of the fermentation.

Sometimes, the isolation of pure cultures is not even necessary. For example, in the fermentation of cucumbers for the manufacture of pickles, conditions are established in the fermentation tanks such that part of the natural microbial flora on the cucumbers is favored. Certain bacteria grow at the practical exclusion of all other micro-organisms and the fermentation goes to completion without resort to a pure culture. The efficacy of pure culture starters in pickle making is being studied.

In the native African fermentation of corn, called magou, conditions of anaerobiosis and temperature are so adjusted that a high-temperature lactic acid fermentation occurs without the use of pure culture starter. Part of a previous batch is used as starter for each new batch.

Another general approach is to isolate a microbial strain from the natural flora of a choice sample of material and to use it to make a uniformly good product. For example, in the United States for the past twelve years a culture of *Pediococcus* has been used to inoculate commercial sausage. According to information supplied by the manufacturer, a search was made of cultures for samples of superior quality sausage. From these sausages, a strain was selected that produced the desired fermentation; that also could be grown under conventional fermentation conditions and then be preserved by freeze drying without undue loss of vitality; and that could initiate growth rapidly in fresh sausage in spite of competing bacteria.

However, these various approaches cannot always be used because we do not know where to look for suitable strains or because the actual nature of the product desired is not known. The search for antibiotics is an example. In this instance, initial searches indicated that *Streptomyces* had excellent possibilities. From earlier studies on soil microbiology, it was known that soil, particularly grassland soil, had a great number and variety of species. The method used in this research was to get many different samples of soil from various geographical and ecological areas. The soils were plated out on media suitable for the growth of *Streptomyces* and bacteria, and myriads of strains of actinomycetes were isolated. Selection of colonies for making pure cultures was often influenced by observation of their inhibition of adjacent bacterial or mold colonies. A multitude of strains so selected were then tested on plates against the target pathogen or, more often, against a harmless micro-organism known to be closely related to the pathogen.

Ultimately, the antibiotic from a particular culture had to be tested against strains of the pathogen either *in vitro* or *in vivo*. As an aside, a search of this sort for new antibiotics now appears to be a useless activity except by possibly the most highly skilled and experienced researchers in industrial laboratories. A better approach is the modification of known antibiotic compounds.

The plating technique can be used to isolate single cells or spores of practically all micro-organisms that grow in laboratory media. This technique involves the dilution and separation of propagules of the micro-organism. These then grow into colonies of sufficient size to be seen with the naked eye or under a dissecting microscope on, or in, the agar medium. Using aseptic techniques, some or all of a colony can be picked off the substrate and a pure culture established. However, with many fungi, a more rapid and efficient technique is the isolation of a few spores from one fructification.

Routinely in our laboratory, mold growth that has fruiting heads on the substrate (this growth may be on medium in Petri dishes) is used to start pure cultures. This technique can readily be done by placing the material on which the mold is growing under a dissecting microscope and selecting a well-isolated fruiting head containing mature spores. A transfer needle with a fine straight wire (a filament from an electric light bulb mounted in a holder works very well) is flame sterilized, cooled, and moistened in sterile agar; the tip is carefully brought into contact with the fruiting head. The adhering spores may then be transferred to suitable nutrient agar and a pure culture established in a matter of a few seconds. This technique requires considerable hand dexterity and practice. It works well for all fungi that produce spores on stalks. We use it routinely for the isolation of Mucorales, *Aspergillus, Penicillium,* and the Fungi Imperfecti — including genera such as *Alternaria, Cladosporium,* and *Gliocladium.*

A modification of this technique is the use of a micromanipulator to isolate single spores from the surface of agar. It is particularly useful when all the ascospores in an ascus or the basidiospores on a basidium are to be isolated for genetic studies. In some yeasts and fungi, spores are forcibly discharged and these can be isolated from an agar surface placed above growing colonies.

For some fungi, especially those of the class Basidiomycetes, spores are only produced in, or on, large macroscopic fruiting structures. Basidiospores may be allowed to discharge on nonnutrient agar, then a few are transferred to nutrient agar, and cultures become established. However, one often encounters basidiospores that fail to germinate. Frequently cultures can be made, if the fruiting body is large enough, by carefully dissecting a small fragment of tissue from inside the sterile fruiting body. The tissue is transferred to an appropriate medium. Since colonies will produce mycelium but no fruiting structure, extreme care must be taken to ensure that one is not isolating mold contaminants. There is no way of positively identifying a culture that produces sterile mycelium.

One last technique often combined with one or more of the general methods described is the use of specific inhibitors in either liquid or solid medium to eliminate other groups of micro-organisms. Routinely, in making yeast and mold counts of cereal products, tetracycline is incorporated into the medium to inhibit almost all bacteria. It has no adverse effect on the growth of molds or yeasts. On the other hand, a second antifungi antibiotic, actidione, can be incorporated into nutrient media to inhibit both yeasts and fungi without affecting the growth of bacteria. We are not aware of any combination of materials that will inhibit all bacteria and fungi but will permit the exclusive growth of actinomycetes.

6. CLASSIFICATION OF MICRO–ORGANISMS USED FOR PRODUCTION OF FERMENTATION PRODUCTS

Since this topic is reviewed in great detail in various texts on fermentation,[11,12,13] no attempt will be made to discuss the micro-organisms involved except to summarize the information in tabular form (Table 2). Even this summary cannot be complete because some fermentation products were, or are, made on a limited custom basis and are not regular articles of commerce. When one considers all the types of fermented food made all over the world, some of which are quite local, it becomes quite impossible to enumerate even a small portion of them. Also, some products once made by fermentation are now made by chemical synthesis especially from petrochemicals. Thus, ethanol for industrial uses is made exclusively via this method in the United States. Other countries with a shortage of oil, but with enormous amounts of molasses, still make ethanol by fermentation.

TABLE 2. ALPHABETICAL LISTING OF COMMERCIAL PRODUCTS
PRODUCED BY MICRO-ORGANISMS

Product	Genus	Type of micro-organism
AMINO ACIDS		
Aspartic acid	*Pseudomonas*	Bacteria
Glutamic acid (monosodium glutamate)	*Bacillus*	Bacteria
	Brevibacterium	Bacteria
	Micrococcus	Bacteria
Isoleucine	*Pseudomonas*	Bacteria

TABLE 2 (Cont.)

Product	Genus	Type of micro-organism
Lysine	*Micrococcus*	Bacteria
Phenylalanine	*Micrococcus*	Bacteria
Threonine	*Bacillus*	Bacteria
Valine	*Micrococcus*	Bacteria
ANTIBIOTICS		
	Aspergillus	Mold
	Bacillus	Bacteria
	Cephalosporium	Mold
	Fusidium	Mold
	Micromonospora	Actinomycetes
	Nocardia	Actinomycetes
	Penicillium	Mold
	Streptomyces	Actinomycetes
BEVERAGES		
Beer	*Saccharomyces*	Yeast
Distilled spirits	*Saccharomyces*	Yeast
Sake	*Aspergillus*	Mold
	Saccharomyces	Yeast
Wine	*Saccharomyces*	Yeast
ENZYMES[6]		
α-Amylase (EC 3.2.1.1)*,†	*Aspergillus*	Mold
	Bacillus	Bacteria
	Endomycopsis	Yeast
	Rhizopus	Mold
(Amyloglucosidase see Glucoamylase)†		
Asparaginase (EC 3.5.1.1)	*Erwinia*	Bacteria
	Escherichia	Bacteria
Catalase (EC 1.11.1.6)	*Aspergillus*	Mold
	Penicillium	Mold
Cellulase (EC 3.2.1.4)	*Aspergillus*	Mold
	Myrothecium	Mold
	Trichoderma	Mold
Dextranase (EC 3.2.1.11)	*Penicillium*	Mold
β-Fructofuranosidase (EC 3.2.1.26)	*Saccharomyces*	Yeast
β-Galactosidase (EC 3.2.1.23)	*Saccharomyces*	Yeast
Glucoamylase (EC 3.2.1.3)	*Aspergillus*	Mold
	Endomycopsis	Yeast
	Rhizopus	Mold
(Glucose isomerase see Xylose isomerase)		
Glucose oxidase (EC 1.1.3.4)	*Aspergillus*	Mold
	Penicillium	Mold
α-Glucosidase (EC 3.2.1.20)	*Aspergillus*	Mold
β-Glucosidase (EC 3.2.1.21)	*Aspergillus*	Mold

TABLE 2 (Cont.)

Product	Genus	Type of micro-organism
(Hemicellulase see Xylanase)		
(Invertase see β-Fructofuranosidase)		
(Lactase see β-Galactosidase)		
(Laundry enzymes see Proteases, alkaline)		
Lipase (EC 3.1.1.3)	Aspergillus	Mold
	Candida	Yeast
	Mucor	Mold
(Milk-clotting enzymes see Rennin)		
(Pectinase see Polygalacturonase)		
Penicillinase (EC 3.5.2.6)	Bacillus	Bacteria
Penicillin amidase (EC 3.5.1.11)	Many micro-organisms	
(Pentosanase see Xylanase)		
Plasmin (EC 3.4.4.14)	Streptococcus	Bacteria
Polygalacturonase (EC 3.2.1.15)	Aspergillus	Mold
Proteases (EC 3.4.4)	Aspergillus	Mold
	Conidiobolus	Mold
	Mucor	Mold
	Streptomyces	Actinomycetes
Proteases, alkaline	Bacillus	Bacteria
Rennin (EC 3.4.4.3)	Endothia	Mold
	Mucor	Mold
(Streptokinase see Plasmin)		
Xylanase (EC 3.2.1.8)	Many micro-organisms	
Xylose isomerase (EC 5.3.1.5)	Streptomyces	Actinomycetes
FOODS		
Ang-kak	Monascus	Mold
Bacterial starters (fermented dairy products and sausage)	Lactobacillus	Bacteria
	Leuconostoc	Bacteria
	Pediococcus	Bacteria
	Propionibacterium	Bacteria
	Streptococcus	Bacteria
Bantu beer	Lactobacillus	Bacteria
	Saccharomyces	Yeast
Blue cheese flavor	Penicillium	Mold
Bread (bakers' yeast)	Saccharomyces	Yeast
Cheese and fermented dairy products	Lactobacillus	Bacteria
	Penicillium	Mold
	Propionibacterium	Bacteria
	Streptococcus	Bacteria
Chinese yeast	Chlamydomucor	Mold
	Hansenula	Yeast
	Rhizopus	Mold
	Saccharomyces	Yeast
Fermented fish	Aspergillus	Mold
	Unidentified	Halophilic bacteria
Hamanatto	Aspergillus	Mold
Koji	Aspergillus	Mold
	Rhizopus	Mold
Magou	Lactobacillus	Bacteria

TABLE 2 (*Cont.*)

Product	Genus	Type of micro-organism
Miso	*Aspergillus*	Mold
	Saccharomyces	Yeast
Nata	*Acetobacter*	Bacteria
Natto	*Bacillus*	Bacteria
Ontjom	*Neurospora*	Mold
Pickles and sauerkraut	*Lactobacillus*	Bacteria
	Streptococcus	Bacteria
Shoyu	*Aspergillus*	Mold
	Saccharomyces	Yeast
	Torulopsis	Yeast
Sufu	*Actinomucor*	Mold
	Mucor	Mold
Tempeh	*Rhizopus*	Mold
Yeast	*Candida*	Yeast
	Saccharomyces	Yeast
INDUSTRIAL SOLVENTS		
Acetone	*Clostridium*	Bacteria
Butanol	*Clostridium*	Bacteria
2,3-Butanediol	*Aerobacter*	Bacteria
	Bacillus	Bacteria
Dihydroxy acetone	*Acetobacter*	Bacteria
Ethanol	*Clostridium*	Bacteria
	Saccharomyces	Yeast
Glycerol	*Saccharomyces*	Yeast
	Torulopsis	Yeast
MISCELLANEOUS		
Alkaloids	*Claviceps*	Mold
Bioinsecticides	*Bacillus*	Bacteria
Dextran	*Leuconostoc*	Bacteria
Gibberellin	*Gibberella (Fusarium)*	Mold
Inosinic acid	*Bacillus*	Bacteria
	Several genera	Yeast
Nucleotides	*Brevibacterium*	Bacteria
	Candida	Yeast
Rhizobium culture	*Rhizobium*	Bacteria
Silage	*Lactobacillus*	Bacteria
Sorbose	*Acetobacter*	Bacteria
Steroid transformations	*Aspergillus*	Mold
	Corynebacterium	Bacteria
	Curvularia	Mold
	Rhizopus	Mold
	Streptomyces	Actinomycetes
ORGANIC ACIDS		
Acetic	*Acetobacter*	Bacteria
Citric	*Aspergillus*	Mold
Erythorbic	*Penicillium*	Mold
Fumaric	*Rhizopus*	Mold
Gluconic	*Aspergillus*	Mold
Itaconic	*Aspergillus*	Mold
Itatartaric	*Aspergillus*	Mold

TABLE (*Cont.*)

Product	Genus	Type of micro-organism
2-Ketogluconic	*Pseudomonas*	Bacteria
5-Ketogluconic	*Acetobacter*	Bacteria
α-Ketoglutaric	*Pseudomonas*	Bacteria
Kojic	*Aspergillus*	Mold
Lactic	*Lactobacillus*	Bacteria
PROTEIN	*Chlorella*	Algae
	Saccharomyces	Yeast
	Torula	Yeast
VITAMINS		
Ascorbic acid (vitamin C) (in part)	*Acetobacter*	Bacteria
B_{12}	*Bacillus*	Bacteria
	Propionibacterium	Bacteria
	Streptomyces	Actinomycetes
β-Carotene	*Blakeslea*	Mold
Riboflavin	*Ashbya*	Yeast
	Eremothecium	Yeast

*. Enzymes are identified by their code numbers, the key to classification as assigned by the Commission on Enzymes, International Union of Biochemistry.[6]

†. Included in this alphabetical list of enzymes are those names familiar to microbiologists but no longer recommended for usage by the Enzyme Commission. Recommended and familiar names are cross-referenced.

7. PROBLEMS IN MAINTAINING STABLE INDUSTRIAL CULTURES

From years of experience in handling cultures of all kinds, we believe certain steps should be taken as soon as a micro-organism is isolated in pure culture to ensure that it remains in a stable state. These steps need to be taken as soon as the culture is brought into the laboratory or isolated, long before its potential or lack of potential is known.

1. The culture is examined under a dissecting microscope to determine (*a*) that the culture is growing uniformly; (*b*) that it is free of other micro-organisms; (*c*) that if it is a mold, mature spores are present; and (*d*) that the culture appears to be the genus and perhaps species isolated or named when received. This examination will then determine the next step to be taken.

2. If the culture appears to be pure, shows vigorous and uniform growth, and has mature spores, three to five ampules of the micro-organism should be lyophilized immediately.

3. If the strain is known to be a member of a species or genus in which lyophilization is always successful, no viability check is needed. However it is good practice, even with such strains, to sacrifice one lyophil tube and to dilute or streak the culture out onto a suitable growth medium. The check on viability will show three things: (*a*) If the propagules have survived in large numbers, (*b*) if the lyophil preparations are free of other micro-organisms, and (*c*) if the regenerated culture is still uniform in growth and sporulation. If the strain being preserved has not been lyophilized before, the viability check is a must because some micro-organisms fail to survive this process. In

case of failure, an alternative method must be found while the first generation culture is still available and in a healthy state.

4. If the culture is recovered successfully from lyophil or from a preparation preserved by another method, such as freezing in liquid nitrogen, records should be kept of the proper medium for growth and sporulation, as well as for any special requirements, such as temperature of incubation, length of incubation, or pH. For example, the mold *Blakeslea trispora* will sporulate rapidly (3 to 4 days at 25°C), but often within 10 days the spores will germinate in place. Lyophilization subsequent to this occurrence will be a complete failure. The process works beautifully if the spores are processed when they have just reached maturity.

5. The lyophil tubes should be stored at 4° to 10°C and, perhaps, checked for viability at the end of each 10 years of storage.

6. At the time of lyophilization, the culture should be examined in the appropriate way to determine its identity. Sometimes this examination may lead to species and variety identification, but other times it leads only to the approximate species and genus. The records should certainly show its approximate identity because (a) it allows the person reviving the culture, perhaps years later, to know what was preserved and (b) it makes the records more complete and, therefore, more useful. In some collections of fungi, a microphotograph is made of the fungus at the time of identification. This is an excellent type of record.

7. At the same time lyophil preparations are made, records should be completed showing the following items: (a) the name of the organism; (b) where obtained — whether it was isolated in a laboratory or received from another microbiologist and, if the latter, his name and address; (c) accession number assigned and any other designation given to it, such as a temporary number or other laboratory or collection number; (d) location and original source of the material (where was the organism found in nature); (e) special requirements — medium of maintenance, optimum temperature, and other conditions; (f) products or unique properties and approximate yields; (g) number of cultures made; and (h) references if strain is cited in a paper or patent. Rarely can all this information be assembled. With time, additional information will be needed for the record. The data can be placed on cards, which should be cross-indexed so that one can find a culture by number, source, product, and name. Some collections are being indexed for computer sorting. With a large collection, such as ours (35,000 + strains), to put this data now onto cards would be a Herculean task.

Records should be kept showing who uses a given strain in the laboratory and to whom it is sent in laboratories at other institutions. This information is useful if it should become necessary to obtain a replacement of a culture which died or degenerated, but which may still be in the original state in someone else's laboratory. Written and dated records are also very essential on industrial strains, which may become involved in patent and legal cases. For this reason, we ask anyone requesting cultures from our collection to put his request in writing. When the culture is sent, a letter to the requester is prepared by the appropriate curator as a matter of written record.

8. If the culture does not meet the requirements set forth in item 2, the following steps are taken: If the culture is pure but shows sectoring, then the nonuniform culture is lyophilized; but also, the various forms are isolated separately, and each type is lyophilized individually. Sometimes a heterogenous culture cannot be separated into its components. The philosophy of lyophilizing the sectoring culture is to try to preserve all the component parts because typically, at this stage, you do not know which part you actually may need later.

If the culture is impure, methods must be used to rid the culture of its contaminants either by dilution and picking an isolated colony or by picking one or a few spores from a fruiting head. Occasionally two or more organisms are associated (mixed culture) as, for instance, in koji starters, where it may be necessary to lyophilize the total starter as well as its components.

9. Frequently, two different methods of preservation should be used at the same time. For instance, we still carry some of our fungus cultures on agar slants with periodic transfers, as well as in lyophil. Oiled and soil cultures are other possibilities. The former consists of agar slant cultures covered with sterile mineral oil. In the latter, spores are placed in sterile soil or sand and allowed to dry. Details of techniques of preservation require too much space to describe them here.

10. If a culture that has degenerated is received, certain steps must be carried out to obtain a better one. Perhaps the ability to sporulate is deteriorating; often a series of dilution plates will produce some colonies that grow more vigorously or sporulate more heavily in a natural manner. In some cultures, especially molds, the isolation of spores from individual heads may lead to better cultures. In others, the fault may be the medium or the growth conditions. Many Aspergilli and Penicillia grow normally and vigorously on a synthetic medium. However even though some cultures grow and fruit on a synthetic medium, they will do much better if the medium contains organic nitrogen and growth factors in the form of malt or yeast extract. This response may represent a better form of nitrogen, or it may reflect a partial vitamin deficiency that has been overcome.

We believe that it is appropriate to list a number of principles which we consider important for the cultivation of micro-organisms in order to ensure vigorous, healthy, and stable starters:

1. For the maintenance of stock cultures a chemically undefined, but reproducible, stock medium is better than a synthetic one. A microorganism, as it occurs in nature, almost always is growing on an undefined substrate. A defined medium will more likely select a certain part of the genetic population. The result may well lower the yield of the desired product.

2. In general, a stock culture medium should be no more nutritionally rich than is required to perpetuate the culture without change. Thus, glucose (or other sugar) is customarily excluded, or if glucose is essential (as it is for lactic acid bacteria), a buffer is incorporated to control the pH. If the pH were allowed to drop, the longevity of cells might be endangered. Worse yet, the population imbalance mentioned in number 1 might reduce or destroy the usefulness of the culture. Appropriate media for use with a variety of

micro-organisms are given in a paper entitled 'Maintenance of cultures of industrially important micro-organisms'.[8]

3. Stock cultures are usually subjected to two different sets of conditions. First, they are encouraged to grow rapidly and vigorously for a relatively short time by incubating them at or near their optimum temperatures and, if they are aerobic, allowing them free access to air. Then they are induced to slow down metabolically by storing them for a comparatively long time in a refrigerator and sometimes also by limiting their access to air by stoppering test tubes and flasks and sealing petri dishes. Stoppering also hinders loss of moisture from the culture. These variations are not too much different from those they encounter in nature. Regardless, they seem not to harm the micro-organisms.

The pH of the medium is also important. Generally, bacteria are grown in neutral media; molds are grown in media that have a pH between 6 and 7; and yeasts, in the vicinity of pH 6.

4. When new cultures are started, inoculum is taken from a mature culture. It consists of a small amount of growth of yeasts or bacteria or, for molds, a few spores without mycelium.

8. PROBLEMS OF STRAIN DEGENERATION AND LOSS

In looking back on our experiences involving the loss of pure cultures, five causes come to mind: (1) contamination by other micro-organisms, (2) infestation by mites, (3) phage infestations, (4) natural selection and mutation, and (5) untrained staff.

8.1. CONTAMINATION BY OTHER MICRO-ORGANISMS

We have encountered many cultures reputedly pure but which carried a second micro-organism never separated from the original culture at the time of isolation. This situation is particularly true when colonies are picked from dilution plates in which an inhibitor was placed in the medium to control bacterial growth. Often colonies growing on the surface of the agar plates with tetracycline as a bacterial inhibitor appear to be well isolated and pure. When the colony is picked off however, bacterial cells that are dormant are removed with it and then when placed on a medium free of the inhibitor grow again. A bacterial culture producing a thin growth may be obscured by the more luxuriant growth of the mold.

A common cause of contamination is the storage of cotton-plugged agar slant cultures in refrigerators. Often in the summer the air is warm and moist. When the refrigerator door is opened, moist air enters and upon cooling condenses upon the labels and cotton plugs. Some Penicillia can grow on the moist cotton and the labels at 4° to 5°C in a matter of a few weeks. When they do, the conidia present on the tubes make pure culture transfers all but impossible. If sufficient time is allowed, mycelial growth of the Penicillia will penetrate the cotton plug where they sporulate, and conidia will then drop onto the surface of the agar and develop new colonies.

8.2. INFESTATION BY MITES

Certain species of mites feed on fungus spores. These mites are extremely small and can barely be seen with the naked eye. They occur in nature in decaying plant material and are worldwide in distribution. When an active program of isolating fungi from soil, humus, or moldy plant material is going on, these animals are often present as adults or eggs. If care is not taken, the mites will travel from the contaminated material into petri dishes and test tube cultures. They invariably carry various mold spores on them, and they appear to be attracted by the odor of certain mold species. Even though cotton plugs are a good barrier to mold spores, the fungus mites traverse the cotton plugs into pure cultures unless the cotton plugs are poisoned. Besides contaminating a culture, they will lay eggs that hatch in a few days and the young will migrate into new cultures. In a short time, hundreds of cultures are contaminated, and a whole collection may be lost. If infestation has not spread too far, the contaminated cultures may be destroyed, but many other cultures that appear to be pure will contain mites, and the contamination will reoccur.

Once mites are introduced from natural material or have been introduced from cultures deposited in the collection, they are difficult to control. The best solution is to prevent mites from becoming free in the laboratory by quarantining all suspected, contaminated material in a location away from stock cultures. The same precaution should be taken with respect to cultures received from outside the culture collection. The further precaution of poisoning all cotton plugs of stock cultures should be observed.

8.3. PHAGE INFESTATIONS

Some strains of bacteria and actinomycetes carry phage in one form or another. Generally, these are difficult to detect, and it is even harder to free the culture of them. At one time, yeasts and mold species were believed never to be infected with phage, but this belief is now known not to be true.

8.4. NATURAL SELECTION AND MUTATION

Changes in the genetic population in a culture will occur in all micro-organisms. It is our personal belief that some of these changes may be prevented by the use of more natural media. For example, in our collection of Mucorales no change appears to occur if the cultures are carried on a potato-dextrose-salts medium. Some of the Mucorales will develop sectoring and sterile growth if cultures are carried on a synthetic medium. Culture rundown frequently occurs in some of the species of *Penicillium* and *Aspergillus*. Once this process has gone to a certain phase, it is impossible to regain the original form. For example, some three culture lines of the type strain of *A. parasiticus* NRRL 502 are poor aflatoxin producers, but the same strain carried under different cultural conditions in two other laboratories over many years is still a good producer of the mycotoxin.

8.5. UNTRAINED STAFF

Probably as serious as any cause of culture failure is the handling of cultures by untrained persons. Often media is improperly sterilized by people who do not comprehend that some complex media require more heat than others and that the larger the amount of media, the more sterilization is needed. Sometimes inadequately trained people just do not know how to transfer cultures so as to avoid contamination. Often they transfer a large mass of spores and mycelium of a mold causing the whole work area in the transfer room to be filled with spores suspended in the air and on the table tops. Only a few spores attached to sterile agar on a transfer needle are needed to start new stock cultures. Some microbiologists do not recognize even the micro-organisms they are working with. One prominent microbiologist has estimated that from one-third to one-half of all the work published on bacterial physiology has been done with contaminated cultures or with the wrong species!

9. PHYSICAL CONDITIONS AFFECTING MICRO-ORGANISMS

The physical conditions that affect the growth and longevity of micro-organisms are the same as those that influence other forms of life; viz, pH, temperature, light, humidity, pressure, oxidation/reduction potential, surface tension, and radiations. In the context of this discussion, we are interested in the effect of these factors on the survival of microbes in culture collections and while the organisms are *in transit*.

In most modern culture collections, stock strains are carried as lyophilized (freeze-dried) cultures. Essentially to lyophilize a culture, microbial cells, spores, or, sometimes, portions of mycelium are suspended in a protective colloid — such as blood serum or skim milk, quickly frozen at about −40°C, and dehydrated by allowing sublimation of moisture *in vacuo*. The dried preparation is sealed under vacuum and stored, usually in a refrigerator at 5° to 10°. In the lyophilized state, microbes take on some of the properties of bacterial endospores becoming less susceptible to extremes of temperature, dryness, and radiations. They are safe from contamination, changes in pressure, pH, humidity, oxidation/reduction potential, and surface tension. They can be shipped by land, sea, and air in temperate, tropical, or frigid climes without loss of life or change in character. (See reference[5] for a review of methods.)

Some micro-organisms cannot withstand lyophilization and must be maintained by other less convenient means. One that has come into use in recent years, and that has some of the advantages of lyophilization, is preservation by freezing and storage in and over liquid nitrogen (−165° to

−195°C). The full range of microbial types that can be preserved in this manner is not yet known, but many fastidious forms that fail to survive lyophilization have remained viable for long periods in liquid nitrogen refrigerators. For instance, some fungus cultures are reported still viable and apparently unchanged after 5 years in a liquid-nitrogen refrigerator.[10] Ultra-low temperature frozen cultures are sealed in glass vials or ampules so they have essentially the same protection as lyophilized preparations against contamination and changes in the physical environment. However, they must be shipped in special trucks, freight cars, or in portable liquid nitrogen refrigerators because it is only while they are kept at −165° to 195° that they are guarded against damage. Although the method is less convenient than lyophilization, it is becoming ever more common because it is still better than alternative methods.

Use of alternative methods is still inescapable for microbial cells that cannot be preserved by either of the two techniques already discussed. These alternatives have been in use for many years although more time consuming and subject to hazards that are minimal or absent in the other two methods. Basically, there is a single technique but with modifications. It is the serial transfer method by which some growth (vegetative cells, spores, mycelium, tissue) is transferred from one culture (agar slant, agar stab, agar plate, broth, tissue culture) to fresh medium, allowing the new culture to grow under optimum conditions to maturity, storing the new stock culture for a time, and then repeating the cycle.

Storage usually is in a refrigerator (5° to 10°C) but sometimes is at room or some other temperature. The length of time between transfers varies, depending upon the nature of the strain, from one or a few days to several months or even years. Often the interval may be lengthened by preventing dehydration of stock cultures by covering them with mineral oil (oiled cultures) or by closing the cultures with rubber stoppers, corks, or by impregnating the cotton plugs with paraffin. The rate of growth is slowed by refrigerating the cultures, a step minimizing changes in pH, and oxidation/reduction potential and reducing the danger that one or more cells mutated by stray radiations (cosmic rays) will gain predominance in the population.

These are the principal methods of maintaining and preserving cultures, and they all succeed to some degree in minimizing damage to, or loss of, life of cultures by inimical physical conditions.

10. REGULATIONS REGARDING DEPOSIT OF CULTURES FOR PATENT PURPOSES

One activity in which many of the larger culture collections become involved is the handling of cultures of micro-organisms deposited in connection with patent applications. In some countries it is desirable to deposit a culture, not necessarily a high producer of a product, with a recognized culture collection. This deposit is to ensure that a process being patented is fully disclosed to the public. In other words, a fermentation process is not considered fully operable until a culture is available for use in the process.

Over the years, we, at the Northern Regional Research Laboratory, have developed guidelines regarding this culture collection activity. They are based on considerable experience and also on consultations with inventors, companies, and patent lawyers. These guidelines are updated from time to time and are not to be construed as being final. The latest revision was on September 1, 1969.

11. PROCEDURES AND POLICIES FOR DEPOSITION OF CULTURES FOR PATENT PURPOSES IN THE ARS CULTURE COLLECTION

The ARS Culture Collection serves as a depository for cultures that are involved in fermentation patents and, therefore, will be glad to receive such a culture in connection with a patent application. When such a culture is received, it is assigned a number in the collection and is maintained thereafter in a living state. Immediately after receipt, a letter is written to the depositor advising of the number assigned and including the following statement:

> Furthermore, insofar as is practicable in carrying out the business of the Department of Agriculture, we shall refrain from distributing this culture pending the issuance of the United States patent to your Company, with the exception, however, that access to this culture by other parties will be granted upon receipt of written authorization from your Company specifying the name and the ARS Culture Collection designation (NRRL number) of the culture and identifying the party who is to receive it.

More recently some depositors have requested replacement of the paragraph above by a simple statement such as:

> As of this date, the subject culture(s) will be made available to anyone who requests the same.

It is suggested that you seek advice from your attorney as to which type of statement you should use. Either one of these statements will be written depending upon your wishes. The ARS Culture Collection letter then can be attached to the patent application for the Patent Examiner.

Curators in the ARS Culture Collection do not attempt to make an identification or to name any organism that has been deposited in connection with a patent application, nor do they carry out research work with such deposits until a U.S. patent issues. It is not necessary, of course, to provide a precise identification, but the microbiologist concerned should at least state to what genus the micro-organism belongs. Also, if special media are required for its maintenance, the curators need to know this. Ordinarily, one or two agar slant cultures, one lyophilized preparation, or both are received from depositors. Depositors also are responsible for resupplying material should the need ever arise.

The depositor has the option of sending cultures for deposit in the ARS Culture Collection in three ways:

1. Thirty lyophilized preparations, clearly labeled with the depositor's original strain designation and preferably in tubes no longer than 2 inches. One of these is checked for viability, the NRRL designation is placed on each tube, and the supply of tubes is stored at 3° to 5°C. *Bona fide* letter requests for the culture would be shipped from this stock.

2. One lyophilized preparation, clearly labeled with the depositor's original strain designation. On receipt, the micro-organism is cultivated on appropriate agar media and 30 lyophilized preparations made. One of these is checked for viability, the remainder handled as in option 1.

3. One, or preferably two, agar slant cultures of the micro-organism growing on an appropriate medium. Sufficient material is prepared by our curators to make 30 lyophilized preparations; one is checked for viability and the remainder are handled as in options 1 and 2. When the initial agar slant cultures deposited appear suitable, lyophilizations often are made from that material.

There is no charge for the deposit or maintenance of cultures.

Cultures deposited in connection with patent applications may be obtained, free of charge, by letter request stating the name of the micro-organism and the ARS Culture Collection strain designation (NRRL number).

The ARS Culture Collection does not issue a catalog or list. It has no regulations imposing restrictions on the use of such cultures deposited for patent purposes. Such materials are distributed according to the depositor's wishes which, in turn, generally are based on his interpretation of Patent Office requirements. Use of such materials, once distributed, are the responsibility of the requestor. Cultures are automatically removed from any restrictive category, once a U.S. patent issues wherein the particular micro-organism is involved.

12. SHIPMENT OF MICRO-ORGANISMS

Living cultures of micro-organisms are items of international commerce. Every year many thousands of strains are transported by land, sea, and air to scientists on every continent, with the possible exception of Antarctica. Most cultures are dispatched from large culture collections, such as the American Type Culture Collection in the United States, the various national collections in England, and the Institute for Fermentation in Japan. Many strains are also distributed by small, specialized collections and by individual scientists who maintain a few micro-organisms, primarily for their own research. In the course of time the large collections have learned to solve the problems associated with packaging and shipping living cultures so that they arrive at their destinations whole, alive, and unchanged.

Except for micro-organisms used in the pharmaceutical industry to produce vaccines and antisera, very few of the microbes used in the food, feed, and fermentation industries and carried in the mails endanger public health or agriculture. In the United States, and presumably in other nations, nonpathogenic cultures are virtually immune from legal restrictions on their movement from laboratory to laboratory. However in the United States, and very likely in other countries, a number of laws apply to the import, export and internal transport of 'etiological agents and vectors'. These are discussed in a brochure published in 1970[1] by the American Type Culture Collection.

Six Departments of the U.S. Government — Agriculture; Commerce; Health, Education, and Welfare; Transportation; Treasury; and the U.S. Postal Service are listed as those concerned with regulating potentially dangerous micro-organisms.

In the U.S. Department of Agriculture the Animal Plant Health and Inspection Services' (APHIS), Animal Health Programs are charged with responsibility to see that 'no organisms (which may introduce or disseminate any contagious or infectious diseases of animals, including poultry) ... shall be imported into the United States or transported from one State ... to another State ... without a permit issued by the Secretary and in compliance with the terms thereof'. APHIS' Quarantine Inspection Programs require a permit for the movement of any plant pest into or through the United States or any of its territories and possessions. 'Plant pest', as defined in the Federal Plant Pest Act, includes microbial cultures which can directly or indirectly injure or cause disease or damage in plants.

The primary concerns of the Department of Commerce, as it relates to distribution of living cultures, are safeguarding our national security and furthering our foreign policies. It has published a 'general license' that authorizes the export of living cultures to most destinations. Special 'validated' licenses are required for the export of living cultures to a few nations.

The Public Health Service of the U.S. Department of Health, Education and Welfare, through its Foreign Quarantine Program (Center for Disease Control), promulgates and enforces regulations to prevent the introduction and spread of communicable disease from foreign countries into the United States ..., or from one state to another'. It operates under a law which states, 'A person shall not import into any place under the control of the United States, nor distribute after importation, any etiologic agent ... of human disease ... unless accompanied by a permit issued by the Surgeon General'.

Shipments of living cultures are free of customs duty. The principal reasons that the Bureau of Customs of the United States Department of the Treasury is involved are:
1. To determine from documentation and inspection at ports and airports of entry if shipments are controlled by Federal law and regulations enforced by other Government agencies.
2. To select required samples for referral to enforcement agencies.
3. To withhold release of shipments during examination of referred samples and pending issuance of permits or licenses that are required as a condition of release.

Regulations of the U.S. Postal Service and the Department of Transportation specify how pathogenic micro-organisms shall be packaged. The intent of these instructions is the same as those governing permits and licenses; that is, to ensure that etiological agents do not escape and endanger public health and agriculture.

The reasons we have detailed the United States agencies and their requirements are that we suspect other nations either have similar regulations or will ultimately pattern theirs after those of the United States. It behooves culture-collection curators to be cognizant of such laws, if they wish to avoid difficulty and delay in receiving and sending cultures.

In perusing culture catalogs from several of the larger collections, we found only three that make any mention of possible need for licenses and permits. One is the Catalogue of Strains of the American Type Culture Collection.[2] Another is the Catalogue of the Culture Collection of the Commonwealth Mycological Institute.[4] The third is the List of Cultures of the Centraalbureau voor Schimmelcultures.[3]

Inasmuch as no mention is made in catalogs of other collections (Argentina, Czechoslovakia, England, Germany, India, Indonesia, Japan, Netherlands, Scotland, U.S.S.R.) about the need for licenses, permits, or customs arrangements, we suspect that either such requirements in countries other than the United States and Canada do not exist or else enforcement is ineffective. If true, curators need concern themselves only with safe packaging and labeling of cultures destined for most nations.

Problems to be overcome in packaging are the selection of a sturdy mailing tube or carton that will remain intact, despite rough handling and possible exposure to moisture, and that will protect the enclosed cultures from breakage. Additionally, the container must be so made that, if despite all precautions breakage does occur, the released micro-organisms cannot escape to the outside.

Although the requirements delineated may seem formidable to the uninitiated, curators ordinarily have little or no trouble in obtaining cultures from anywhere in the world or in sending cultures to anyone who has a legitimate need for them.

This review was presented at the United Nations Organization Expert Working Group Meeting on the Manufacture of Chemicals by Fermentation, December 1–7, 1969, Vienna, Austria.

Trade names are used in this publication solely for the purpose of providing specific information. Mention of a commercial product or company does not constitute a guarantee or warranty of the product by the U.S. Department of Agriculture or an endorsement by the Department over other products not mentioned.

LITERATURE CITED

1 ALEXANDER, M. T., and BRANDON, B. A. (1970) *The packaging and shipping of living reference cultures.* ATCC Pub. No. 1, 26 pp., American Type Culture Collection, Rockville, Md. 20852.
2 AMERICAN TYPE CULTURE COLLECTION. (1972) *Catalogue of strains.* Ed. 10, 312 pp. Rockville, Md. 20852.

3. CENTRAALBUREAU VOOR SCHIMMELCULTURES. (1968) *List of cultures.* Ed. 27, 262 pp. Baarn, Netherlands.
4. COMMONWEALTH MYCOLOGICAL INSTITUTE (1968) *Catalogue of the culture collection of the commonwealth mycological institute.* Ed. 5, 162 pp. Kew, Surrey, England.
5. FENNELL, D. I. (1960) *Bot. Rev.* 26, 79.
6. FLORKIN, M., and STOTZ, E. H., eds. (1965) *Enzyme nomenclature.* Ed. 2, Vol. 13. In *Comprehensive Biochemistry*, New York.
7. HAYNES, W. C. (1963) *Discussion ii. (On the organization of a type culture collection* by Shewan, J. M., Torry Research Station, Aberdeen, Scotland.) In Martin, S. M. (ed.), *Culture Collections: Perspectives and Problems. Proc. Specialists' Conf. on Culture Collections,* August 1962, p. 36. Ottawa. University of Toronto Press, Toronto, Canada.
8. HAYNES, W. C., WICKERHAM, L. J., and HESSELTINE, C. W. (1955) *Appl. Microbiol.* 3, 361-368.
9. HESSELTINE, C. W., HAYNES, W. C., WICKERHAM, L. J., and ELLIS, J. J. (1970) *History, policy, and significance of the Ars Culture Collection.* In Iizuka, H. I., and Hasegawa, T. (eds.), *Culture Collections of Microorganisms. Proc. Internatl. Conf. on Culture Collections,* October 7–12, 1968, p. 21. University of Tokyo Press, Tokyo, Japan.
10. HWANG, S. H. (1966) *Appl. Microbiol.* 14, 784.
11. PEPPLER, H. J., (1967) *Microbial technology.* 454, Reinhold Publishing Co., New York.
12. PRESCOTT, S. C., and DUNN, C. G. (1959) *Industrial microbiology.* Ed. 3. p. 945 McGraw-Hill, New York.
13. SMITH, G. (1969) *An introduction to industrial mycology.* Ed. 6. p. 390, St Martin's Press, New York.

Part III

STRAIN DEVELOPMENT

Editor's Comments
on Papers 4, 5, and 6

4   BACKUS and STAUFFER
    The Production and Selection of a Family of Strains in Penicillium chrysogenum

5   THOMA
    Use of Mutagens in the Improvement of Production Strains of Microorganisms

6   PONTECORVO and SERMONTI
    Parasexual Recombination in Penicillium chrysogenum

The Backus and Stauffer paper (Paper 4) is a unique account of the development of a family of strains of unparalleled importance to the fermentation-based industries. Microbiologists generally acknowledge that all penicillin-producing cultures used in this country, and possibly throughout the world, are derived from Wisconsin strains or have a common ancestor, namely, *P. chrysogenum* NRRL 1951 (Roper and Alexander 1945). Backus and Stauffer not only present the history of the Wisconsin family of strains but describe methods not formally published elsewhere, although the methods were largely disclosed to and exploited by industry prior to the publication of this paper in 1955.

The paper by Thoma (Paper 5) presents a brief summary of the results of a number of strain development programs carried out in a large fermentation-based pharmaceutical house over a period of about 15 years. The experience cited is probably typical of that of many other pharmaceutical manufacturers throughout the world. The paper was published at a time when many research directors were questioning the value of initiating or continuing intensive strain development programs.

The paper by Pontecorvo and Sermonti (Paper 6) followed their earlier brief report of work with a penicillin-producing culture (Pontecorvo and Sermonti 1953). The techniques used were developed still earlier with *Aspergilli* (Pontecorvo et al. 1953, and Pontecorvo, Roper, and Forbes 1953). Later, a parasexual cycle was demonstrated with an actinomycete (Sermonti and Spada-Sermonti 1956). Subsequently, attempts to exploit the possibility of effecting genetic recombination by the parasexual mechanism with a variety of antibiotic-producing organisms were widespread in industry and, in some laboratories, intense. However, only a few modest successes have been reported (Bryson, Thoma, and Nimeck 1968, Mindlin *et al.* 1961, and Elander *et al.* 1973).

Several claims for successful transfer of genetic information by the mechanisms of transduction and transformation between antibiotic-producing streptomycetes have been made, some as early as 1959 (Elander 1967). However, rigorous proof for genetic recombination via transduction or transformation among streptomycetes is still lacking (Sermonti 1969).

For an up-to-date overview of mutation in relation to coordination of metabolism in some important industrially useful microorganisms, see the paper on mutation and the production of secondary metabolites by Demain (1973).

REFERENCES

Bryson, V., R. W. Thoma, and M. W. Nimeck. 1968. Application of microbial genetics in industry. Page 250 in *Proc. 12th Int. Congr. Genet., Vol. 2.* Science Council of Japan, Tokyo.

Demain, A. L. 1973. Mutation and the production of secondary metabolites. *Adv. Appl. Microbiol.,* **16**:177–202.

Elander, R. P. 1967. Potential application of microbial genetics in antibiotic-producing microorganisms. Pages 395–402 in *Induced Mutation and Their Utilization.* H. Stubbe, ed. Abhandlungen Deutsche Akademie der Wissenschaften, Berlin.

———, M. A. Espenshade, S. G. Pathak, and C. H. Pan. 1973. The use of parasexual genetics in an industrial strain improvement program with *Penicillium chrysogenum.* Pages 239–253 in Z. Vanek *et al.,* eds., *Genetics of Industrial Microorganisms.* Academia, Prague.

Mindlin, S. Z., S. I. Alikhanian, A. V. Vladinirov, and G. R. Mikhailova. 1961. A new hybrid strain of an oxytetracycline-producing organism, *Streptomyces rimosus. Appl. Microbiol.,* **9**:349–353.

Pontecorvo, G., and G. Sermonti. 1953. Recombination without sexual reproduction in *Penicillium chrysogenum. Nature,* **172**:126–127.

———, J. A. Roper, L. M. Hemmons, D. M. MacDonald, and A. W. J. Bufton. 1953. The genetics of *Aspergillus nidulans. Adv. Genet.,* **5**:141–238.

———, ———, and E. Forbes. 1953. Genetic recombination without sexual reproduction in *Aspergillus niger. J. Gen. Microbiol.,* **8**:198–210.

Roper, J. A., and D. F. Alexander. 1945. Penicillin. V. Mycological aspects of penicillin production. *J. Elisha Mitch. Sci. Soc.,* **61**:74–113.

Sermonti, G. 1969. Genetic recombination of streptomycetes. Page 312 in *Genetics of Antibiotic-Producing Microorganisms.* Wiley-Interscience, New York.

———, and I. Spada-Sermonti. 1956. Gene recombination in *Streptomyces coelicolor. J. Gen. Microbiol.,* **15**:609–616.

Copyright © 1955 by The New York Botanical Garden
Reprinted from *Mycologia*, 47(4), 429–463 (1955)

THE PRODUCTION AND SELECTION OF A FAMILY OF STRAINS IN PENICILLIUM CHRYSOGENUM [1]

M. P. BACKUS AND J. F. STAUFFER

(WITH 4 FIGURES)

A program of research dealing with variation in *Penicillium notatum* Westling and *P. chrysogenum* Thom has now been in progress in the writers' laboratories for over ten years. Begun as one of several projects set up by the War Production Board in its monumental effort to increase production of the then new drug, penicillin, to meet war needs, the program was continued for one year (1944–45) under a grant obtained from the Research Committee of the University of Wisconsin Graduate School, and thereafter with support from various segments of the American antibiotics industry. Initially geared to the sole objective of securing strains of the mold capable of higher yields of the drug, the program was soon modified to emphasize fundamental studies on variability in the fungi in question. Strain development work has at no time been completely discontinued, but at the termination of the government contract it was made subordinate to the broader purposes of an academic research enterprise.

[1] Supported in part by the Research Committee of the Graduate School from funds supplied by the Wisconsin Alumni Research Foundation, and by grants from Eli Lilly and Company, Commercial Solvents Corporation, American Cyanamid Company, Cutter Laboratories, Wyeth Laboratories, Inc., and E. R. Squibb & Sons.

Beginning during the war period and continuing into the two later phases through which the project has progressed, the writers have worked in close conjunction with other groups conducting penicillin research at the University of Wisconsin. The relationship with the Department of Biochemistry has been particularly close, and the writers are especially indebted to Professors W. H. Peterson, M. J. Johnson, and M. A. Stahmann of that Department for their invaluable cooperation. A succession of graduate students—O. H. Calvert, Eugene Dulaney, Sally Kelly, Bruce Churchill, F. Roegner, T. H. Campbell, Roy Curtis, W. F. Whittingham, and James Grosklags—have assisted in the work; and each of these has played an important part in carrying out certain phases of the investigation. Dr. H. C. Greene of the Botany Department also participated in the project briefly during the time when operations were being carried on under the government contract.

Accounts of the whole general wartime penicillin research program in America, including a consideration of the strains of the mold developed and used during that period, have been published by Coghill and Koch (12), Raper (39, 40), Raper and Alexander (41), Peterson (36), Thom (54), Florey *et al.* (15), Perlman (35), and others. The details need not be repeated here. Suffice it to state that although a variety of new strains were secured in this laboratory during the writers' brief participation in the War Production Board program, none was of sufficient merit to be ranked among the best strains in circulation at that time.

In 1945, however, a very superior strain of *P. chrysogenum* was developed at the University of Wisconsin (4) and in the further course of operations an outstanding "family" of strains, stemming from that notable variant, has emerged. It is the purpose of the present communication to give an account of this "Wisconsin Family," with special emphasis on key strains and how they were obtained, as background for a discussion of some of the more specialized phases of the study of variability to be reported later. Reports dealing with certain aspects of the general program have already been published (**1, 45, 49, 50**).

THE ANCESTRAL STOCK

The first penicillin produced on an industrial scale was obtained by surface culture methods, and the organism most extensively employed in these pioneer days was a strain of *P. notatum* known as *NRRL 1249·B21* (**41, 42**). This was a derivative of the original culture contributed by Fleming. However, by the time the writers began their work on variability in penicillin-producing molds, industry in America was shifting to production of the antibiotic in submerged culture. At the U.S.D.A. Northern Regional Research Laboratory, which took the lead in penicillin research in this country when the problem of large-scale production of the drug was first brought to America in 1941, a notable strain of *P. notatum, NRRL 832,* had been brought to light and shown to be capable of giving reasonably good yields of penicillin when grown submerged in aerated containers. Thus strain *NRRL 832* opened the way for development of the tank fermentation method of producing the antibiotic. *NRRL 832* was one of the strains on which the authors began their studies. Since, however, the Wisconsin family of strains to be described in this communication sprang from other stock, strain *NRRL 832* need not be considered further here.

Strains NRRL 1951 and NRRL 1951·B25

In the search for better stocks of penicillin-producing mold, an outstanding step forward was taken in 1943 when Raper and Alexander isolated a strain of *P. chrysogenum* which they designated as *NRRL 1951* (**41**). This strain was selected for further study by its discoverers because in initial tests it gave, in submerged culture, yields somewhat better than those obtained from *NRRL 832*. It is notable, chiefly, however, as the progenitor of a long line of interesting and important descendants, including the entire "Wisconsin Series." Since the middle of 1944 it would appear that descendants of this famous isolate have been the chief source of the world's supply of penicillin.

A large number of spontaneously-occurring variants derived from *NRRL 1951* were selected and studied by Raper and Alexander. One outstanding derivative, *NRRL 1951·B25,* was obtained in two steps. The exact mode of origin of this variant, also its cultural and microscopic characteristics and fermentation behavior are recorded in a series of publications (**17, 30, 31, 41, 42**). In submerged culture, strain *NRRL 1951·B25* gave penicillin yields up to 250 O.U./ml—two to three times the amount obtainable from its wild-type ancestor or from *P. notatum 832*—and quickly began to supplant strain *832* for the production of penicillin when released for industrial use. Although Raper and Alexander endeavored to obtain still higher-yielding strains by isolating variants from *1951·B25,* they were unsuccessful in this attempt (**40, 41**).

Strain X–1612

The crowning event in the wartime effort to secure improved strains of penicillin-producing molds was the emergence of "super-strain"

X–1612, which was obtained through mutagenic (X-ray) treatment of conidia of *NRRL 1951·B25.* As Raper (**39**) has pointed out, "The production of this culture should be regarded as a joint endeavor. The stock was supplied by the Fermentation Division, Northern Regional Research Laboratory; the irradiation was performed by Dr. Demerec and associates at the Carnegie Institution, in Cold Spring Harbor; the initial and indicative production tests were made at the University of Minnesota, by Drs. Christensen and Ehrlich; and the real magnitude of its superiority was demonstrated, at the University of Wisconsin, by Professors Peterson and Johnson in 80-gallon fermenters." This "super strain" was shown to yield approximately twice as much penicillin as its parent, and during 1945 it became the principal race used in commercial production of the drug. Cultural characteristics and the metabolism of this variant are described in the literature (**5, 17, 28, 30, 32, 33, 39, 42, 52, 53**).

In relation to the present communication, special interest centers on strain *X–1612,* since it is the immediate ancestor of the entire "Wisconsin Series" of *P. chrysogenum* strains. Strain *X–1612* first came into the hands of the writers early in the fall of 1944, before its true merit had been established. It was nevertheless almost immediately adopted by the authors as the base stock to be used in a new strain development program being launched at that time, for, shortly after being received, it was tested in shake-flasks in the writers' laboratory along with several strains which had been rated among the best available at that date, and it gave better penicillin yields than any of the competing cultures.

CULTURAL METHODS

The principal solid substrate used for the cultivation of the fungus in the strain development program carried out was honey-peptone agar, a Sabouraud-type medium consisting of 6% honey, 1% Difco bacto-peptone, and 2% agar. To minimize caramelization of the sugar in the preparation of this substrate, the agar is dispersed in the water before the other ingredients are added. The autoclaved mixture has, without adjustment, a pH favorable to fungal growth and inhibitory to most bacteria. This medium was used to prepare tube and bottle slants and Petri plates.

Conidial suspensions, as needed, were made up over agar slant cultures of appropriate age—mainly cultures which had been incubated for 7–10 days at 24–25° C. Sterile distilled water, sometimes with a trace of wetting agent, was introduced into the culture vessel to fill the latter approximately half full. Agitation of the liquid, accompanied when necessary by gentle scraping of the surface of the culture with a transfer loop or serological pipette, served to bring the spores into suspension. Sterile pipettes and water blanks were employed to make dilutions. When suspensions of a particular spore concentration were required, counts were made with a haemacytometer and adjustment of the spore load accomplished by dilution procedures. In some instances spore suspensions were passed through sterile Whatman No. 2 filter paper to remove mycelial fragments and eliminate spore clumps.

For plating, a standardized procedure has been followed, although spore suspensions of various histories have been handled. Plates are poured in advance and allowed to solidify on a leveled table top in a culture room. Then onto the surface of each plate of agar one milliliter of a suitably diluted spore suspension is introduced with a sterile pipette. The plate is rotated to spread the suspension over the agar, after which it is returned to its original position and left to incubate at room temperature.

In a limited number of instances, individual sporelings were isolated through a modification of the single-spore isolation technique described by Thom and Raper (**55**). In this procedure, a diluted spore suspension is spread onto hard (4%) cleared agar containing 0.5% sucrose as the only nutrient. After approximately 15–20 hours (the time varying with the strain) at room temperature, germination has usually progressed to the point where isolation can be begun. The operator, working under a high-powered stereoscopic microscope, then proceeds to remove selected sporelings together with a small square of the underlying agar, utilizing a microscalpel fashioned from a fine sewing needle which has been fused into a glass rod and ground flat on a carborundum stone. This tool is sterilized by dipping it into a vial of alcohol and then quickly running it through a flame. The same equipment has been employed to remove and transfer hyphal tips from the margins of colonies.

A long succession of populations, each arbitrarily limited in size to about 250 isolates, has been studied. Almost invariably these populations have been obtained by seeding plates with a dilute spore suspension and then transferring hyphal tips from the margin of each young colony after about three days incubation. For study of the cultural characteristics of the individuals in such a population, hyphal tip transfers are made to honey-peptone agar plates containing measured amounts of the medium. Four such transfers, evenly spaced, are made onto the agar surface of each culture dish (FIG. 1, B), and the transfers are numbered. The 60–65 plates required to accommodate a given population are then incubated at 24–25° C for a week to ten days. At the same time that the hyphal tip transfers are made to the agar plates, duplicate transfers are also made to correspondingly numbered tube slants. After a suitable incubation period, a second tube slant is seeded from each original tube culture by mass-spore transfer. This subculture, which matures rapidly

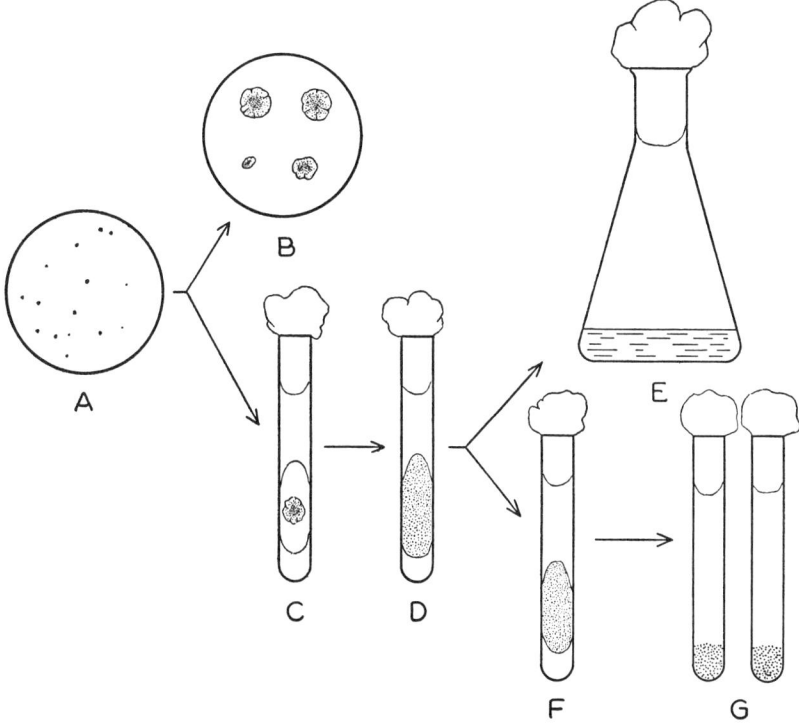

FIG. 1. Diagram illustrating sequence of cultures prepared in processing a population. A. Isolation plate. B. One of a group of plates prepared for study of the cultural characteristics of the isolates. C. Original tube culture of a given isolate. D. Sub-culture No. 1, inoculated by mass-spore transfer from the original tube culture. E. Fermentation flask (for screening test). F. Sub-culture No. 2. G. Master soil preparations.

because of the heavy seeding, furnishes inoculum for an antibiotic screening test. A conidial suspension is made up over this culture and one ml of it is used to seed the broth in a fermentation flask. A loopful of the same suspension is used as inoculum to start tube subculture No. 2. This second sub-culture is eventually either discarded or is employed to set up soil preparations, depending upon the quality of the isolate, as revealed by the outcome of the fermentaion test, etc. For each isolate selected for further study, two duplicate soil preparations are set up from the source indicated above. This pair of soil preparations constitute the "master" cultures for the given isolate and are carefully preserved in the permanent collection. All further operations involving any particular isolate are based upon these master stocks. This basic sequence of cultures, ending with the master stocks, is shown schematically in Fig. 1.

In setting up the soil preparations a modification of the technique described by Greene and Fred (19) has been followed. This involves pipetting 1 ml of a heavy conidial suspension of a given race onto 5 gms of dry, thoroughly sterilized soil (a mixture of equal parts of quartz sand and well-screened and pulverized garden loam) in a culture tube stoppered with a gauze-wrapped cotton plug. The inoculated soil tube is then allowed to dry out slowly at room temperature, covered by a porous paper cap. Once prepared and dried, the soil stocks require no further attention. In the writers' laboratories they are not refrigerated and they are given no special care except to keep them protected from dust. Such soil stocks are not only very easy to set up but serve to maintain the various strains in an unchanging condition over a long period of time. Cultures prepared in this laboratory as long ago as 1944 are still viable and apparently have undergone no degenerative changes. To recover the fungus in active condition it is only necessary to remove aseptically a small quantity of the soil and dust it over the surface of a fresh agar slant. For this purpose a tiny dipper, fashioned from the end of a hammered inoculating needle, is convenient. The same soil tube can be used repeatedly.

Secondary soil preparations in lots of thirty to fifty tubes have been set up for some isolates, and to provide the necessary amount of spore suspension in these cases large bottle slants seeded with soil from the master stocks have been used. Since this laboratory has attempted to supply stock cultures of certain key strains to all penicillin producers and research laboratories requesting them, several successive lots of this magnitude have had to be prepared to meet the demand in a number of instances. In setting up these lots, the unvarying procedure has been to go back to one of the original "master" soil preparations for inoculum. Furthermore, sample tubes from each new lot have been checked against the master culture by fermentation test. These practices have resulted in a high degree of uniformity among successively prepared lots of stock cultures of a given strain.

MUTAGENIC TREATMENTS

Although a substantial amount of spontaneous variation has been encountered in the *P. chrysogenum* lines involved in this study, it has been found that through mutagenic treatments of spores the amount of variation can be increased. Furthermore, through mutagenic treatments there are obtained types of variants which either do not occur spontaneously at all or which appear so rarely that none of them has yet been detected in this laboratory (50). Two types of mutagenic agents have been employed in the program out of which the Wisconsin family of strains has emerged: ultraviolet radiation and the nitrogen mustard, methyl-bis(β-chloroethyl)amine.

The irradiation treatments have been carried out in a special vessel, similar to one used by Duggar and Hollaender (13), of the design illustrated in Fig. 2. This vessel is made of glass but has quartz windows. Through the side arm a spore suspension to be treated is introduced and treated samples can be removed. A sterile stirring rod, driven by a motor, fits through the vertical upper arm and its tip rotates to one side of the center of the main chamber where it effectively circulates the suspension but is not in the path of the UV rays entering the window. Prior to an irradiation treatment the vessel is sterilized by flushing it with 70% ethyl alcohol followed by rinsing with sterile distilled water. While a treatment is in progress the side arm is capped with a glass vial and a glass baffle around the top of the stirring device prevents entrance of contaminants from above.

A 1000-watt General Electric AH–6 water-cooled, high pressure mercury vapor lamp has been used to furnish ultraviolet radiation for the treatments. The lamp is housed in a radiation-tight box, and the image of the arc is focused on the front slit of a Bausch and Lomb quartz monochromator by means of a quartz lens affixed to an opening in the lamp house. Radiation from the rear slit of the monochromator passes through a quartz lens, through the irradiation vessel, and on to the receiver of a thermopile. The latter as part of a thermopile-galvanometer system was standardized for absolute energy measurements. For the treatments used in the phases of the project in question here, ultraviolet radiation of 2750.Å and 2534–37.Å has been exclusively employed.

When a treatment is to be carried out, a spore suspension is prepared and adjusted to a concentration of about 100,000 conidia/ml. With a sterile pipette ten ml of this suspension are introduced into the irradiation vessel. Just before the irradiation is begun, a 0.2 ml control sample is removed and placed in a 100 ml sterile water blank. Additional 0.2

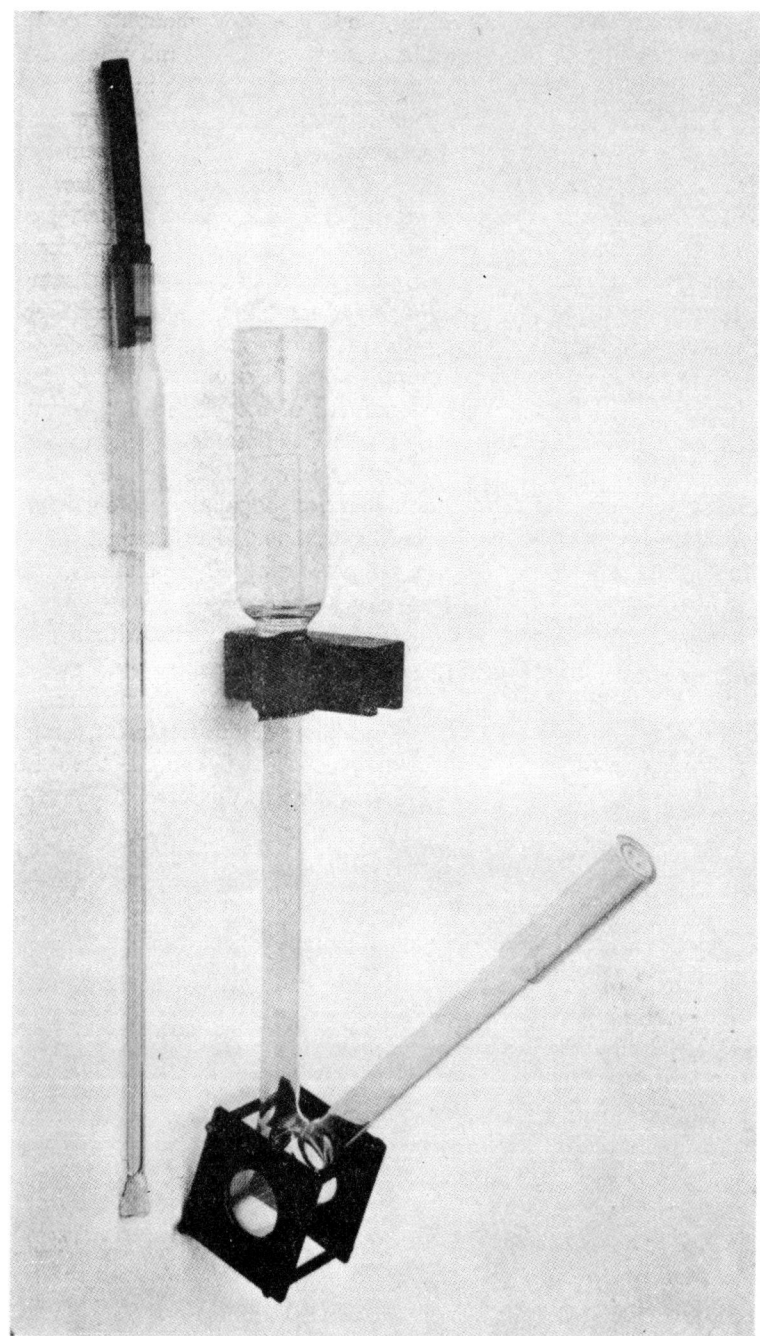

Fig. 2. Irradiation vessel. Capacity, 10 ml.

ml samples are taken at five-minute intervals throughout the progress of the treatment, and these are likewise placed in 100 ml water blanks. The irradiation is usually continued for 1 hour. A few milliliters from each diluted sample are then immediately plated out to determine the number of spores which have survived the treatment. The remainder, representing the bulk of each sample, is refrigerated for a few days until the desired information on killing is available from the test plates. Then the remainder of the one sample showing the desired percentage survival is plated out. During the early part of the program, samples which exhibited up to 99.4% killing were used; in all of the more recent work, however, samples showing a survival of approximately 25% have been chosen, since it had been found that with strain NRRL *1951* this survival level yielded the greatest number of morphological variants—a situation in general agreement with what has been described in the case of *P. notatum* (22, 49) and certain other fungi (8, 23, 24, 25). From the plates prepared, an irradiation-survivor population is obtained in the fashion described in the preceding section. This is then handled according to the plan worked out for processing all populations, whatever their origin.

The methods followed in treating spore suspensions with nitrogen mustard have been described elsewhere (45). Because with nitrogen mustard treatments the percentage of morphological variants increases more or less progressively with the duration of the treatment, samples containing a relatively low percentage of viable conidia have been employed to provide the survivor populations in this case.

FERMENTATION TESTS

In view of the fact that the ability of a given isolate to produce penicillin has been regarded as a matter of major importance and has, indeed, often been the principal basis for selection of key forms in the Wisconsin family of strains, much effort has naturally gone into the task of evaluating the productivity of the cultures.

The fermentations have been carried out in 500 ml Erlenmeyer flasks containing 100 ml of a corn steep-lactose liquid medium. This medium, throughout the study, has consisted of the following basic ingredients: distilled water, corn steep solids 2.0%, crude lactose 4.0%, $NaNO_3$ 0.3%, KH_2PO_4 0.05%, $MgSO_4$ 0.025%, and $CaCO_3$ 0.3 to 0.4%. The amount of calcium carbonate used was varied as indicated to bring the pH of the medium to 5.3–5.6 after sterilization. In essentially all of the tests carried out since 1947 an antifoam agent has been employed, 3 drops of Dow Corning "Antifoam A" emulsion (silicone) being introduced into each flask prior to autoclaving. Also beginning about 1947 a penicillin precursor, β-phenylethylamine (Monsanto), neutralized with acetic acid, has been added, usually at a level of 0.2% or 0.25%. This adjuvant likewise has been introduced into each flask separately prior to sterilization of the broth. The flasks are plugged with cotton, sterilized for 15 minutes at 15 lbs steam pressure, and, when cool, are seeded with spores of the isolate to be tested. One ml of an undiluted conidial suspension, made up over a fresh agar slant culture (usually 7–10 days old), is used as inoculum. The seeded flasks are then placed on a reciprocating shaker operating at 90–95 cycles per minute with a 4-inch stroke, in a fermentation room thermostatically controlled at 24–25° C. At appropriate intervals broth samples are removed with a sterile pipette, diluted with phosphate buffer, and assayed by the cylinder-plate method of Schmidt and Moyer (47).

One of the chief aims of the fermentation tests carried out on a given population was to detect whether there might be in this population variants possessing increased capacity to produce the antibiotic. Since accurate evaluation of the ability of a culture to produce penicillin is not simply accomplished, the procedure adopted may be considered as something of a compromise. Although it was not as thorough as the writers would have liked, it was as intensive as could be arranged for the handling of large numbers of cultures. That the selection program, as carried out, was in some measure effective is attested by the fact that it has yielded improved strains suitable for industrial use.

The initial step in conducting the fermentation studies on a given population has been to run a "screening" test. In this test a single flask is used for each of the approximately 250 isolates to be evaluated, and assays are run at six and seven days. Several flasks inoculated with the parent strain and usually also others inoculated with older strains of known potency are included as controls in each fermentation run. When the screening test is completed, about a dozen of the isolates which have given the highest yields of the antibiotic are selected. A repeat test on these isolates is then set up on a larger scale. Three or four flasks of medium are inoculated with each strain in this trial. Yields are averaged and again those isolates giving the best performance are picked out. In a third test, three or four "finalists" may be put in competition, with up to ten flasks being used for each strain. On the basis of performance in this test a single culture is selected to be used in starting a new "generation"—with or without intervening mutagenic treatment. However, it is not invariably the isolate which comes out best in these fermentation tests that is chosen to form the basis for further operations, since cultural characteristics and other traits may also enter into the picture when the final selection is made. If, as occasionally happened, no isolate in the population gave yields better than the parental form, the usual procedure was, nevertheless, to take what appeared to be one of the best individuals in the population and carry on with it.

Certain key strains emerging in the course of the work have been subjected to additional tests and study in the writers' laboratories, and a few of them have been submitted to the Fermentation Laboratories of the University of Wisconsin Department of Biochemistry for more critical evaluation. Here tests have commonly been run in flasks on a rotary shaker, in 30-liter stirred jars, and occasionally in 80-gallon tanks; the performance of the strains on various media, under varying amounts of aeration, at various precursor levels, etc. has also often been studied by the biochemists. A detailed analysis of chemical changes occurring in the medium as the fermentation progressed has usually been made (1, 6, 17, 26, 29, 36, 53).

STRAIN WIS. Q176

Origin and Productivity

As has been pointed out above, the first significantly-improved strain to emerge in the course of work conducted in the writers' antibiotics research program was obtained in 1945. This variant, designated as Wis. *Q176,* begins the "Wisconsin Family" of strains.

During 1944–45 a series of irradiations had been carried out on a variety of *P. notatum* and *P. chrysogenum* lines. Irradiation "Q" was the second in which spores of strain *X–1612* were treated, and it was colony number *176* in the irradiation-survivor population obtained following this treatment which yielded the notable strain in question.

For the sake of precision, it should be stated here that a special procedure was involved in preparing the *X–1612* stocks employed in these irradiation experiments. Because the writers wished to be sure that they were dealing with a pure line, and because at that time only scanty information was available concerning the history of the *"X–1612"* culture which had been received from the University of Minnesota, a single spore was isolated to start the *X–1612* stocks for this laboratory. However, in view of the fact that this single-spore line agreed closely with the parental culture in growth characteristics and fermentation pattern, it is believed that little significance need be attached to this single-sporing step in tracing the genealogy of the Wisconsin Family.

As has already been mentioned, it was the practice in the early irradiation treatments carried out in this laboratory to prepare the survivor populations from suspensions in which a very large portion of the conidia had been killed. In the suspension used from Irradiation Q less than 1% of the spores were viable.

In the screening test carried out on survivor population Q, many of the individuals yielded appreciably less penicillin than did the *X–1612* control culture, many showed yields of approximately the same magnitude as those from the parent, but for individual *176* the inhibition zone on the assay plate was much larger than any of the other zones at both the six and the seven day assay. The performance of this variant in the larger-scale repeat tests was equally good, and it was soon established that an outstanding new strain had been secured. Tested in 80-gallon fermentors in the University of Wisconsin Biochemistry Department, the new strain gave yields of over 900 O.U./ml in contrast to about 500 O.U./ml obtained with strain *X–1612* run under the same conditions (4, 17).

Released late in 1945, strain Wis. *Q176* was widely sought by commercial producers of penicillin, perhaps as much for use as a "breeding stock" in their own strain development programs as for use in their production plants. Microbiologists at universities, etc. also chose this variant as a base stock for work along similar lines and for a variety of fundamental studies (2, 3, 9, 14, 37, 38, 43, 46, 48, 56, 57, 58). The strain was likewise employed in detailed chemical studies on the fermentation process (6, 7, 16, 18, 20, 26, 27, 34). Over 150 soil preparations of strain *Q176* have been distributed by the writers in response to requests from industrial laboratories, university laboratories, microbiological institutes, culture collection centers, public health organizations etc. in nineteen countries.

What at first appeared to be a weakness of the strain, namely, the fact that it tended to produce a high proportion of penicillin K, was soon overcome when it was discovered that simply by adding to the fermentation broth a suitable precursor (phenylacetic acid or various derivatives of this compound) production could be shifted over almost completely to the desired penicillin G (1, 6, 7, 20, 21). In addition, the use of a precursor was found to increase the total yield of the antibiotics. In contrast to some of the later-derived strains of the Wisconsin Series, which are apparently exacting in their requirements regarding fermentation conditions, strain *Q176* appears to be adapted to perform well under a wide range of circumstances. To judge by reports received, yields surpassing those from *X–1612* seem to have been almost universally

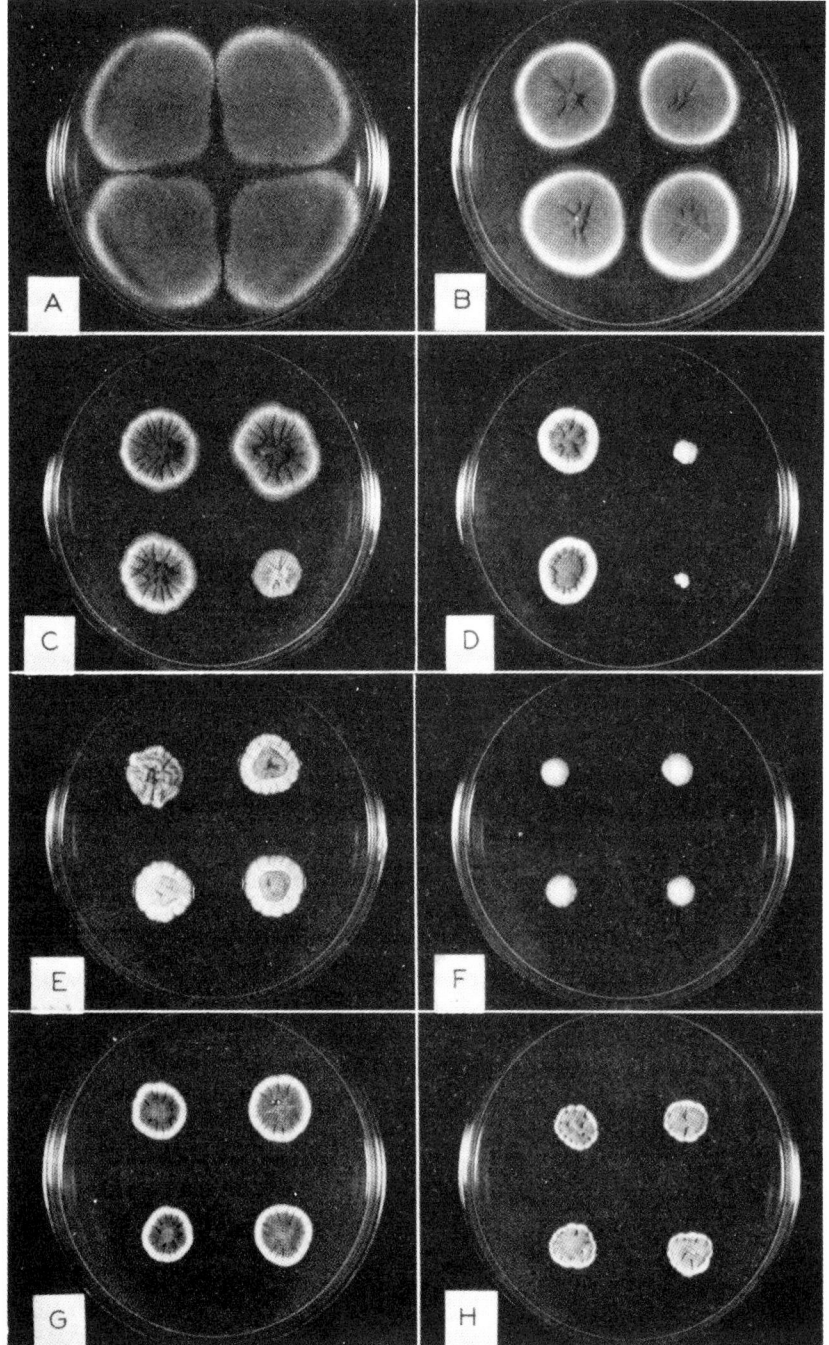

Fig. 3.

obtained with this variant, and it became widely known as a standard culture for penicillin production (29).

Cultural Characteristics

Mass spore transfer cultures of Wis. *Q176* on honey-peptone agar resemble those of *X–1612* but mature somewhat more slowly and sporulate less heavily. There is, indeed, progressive reduction in rate of growth and amount of sporulation following through the entire line of descent from *NRRL–1951* to Wis. *Q176*. This is best observed when a few dozen separate sporelings of each strain are transferred to individual honey-peptone agar plates and the development of the monosporous colonies is studied. After six days incubation at 24–25° C, colonies of *NRRL–1951* have reached an average diameter of about 40 mm, those of *NRRL–1951·B25* approximately 30 mm, and those of *X–1612* about 25 mm, while the average diameter of the *Q176* colonies is less than 20 mm at this age. Colonies of *NRRL–1951* have a flat, spreading growth on honey-peptose agar; in *NRRL–1951·B25* and *X–1612* there is slight radial folding, and such folding becomes very conspicuous in *Q176* (FIG. 3, C). As convolutions increase, the colony outline also tends to become irregular. Conidia are produced in great abundance in the wild-type ancestor, but sporulation is substantially diminished as one moves through the series, and the spores are slower to appear. Raper and Alexander (41) have recorded the occurrence of abnormal penicilli in strain *1951·B25*. Irregularities in penicillus structure are also encountered in *X–1612* and *Q176*. Furthermore, in strain *Q176* there are marked abnormalities in some of the vegetative hyphae. Campbell (10) and Churchill (11) have made a study of the mycelium and reproductive structures in this strain, and some of the microscopic peculiarities have been illustrated in a recent communication from this laboratory (50).

The preparation of populations from untreated spores of Wis. *Q176* first brought to light a remarkable pattern of spontaneous variation which has since been found to extend throughout most of the Wisconsin family of strains and which has been designated "the population pattern phenomenon." Some features of the situation involved here have already been outlined by the writers (50) and the "phenomenon" will be dealt

---

FIG. 3. Plate cultures showing 7-day old colonies from populations of various strains: A. Strain *NRRL 1951;* B. Strain *X–1612;* C. Strain Wis. *Q176;* D. Strain Wis. *48–701;* E. Strain Wis. *49–133;* F. Strain Wis. *51–20;* G. Strain Wis. *51–20A;* H. Strain Wis. *53–844*. Each plate contained 25 ml honey-peptone agar. Inoculation was by hyphal-tip transfers and incubation was at 24–25° C.

with in detail in a later publication. For the purposes of the present report, only those items which bear directly upon the origin and characteristics of the various key Wisconsin strains need be considered.

When sizable populations of strain *Q176* are set up as previously described (see Cultural Methods section above), or if *Q176* sporelings are isolated at random and transferred four to a plate, it is found that the colonies present after 7–10 days incubation are not all alike, as one might expect they would be. Instead, it can be observed that there are mainly five fairly distinct types of colonies present; and a further remarkable fact is that whenever *Q176* populations are prepared, these five colony types always appear in approximately the same proportions. One kind, designated as the "U-type" (usual type), predominates, accounting for roughly 65–75% of the population. "D"- and "C"-type colonies are next most abundant, while "B"- and "A"-types are always

TABLE I

A TYPICAL POPULATION PATTERN IN STRAIN WIS. Q176

Colony type	Av. diameter of colony in mm.*	Occurrence in population per cent
U	31.3	65.0
D	23.4	20.0
C	14.7	8.5
B	12.0	2.5
A	12.5	4.0

* Colonies ten days old, grown singly in Petri dishes on 25 ml honey-peptone agar, at an incubation temperature of 24–25° C.

present in very small numbers. The proportion of colony types found in one typical *Q176* population is recorded in TABLE I. This table further indicates the general difference in colony size which is encountered. The mycelial mats of "C"- and "B"-type colonies are typically strongly raised. Sporulation diminishes progressively from the U-type through the D-, C-, and B-types. The small B-type colonies usually have only a faint tinge of green near the center, where most of the scanty spore production is concentrated. The A-type colony departs most strongly from the usual type. It forms no conidia at all, and it has a distinctive flesh color; the mat is thick and compact but tends to be translucent and may have a water-soaked appearance. A- and B-type colonies show a strong tendency to sector, with the sectors growing much more rapidly than the parent and sporulating. Hyphal-tip transfers from such sectors usually yield typical U-type colonies.

From the situation just related, it follows that one cannot describe the cultural characteristics of a strain like *Q176* in simple terms. Since a spore suspension or soil preparation of this variant contains conidia of five or more definitely different potentials, it also follows that the usual concept of a strain must be modified in applying the term to the entity at hand. Strain *Q176* is not to be thought of as a homogeneous entity but rather as a balanced mixture of several components, each with its own distinctive pattern of behavior. *Q176* may be defined, in part, in terms of its population pattern; and, because of the predominance of the U-type colony, it may properly be referred to as a U-type strain.

STRAIN BL3–D10 AND THE BEGINNING OF THE WISCONSIN PIGMENTLESS SERIES

About the time that strain *Q176* was obtained, it began to appear that in addition to increased ability to produce penicillin there were other traits to be encountered among variants which might make a stock more desirable for commercial use. One such trait concerned the matter of pigment secretion by the fungus.

As is well known, most isolates of *P. chrysogenum* obtained from nature tend to produce relatively large quantities of water-soluble yellow pigment. This passes out into the broth in a liquid culture, and in cultures on Czapek's agar it not only diffuses into the substrate but colors the droplets of liquid which characteristically accumulate over the surface of the colony. Indeed, so consistently is this trait encountered that Thom, in naming the species, chose the specific epithet "*chrysogenum*" as a descriptive term. The variant strains *NRRL–1951, B25, X–1612,* and Wis. *Q176,* like their wild-type ancestor, *NRRL–1951,* produce yellow pigment in abundance, and commercial producers using such strains were forced to remove the pigment if they desired to market a white product. By 1946 it was general practice to completely remove the pigment although the necessary extraction procedures usually resulted in loss of some of the antibiotic. In earlier years the pigment was not all removed, and the penicillin of commerce at that time had a golden color.

It was suggested to the writers that a mutant strain which did not form any pigment in the medium would be industrially desirable, particularly if such a strain also had potency in penicillin production equal to that of Wis. *Q176.* That it might not be easy to secure such a strain, however, was indicated by certain observations which had been made in this laboratory, not only in conection with the improved *P. chrysogenum* lines being studied at that time but also in connection with the various *P. notatum* strains worked with previously: (1) even in populations grown from conidia surviving mutagenic treatment almost all the

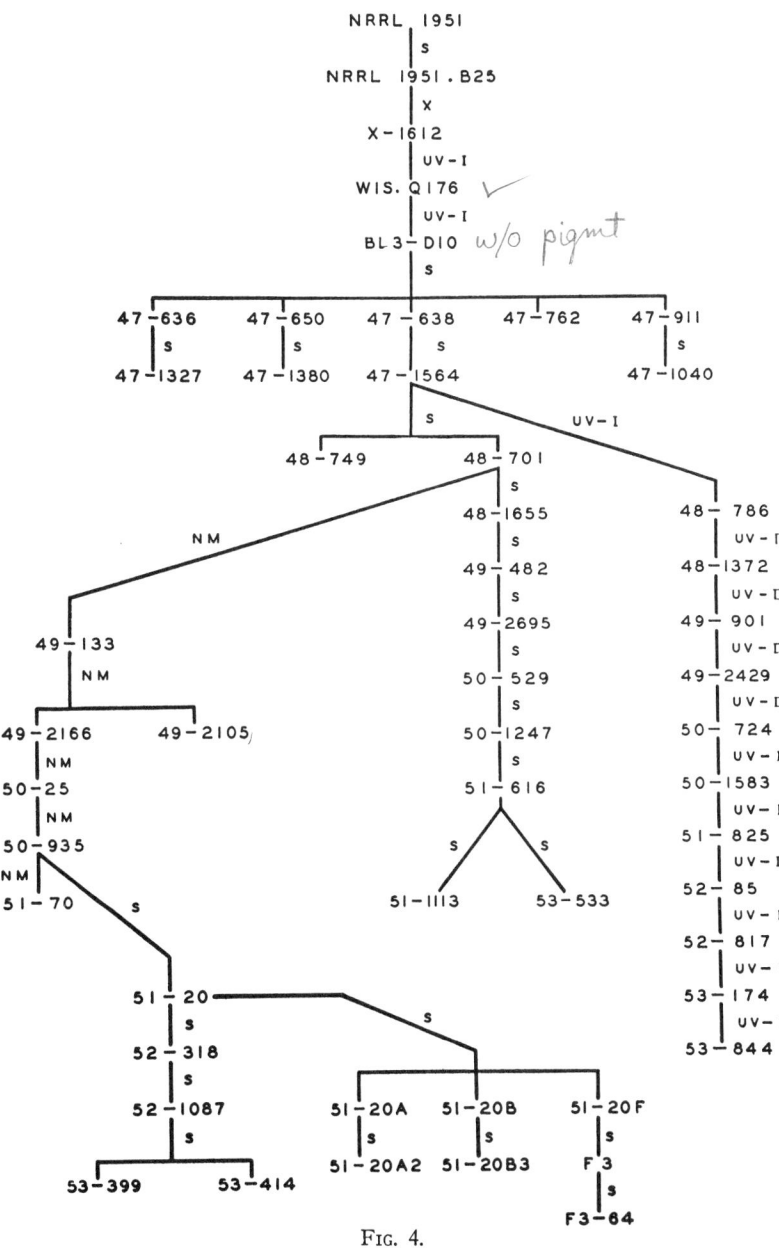

FIG. 4.

FIG. 4. Genealogy of the Wisconsin Family of strains of *Penicillium chrysogenum*. S indicates selection without mutagenic treatment; X indicates selection following X-irradiation; UV–I indicates selection following 2750Å UV-irradiation; UV–II indicates selection following 2534–37Å UV-irradiation; NM indicates selection following nitrogen mustard treatment.

isolates produced pigment in some amount, (2) isolates secreting a reduced amount of pigment almost invariably were also found to yield reduced amounts of the antibiotic, and (3) the rare isolates which seemed to form no yellow pigment at all were variants which had been severely damaged by radiation.

Nevertheless, during the spring of 1947, in studying populations grown from spores which had been subjected to various amounts of UV radiation, a very promising completely pigmentless strain, *BL3–D10,* was encountered. Isolate *BL3–D10* was found in a population grown from *Q176* conidia which had received a very low dosage of radiation. The type of population involved has been termed an "ultra-violet-stimulated population" (51), since conidial suspensions of *Q176* so treated actually yield more colonies than do the unirradiated controls. Although *BL3–D10* produced only about 75% as much penicillin as *Q176* in the fermentation tests conducted, this was by far the highest yield that the writers had encountered in any variant which formed no yellow pigment. Furthermore, this was a vigorous race compared with the only other completely pigmentless variants that had been seen heretofore in this laboratory. It was therefore decided to utilize this as a "breeding stock" in attempting to obtain additional pigmentless variants which might be better producers of penicillin than *BL3–D10* itself appeared to be. This proved to be a wise decision, for among the derivatives eventually obtained from this strain there were many which surpassed *Q176* in ability to produce the antibiotic, while maintaining the pigmentless quality; and some had further advantages as well. As a matter of fact, all the remaining members of the Wisconsin family of *P. chrysogenum* strains are descendants of strain *BL3–D10* and belong to the Wisconsin "pigmentless series." Once arrived at, pigmentlessness proved to be an enduring character and has been found to persist almost without exception through all the many thousands of descendants of *BL3–D10* which have been studied.

As indicated in the genealogy chart (FIG. 4), the next step in the development of the Wisconsin pigmentless series involved a program of selection without further mutagenic treatment. During 1947 a series of populations were prepared and over 2000 cultures were screened. From among these, nine key strains were selected, and four of them, *47–911, 47–1040, 47–1380,* and *47–1564,* in repeated tests gave yields at least as good as those obtained from *Q176* under the fermentation conditions employed. On August 1, 1948, all four strains were made available for general use. Because further tests appeared to indicate that strain *47–1564* was probably slightly superior to the other three, this strain was chosen as a parental stock from which to prepare additional untreated populations in 1948. Two more key strains, *48–701* and *48–749,* were selected when the testing work had been completed, and the release of these strains followed soon afterwards. Yields obtained with these strains in this laboratory have averaged 15%–25% above those secured with the parent, *47–1564,* when the latter was used as a control in the tests. Strain *48–701* has commonly outperformed *48–749* by small margins in fermentations conducted on this campus, and the two strains have been shown to have a somewhat different population pattern (TABLE II); in most respects, however, they are very similar.

In contrast with strain *Q176*, all of the pigmentless strains which have been mentioned above are D-type races. That is, the D-type colony predominates in populations grown from untreated spores; and the U-type

TABLE II

TYPICAL POPULATION PATTERNS IN KEY STRAINS OF THE WISCONSIN PIGMENTLESS SERIES COMPARED WITH THE PATTERN IN Q176

Colony type	Percent						
	Strain Q176	Strain 47–911	Strain 47–1564	Strain 48–701	Strain 48–749	Strain 49–133	Strain 51–20
A	3.7	3.6	7.1	7.6	0.4	3.6	0.5
B	2.6	5.0	2.9	8.4	1.3	0.7	5.4
C	8.5	13.3	5.4	9.3	7.6	4.3	94.1
D	20.2	76.3	84.2	74.7	90.8	91.3	0.0
U	64.8	0.0	0.0	0.0	0.0	0.0	0.0
All others	0.4	1.8	0.4	0.0	0.0	0.0	0.0

colony, which is so characteristic of *Q176,* has almost completely disappeared. Typical population patterns in four of the pigmentless strains under discussion are included in TABLE II. It must be pointed out, however, that the D-type colonies of strains *48–701* and *48–749* differ appreciably from those encountered in strains of the "*47*" series. Roegner (**44**) has distinguished several sub-types in the "D-group" and points out that most of the D-type colonies of strains *48–701* and *48–749* fall in subgroup IV, whereas in *47–1564,* their parent, they are predominantly of the "D–II" type. The chief distinguishing feature of the D–IV colony in contrast with other sub-types is a pale color, resulting apparently from the presence of short sterile aërial hyphae which are intermingled with the conidiophores and in part overgrow them. Because all of the Wisconsin pigmentless strains thus far considered have D-type population patterns, it follows that in all of them the growth on agar is slower and sporulation is somewhat less than in strain *Q176.* The decreased rate of growth of variants *47–1564* and *48–701* is evident in TABLE III. Interestingly enough, however, the mycelial development in aerated liquid cultures on corn steep-lactose medium appears to be about as rapid as it is in the case of older strains in the family tree.

The various pigmentless strains differ from the grandparental strain *Q176* and to some extent from one another in a number of additional features. Of special interest is the fact that the pigmentless strains, in general, seem to utilize precursor more efficiently, tend to have lower oxygen requirements, and tend to oxidize lactose more rapidly (29). Especially rapid oxidation of lactose, with no initial lag phase in the utilization of this sugar, was reported in the case of strain *48–749* (1).

TABLE III

COLONY SIZE IN FOUR WISCONSIN VARIANTS AND IN THREE ANCESTRAL STRAINS *

Strain	Number of colonies		Average diameter (mm.)			
			7 Days		9 Days	
	Honey-peptone agar	Czapek-Dox agar	Honey-peptone	Czapek-Dox	Honey-peptone	Czapek-Dox
NRRL *1951*	49	11	53.6	35.3	—	—
NRRL *1951.B25*	49	11	39.3	24.9	49.8	30.7
X-1612	44	10	25.2	24.1	32.6	31.3
Wis. *Q176*	34	7	21.0	19.0	26.5	24.7
Wis. *47–1564*	42	8	15.9	12.9	19.8	16.9
Wis. *48–701*	25	7	15.0	14.0	18.5	18.0
Wis. *49–133*	43	9	14.0	11.3	18.8	16.0

* Colonies grown at 24–25° C in Petri dishes containing 25 ml medium. Each plate was inoculated with one 1-day-old sporeling chosen at random.

In addition to lack of pigment, there are a number of other features about some of the pigmentless strains which have made them attractive to certain commercial producers of the drug.

A THREE-WAY SELECTION PROGRAM IN THE WISCONSIN PIGMENTLESS SERIES

Results obtained in studies carried out on the mutagenic effects of nitrogen mustard on *P. notatum,* strain *NRRL 832* (49) were primarily responsible for the adoption, in 1948–49, of a new scheme of operations in carrying forward the strain development program in the Wisconsin Pigmentless Series. Beginning at that time, three separate lines of descent were followed, as indicated by the trifurcation in the family tree (FIG. 4). Since one of the principal aims involved in the new plan was to determine the effects of repeated mutagenic treatments over a long series of generations, the approach was academic. However, a number of outstanding strains have emerged in the course of the work.

Selection without Mutagenic Treatment

The control branch of the three-way line of descent was anchored on strain *48–701,* although, as inspection of the genealogy chart will make clear, it actually extends back to strain *BL3–D10* through strains *47–1564* and *47–638*. Each step in the progression has involved the preparation of a population from untreated spores and then carrying through the cultural and fermentation tests upon which were based the selection of the key strain which was to serve as the parent for the next generation. In this manner the selections have been carried on slowly through a total of eight generations beyond Wis. *48–701*. Although no spectacular changes in ability to produce the antibiotic have occurred at any point, a gradual rise in productivity can be traced up through strain *50–1247* —beyond which no further improvement could be effected.

Repeated Ultraviolet Treatments

In the second line of descent, a similar routine of testing and selection has been followed, but in this case each population dealt with was a survivor population prepared from a spore suspension which had been subjected to sufficient ultraviolet radiation to kill approximately three-fourths of the spores. Ultraviolet of two different wavelengths, 2750Å and 2534–37Å, has been employed in the treatments. A survivor population grown from irradiated spores of strain *47–1564* provided the material upon which operations were begun in this portion of the three-way program, and the "ultraviolet line" has been continued through twelve generations (FIG. 4). As in the case of the simple selection (control) line, no strains in the UV series have become widely known, since at the time of their emergence they were eclipsed by still better strains from the "nitrogen mustard branch" of the family tree. Nevertheless, there have been substantial gains in ability to produce the antibiotic. Through the first few generations there was no significant improvement over the ancestral form *47–1564*. However, gains of varying magnitude have been recorded at various points along the line of descent between the fourth and twelfth generations. Thus the key strains in the latter part of this series are, in general, very good producers of penicillin. Not only are they clearly superior to *47–1564,* from which they sprang, but also they are better than any of the strains in the parallel control series in which no irradiation was involved. All key strains obtained up to this time in the UV series are of the D- or C–D-type (FIG. 3, H).

The Nitrogen Mustard Line

Although it was originally intended that the third line of descent in the three-way selection program should exactly parallel the other two, differing only in that repeated nitrogen mustard treatments were to be employed in obtaining the populations from which key strains would be selected, complications were introduced by the occurrence of two noteworthy strains falling outside of the regular sequence, as will be explained below.

The nitrogen mustard line is of special interest because the best strains of *P. chrysogenum* to be developed in this program have emerged here. A spore suspension of strain *48–701* which had been treated with the mutagenic chemical provided the initial survivor population in this series, and the high quality of one individual in this population is primarily responsible for the general superiority of the nitrogen mustard branch of the family tree. The key strain selected from this first group of survivors was Wis. *49–133* (FIG. 3, E and TABLE II), which embodies the greatest single improvement effected in the Wisconsin Pigmentless Series. Penicillin yields 50% to 100% above those obtained from the parent, *48–701*, have commonly been recorded for strain *49–133* when the two variants have been included in the same fermentation run. In early tests the superiority of the nitrogen mustard-derived variant over the parental and grandparental strains was repeatedly exhibited and was most marked at a high precursor level (TABLE IV).

In the second generation, Wis. *49–2105* was the highest-yielding strain encountered in the survivor population; this, however, could not be selected as the key strain to continue the main line of descent because of an interesting defect: it produced a pigment in the medium. The appearance of a pigment-producing variant in a long-established pigmentless stock was of course an item of considerable academic interest, particularly since this was the only strain of that character to be encountered among the thousands of descendants of *BL3–D10* studied in this laboratory. However, since the pigment formed by strain *49–2105* is reddish brown and clearly different from the golden yellow pigment characteristic of the species, it is evident that the phenomenon in question cannot be interpreted as reversion.

Studies conducted in the University of Wisconsin Department of Biochemistry (1) have revealed that strains *49–133* and *49–2105* are more efficient producers of penicillin than older members of the Wisconsin Pigmentless Series. In certain tests these nitrogen-mustard-induced variants were found to produce only about 50% as much mycelium as strains such as *47–1564* and *48–701,* yet gave higher penicillin titers. In the same studies exceptionally efficient utilization of the penicillin precursor, phenylacetic acid, by strain *49–133* was noted.

Between 1949 and 1951 five generations of nitrogen mustard treatment survivors were processed; and four key strains, in addition to

TABLE IV

PENICILLIN PRODUCTION IN SHAKE-FLASKS BY STRAIN WIS. Q176 AND CERTAIN OF ITS PIGMENTLESS DESCENDANTS *

Run no.	Concentration of precursor (β-phenylethylamine)	Strain	No. flasks	Av. yields (O.U./ml)	
				6 Days	7 Days
137	0.0%	Q176	3	293	298
		47–1564	3	291	362
		48–701	3	258	354
		49–133	3	218	435
137	0.1%	Q176	3	420	452
		47–1564	3	738	705
		48–701	3	823	870
		49–133	3	1060	1223
137	0.2%	Q176	3	310	402
		47–1564	3	505	695
		48–701	3	807	863
		49–133	3	1147	1290
155	0.25%	Q176	5	346	519
		48–701	5	930	933
		49–133	5	1472	1734

* Composition of the basal fermentation medium and other fermentation conditions are given in the text.

49–133, were selected in the main line of descent. Strains *49–2166* and *50–25* from the second and third generations showed no significant improvement over *49–133* in ability to produce the antibiotic. In the case of the fourth generation key strain, *50–935,* however, somewhat better yields were repeatedly secured. This strain is of further interest since it arose from a C-type colony and exhibits a C-type population pattern—the first such variant to be selected as a key strain in this program of studies. No further improvement in strain quality was effected through use of the mutagenic chemical.

In 1951 attention was diverted from the main line of descent in the nitrogen mustard series by the appearance of strain Wis. *51–20,* obtained from a population prepared from untreated conidia of *50–935*. This strain arose from a small C-type colony and yields a population best characterized as a "weak C"- or "C–B"-type (TABLE II). This means that all of the colonies in the population exhibit very slow growth and weak sporulation on honey-peptone agar (FIG. 3, F). Mass-spore transfers yield extremely poorly sporulating agar slant cultures. However, in shake-flask fermentation tests conducted in this laboratory, penicillin yields averaging up to 2500 O.U./ml were recorded.

The one undesirable trait of strain *51–20* appeared to be its limited capacity to form spores. Since this feature made the strain difficult to work with in laboratory studies and because it appeared that such a defect might also impair its usefulness in industry, an attempt was made to obtain from it strains of better cultural quality which would still retain this variant's potency. The results of fundamental studies on the behavior of various kinds of colonies suggested that this might best be accomplished by allowing sectoring to occur. A population of *51–20* was accordingly prepared and incubated until sectors appeared on some of the colonies. As expected, these were mainly faster-growing and more heavily sporulating. There were thus obtained various sub-strains, designated as "*51–20A*," "*51–20B*," etc., each originating as a hyphal tip transfer from a different sector. The results of the fermentation tests were disappointing, however. In general, penicillin production was found to decline as sporulation and vigor were increased over that of the parental strain, *51–20*. Nevertheless, three of the variants were retained for further study: *51–20A*, *51–20B* and *51–20F*. The first of these was markedly improved over the parent in capacity to sporulate (FIG. 3, G) but far inferior to it in production of the antibiotic. Strain *51–20B* showed only slight improvement over *51–20* in sporulation but gave yields about on a par with it. *51–20F* was intermediate between the other two sub-strains in sporulation and in penicillin production. Single-spore isolates from these three sub-strains were next tested. All of the derivatives of *51–20A* gave inferior yields; some of the derivatives of *51–20B* equalled or slightly surpassed the yields obtained from strain *51–20* used as a control; while those from *51–20F* were again intermediate. Strain *51–20F3* (= Wis. *F3*) was selected as the best in the latter group, but yields as good as those from the parental form, *51–20*, were never obtained from it. Interestingly enough, results quite at variance with these were reported when certain of the strains were tested in the University of Wisconsin Department of Biochemistry under different fermentation conditions. Here high yields were obtained with strain *51–20F3* and *51–20A* but poorer yields with *51–20B*. Mainly on the strength of the performance of strain *51–20F3* in these latter tests, a population was prepared from it by isolating single spores. The key strain selected from this population, *51–20F3–64* (= Wis. *F3–64*) has usually equalled or surpassed strain *51–20* in penicillin production tests conducted in this laboratory, and its capacity to sporulate is substantially better than that of the latter strain.

One additional attempt to improve the *51–20* stock has succeeded in some measure. Although the surest way to obtain increased vegetative

TABLE V

PENICILLIN PRODUCTION BY REPRESENTATIVE WISCONSIN STRAINS OF PENICILLIUM CHRYSOGENUM TESTED SIMULTANEOUSLY

Strain	Sporulation rating*	Average yield (O.U./ml)**	
		6 Days	7 Days
Q176	4.0	605	640
47–638	4.0	740	980
48–1564	4.0	930	1357
48–701	3.8	1107	1365
50–1247	4.0	1257	1506
48–1372	4.0	904	1343
53–844	2.6	1501	1846
49–133	2.2	1977	2230
49–2105	2.1	1798	2266
51–20	1.5	2140	2521
F–3	3.4	1767	2140
F3–64	3.1	2140	2493
53–414	2.5	1990	2580
53–399	3.1	2018	2658

* Scale 1.0 to 10.0: 1.0 signifies extremely weak sporulation, the cultures being almost white; 4.0 signifies "good" sporulation (Q176 level); 10.0 signifies very heavy, early sporulation such as is encountered in the wild-type ancestor. Figures given are averages of four separate ratings of 7-day-old honey-peptone agar slant cultures by two individuals.
** In each case, the yield given is the average of 12 flask cultures, 4 each from Runs 433, 435, and 442. In each of these runs the fermentation broth contained 0.25% β-phenylethylamine; other fermentation conditions as described in text.

and reproductive vigor in a C- or B-type race is through the sectoring phenomenon, occasional spores will yield more vigorous colonies; and the possibilities of this approach were here explored. Two strains showing the desired combinations of high penicillin yield and at least moderate sporulation were secured after selections had been carried out over three generations. Strains *53–414* and *53–399* have exhibited high productivity in shake-flask tests conducted in this laboratory. Average yields over 2500 O.U./ml have commonly been obtained (TABLE V) and titers have occasionally exceeded 3000 O.U./ml.

Unfortunately the most recent strains have not yet been extensively tested outside of the laboratory of their origin. According to present evaluations, however, the *51–20* side branch of the nitrogen mustard line constitutes the most superior stock thus far to emerge in the Wisconsin Family of *P. chrysogenum* strains.

DISCUSSION

As has already been intimated, the "Wisconsin Family" of *P. chrysogenum* strains by no means comprehends all the strains of penicillin-producing mold which have been developed and studied in this laboratory. According to the writers' definition, the "Family" includes only strain Wis. *Q176* and the select Wisconsin stocks derived therefrom. Excluded are all of the strains obtained in the early days of the Wisconsin project, as well as the thousands of variants secured in numerous experiments on mutation induction for which strain *NRRL 1951* has been used as a parental form.

In view of the fact that penicillin production has been one of the principal characters dealt with in the selection program out of which the Wisconsin Family of strains emerged, certain matters pertaining to the evaluation of strain potency must now be made clear. As is well known to those familiar with penicillin fermentation work, yields obtained with a given strain vary greatly under different fermentation conditions. The composition of the medium, the type of inoculum used, temperature, pH, the amount of aeration, and the type of precursor employed (if any) as well as the manner in which it is added are among the many factors which may profoundly influence the expression of a strain's basic capacity to produce the antibiotic. After extensive experimentation with fermentation conditions, Brown and Peterson (7) were able to secure yields in excess of 2000 O.U./ml from strain Wis. *Q176* —more than double the amount obtained during earlier tests in the same laboratory and three to four times as much as this strain usually produces in the laboratory of its origin. However, it is possible that even these high yields do not represent the maximum of which *Q176* is capable. Yields from a given strain may also vary appreciably from run to run, although every effort may have been made to keep fermentation conditions constant; and, in shake-flask tests, yields vary even among replicate flasks in a single run. Apparently such a minor detail as the relative tightness of a cotton plug may, by affecting aeration, influence the yield obtained. Variability among lots of corn steep liquor, used in the preparation of the medium, is believed to account for some of the yield differences encountered over a period of time. It is also now well established that strains differ in their requirements for effective synthesis of the antibiotic and in their response to different fermentation conditions. Thus a given variant may perform in an outstanding fashion in one situation but appear to be mediocre in another, even though the latter may have been found favorable to some other strain. All of these considerations mean that evaluation of the abilities of a group of variants to produce penicillin is fraught with difficulties, that absolute potencies are rarely determined, and that ratings of cultures must to some extent at least be qualified in terms of fermentation conditions.

As has previously been explained, in the selection program in question here the evaluation of potency has been based upon a relatively uniform set of fermentation conditions—conditions, incidentally, which were probably far from optimum for most of the strains tested. As far as penicillin production has been concerned, then, this program has dealt with changes in potency as measured according to an arbitrary yardstick. In relation to the primary objectives of the study, that does not appear to have been a serious liability, but it is one which the reader should keep in mind.

Limited information on the performance of certain key strains of the Wisconsin Family in tests carried out elsewhere suggests that two interesting situations have developed simultaneously in connection with the emergence of this group of variants. It is evident, on the one hand, that the selection procedures followed have yielded a group of strains especially adapted to perform well under the fermentation conditions here employed, with the result that increases in potency registered by this laboratory's yardstick may not reflect a proportional change in basic potency or in ability to produce the antibiotic under other conditions (**29**). Thus, as has previously been noted, yields in the range of 2000–3000 O.U./ml have been secured in this laboratory with Wis. *51–20* and some other strains near the present end of the selection series, while the output of *Q176,* run as a control, has commonly been in the range of 400–650 O.U./ml—a situation which undoubtedly exaggerates the extent of the improvement in the newer strains. On the other hand, tests conducted outside of this laboratory have also indicated that in the Wisconsin Family of strains there have been realized some substantial gains in absolute potency, or at least increased productivity not confined to any narrowly restricted set of circumstances. There is far from perfect agreement in the rating of strains by this laboratory and by antibiotics workers elsewhere, but in the light of the situations discussed above this is hardly surprising. Perhaps the remarkable thing is that the system of evaluation employed here has successfully picked out so many strains which later have come to be generally regarded as superior.

Although *P. chrysogenum* has now been studied intensively and by many individuals for over a decade, no way has yet been found to tell with certainty from the appearance of a culture what its capacities in regard to production of the antibiotic may be. Nevertheless, an extremely interesting and suggestive general correlation between cultural characteristics and potency in antibiotic production can be traced among the descendants of strain *NRRL 1951*. Beginning with strain *NRRL 1951·B25* and continuing through strains such as Wis. *50–935* and Wis. *51–20*, progressive increase in ability to produce the antibiotic, at least as measured by test conditions in this laboratory, has been paralleled by a more or less progressive reduction in vegetative and reproductive vigor. Strains rated highest are all extreme degenerates and weaklings compared with wild-type strains. Similar degenerative changes were reported by Farrell (**14**) in relation to her most productive strains. While it is unlikely that these relationships are entirely accidental or meaningless, it must be emphasized that it by no means follows that any degenerate strain will be found to be a good producer. Certainly among the survivors of mutagenic treatments one encounters many slow-growing and weakly-sporulating variants which have very poor capacity to synthesize the antibiotic. In this same connection it is important to bear in mind also the several instances in which substantial improvement in capacity to sporulate was secured in the Wis. *51–20* line without any loss of ability to produce penicillin.

The history of the Wisconsin Family of strains taken in conjunction with that of the ancestral stock would seem to indicate that, as far as the type of change which makes strains more attractive commercially is concerned, there is no one method of securing variants which is overwhelmingly superior. In this case each of the various techniques employed to obtain new strains has, in fact, made significant contributions to the building up of the stock. As will be recalled, the first improvements over the wild-type ancestor came with the selection of spontaneous variants by Raper and Alexander. The first significant improvements in the Wisconsin pigmentless stock were also obtained through spontaneous variation. Strains *47–638, 47–1564,* and *48–701* represented progressive steps of improvement over strain *BL3–D10* and gave us the first industrially-attractive material in the pigmentless line. Later a spontaneous variant, *51–20*, initiated an attractive series as an offshoot from the nitrogen mustard branch of the family tree. To the credit of ultraviolet irradiation may be listed the following: the great improvement represented by strain Wis. *Q176,* the initiation of the Wisconsin pigmentless series through production of race *BL3–D10,* and a substantial rise in potency in the latter part of the UV line of the three-way selection program. Interestingly enough, the one major contribution of the nitrogen mustard treatments came in the very first generation of the nitrogen mustard line. Unfortunately fewer generations have been followed through in this series than in the others. X-irradiation was not among the types of mutagenic treatments employed in securing variants in connection with the Wisconsin Family of strains; but strain *X–1612,* a product of X-irradiation, unquestionably embodies one of the outstanding improvements in the general stock to which this Family belongs. Possibly there is merit in integrating various types of mutagenic treatments and simple selection, as was somewhat inadvertently done in the case of the present *P. chrysogenum* stock. It will be recalled that Raper and Alexander (**41**), operating solely by selecting naturally-occurring variants, were unable to secure improvements beyond those represented in strain *NRRL 1951·B25.* However, after X-ray and UV treatments had established a new line of departure it was again possible to obtain advantageous changes by simple selection.

The outcome of the various UV irradiation treatments involved in the emergence of the Wisconsin Family of strains also fails to mark either of the two UV wave lengths employed or any particular dosage as especially advantageous. A wavelength of 2750Å was employed in the treatment which yielded *Q176,* while some of the rises in potency in the UV line of the three-way selection program were secured with radiation of 2534–37Å. As previously noted, studies on irradiation effects in strain *NRRL 1951* have indicated that the maximum number of morphological variants are obtained at a survival level of about 25%, and this general level gave satisfactory results in the three-way selection series. Yet it must be remembered that strain *Q176,* representing undoubtedly the greatest single step of improvement obtained through irradiation in this program and probably the greatest single step of progress in the development of the entire Wisconsin Family of strains, emerged following a severe treatment which left less than 1% of the spores able to form colonies. It should likewise be recalled that another highly important strain, *BL3–D10,* came into being followed an irradiation so gentle that no killing at all could be detected.

Finally it is to be emphasized that, even though a number of variants from the Wisconsin Family of strains have won important roles in the antibiotics industry, the selection program in which they emerged was designed primarily to serve the interests of pure science. It is believed that much of fundamental interest has been learned by tracing the behavior of the organism through the long series of generations and treatments which have been involved. It seems likely that a knowledge of the evolution of this particular *P. chrysogenum* stock may be of value in appraising problems of variability in other imperfect fungi and provide perspective and tools for future workers desiring to obtain "tailor-made" strains of other organisms.

SUMMARY

A ten-year study on spontaneous and induced variation in penicillin-producing molds has yielded, in the writers' laboratories, an assemblage of superior variants which have become well known in industrial circles and which are referred to as the "Wisconsin Family" of *P. chrysogenum* strains. The origin, inter-relationships, and characteristics of key strains in this "Family" are here described.

The methods employed in obtaining, testing, and maintaining the improved stocks are considered. Strains embodying various advantageous changes have been secured through treatment of spores with ultraviolet radiation and with a nitrogen mustard. In addition, spontaneous changes have provided some improvements in the stock. Of the two wave lengths of ultraviolet employed in mutagenic treatments, neither has appeared to have any clear advantage over the other; and desirable changes were secured at high, moderate, and low UV dosage levels.

The Wisconsin Family begins with strain Wis. *Q176*. This strain, obtained in 1945, gave yields approximately double those obtainable from the best strain previously known. From this outstanding variant all other members of the Wisconsin Family trace their descent. The "Pigmentless Series," consisting of strains which secrete no yellow pigment, was inaugurated in 1947. In the "nitrogen mustard branch" of this series have emerged the best strains obtained in this selection program to date. Yields as high as 3000 O.U/ml have been secured from certain variants in this group under fermentation conditions employed in this laboratory. This is more than ten times the amount of penicillin produced under the same conditions by the improved strain from which the "Wisconsin Family" arose and about forty times the amount produced by the wild-type ancestor.

A more or less progressive decline in the vegetative and reproductive vigor of the strains has accompanied the increase in capacity to produce the antibiotic, not only in the Wisconsin Family itself but also in the succession of improved stocks which were ancestral to the Wisconsin

Family. Feeble sporulation and slow growth of the mycelium on agar are characteristic of all of the top-ranking strains. It is emphasized, however, that reduced vigor is not an infallible criterion for the selection of superior penicillin-producing variants. It is further emphasized that there is no close correlation between rate of colony growth on agar and rate of development of mycelium in aerated liquid cultures. In both small and large scale fermentations the best lines regularly produce their high yields about as quickly as maximum yields are obtained from wild-type or inferior strains.

DEPARTMENT OF BOTANY
UNIVERSITY OF WISCONSIN
MADISON, WISCONSIN

LITERATURE CITED

1. **Anderson, R. F., L. M. Whitmore, Jr., W. E. Brown, W. H. Peterson, B. W. Churchill, F. R. Roegner, T. H. Campbell, M. P. Backus and J. F. Stauffer.** Production of penicillin by some pigmentless mutants of the mold, *Penicillium chrysogenum Q176*. Ind. and Eng. Chem. **45**: 768–773. 1953.
2. **Arima, Kei.** Microbiological studies of penicillin production, VIII. The artificial mutation of penicillin producing molds; X ray induced mutations in Penicillium. Jour. Antibiotics **4**: 277–280. 1951.
3. ———. Microbiological studies of penicillin production, IX. On the pigmentless saltant *P. chrysogenum Q176* Arima et Ogasawara. Jour. Antibiotics **4**: 281–284. 1951.
4. **Backus, M. P., J. F. Stauffer and M. J. Johnson.** Penicillin yields from new mold strains. Jour. Am. Chem. Soc. **68**: 152–153. 1946.
5. **Bowden, J. P. and W. H. Peterson.** The role of corn steep liquor in the production of penicillin. Arch. Biochem. **9**: 387–399. 1946.
6. **Brown, W. E. and W. H. Peterson.** Factors affecting production of penicillin in semi-pilot-plant equipment. Ind. and Eng. Chem. **42**: 1769–1774. 1950.
7. ———. Penicillin fermentation in a laboratory type Waldhof fermenter. Ind. and Eng. Chem. **42**: 1823–1826. 1950.
8. **Calvert, O. H., G. S. Pound, J. C. Walker, M. A. Stahmann and J. F. Stauffer.** Induced variability in *Phoma lingam*. Jour. Agric. Res. **78**: 571–588. 1949.
9. **Camici, L., G. Sermonti and E. Chain.** Osservazioni sul *Penicillium chrysogenum* in cultura sommersa. I. Accrescimento miceliale e autolisi. Rendiconti Instituto Superiore di Sanità (Roma) **16**: Special fascicle: Primo Symposium Internazionale Di Chimica Microbiologica (1951), pp. 330–354. 1953.
10. **Campbell, T. H.** Morphological, cytological, and cultural studies on *Penicillium chrysogenum* Thom, with special reference to the population pattern phenomenon in strain Wis. Q176. Ph.D. thesis, 202 pp., University of Wisconsin. 1952.

11. **Churchill, B. W.** Cultural and nutritional studies of "A-type" variants of *Penicillium chrysogenum*. Ph.D. thesis, 111 pp., University of Wisconsin. 1949.
12. **Coghill, R. D. and R. S. Koch.** Penicillin—a wartime accomplishment. Chem. Eng. News. **23**: 2310–2316. 1945.
13. **Duggar, B. M. and A. Hollaender.** Irradiation of plant viruses and of microorganisms with monochromatic light. II. Resistance to ultraviolet radiation of a plant virus as contrasted with vegetative and spore stages of certain bacteria. Jour. Bact. **27**: 241–256. 1934.
14. **Farrell, Leone.** Induced variation and strain selection of *Penicillium chrysogenum* in relation to titer of natural penicillins. Canad. Jour. Med. Sci. **31**: 512–522. 1953.
15. **Florey, H. W., E. Chain, N. G. Heatley, M. A. Jennings, A. G. Sanders, E. P. Abraham and M. E. Florey.** Antibiotics. 2 vols. Oxford Univ. Press, London. 1949.
16. **Foster, J. W., H. B. Woodruff, D. Perlman, L. E. McDaniel, B. L. Wilker and D. Hendlin.** Microbiological aspects of penicillin. IX. Cottonseed meal as a substitute for corn steep liquor in penicillin production. Jour. Bact. **51**: 695–698. 1946.
17. **Gailey, F. B., J. J. Stefaniak, B. H. Olson and M. J. Johnson.** A comparison of penicillin-producing strains of *Penicillium notatum-chrysogenum*. Jour. Bact. **52**: 129–140. 1946.
18. **Goldschmidt, M. C. and H. Koffler.** Effect of surface-active agents on penicillin yields. Ind. Eng. Chem. **42**: 1819–1823. 1950.
19. **Greene, H. C. and E. B. Fred.** Maintenance of vigorous mold stock cultures. Ind. Eng. Chem. **26**: 1297–1299. 1934.
20. **Higuchi, K., F. G. Jarvis, W. H. Peterson and M. J. Johnson.** Effect of phenylacetic acid derivatives on the types of penicillin produced by *Penicillium chrysogenum* Q176. Jour. Am. Chem. Soc. **68**: 1669. 1946.
21. —— **and W. H. Peterson.** Penicillin in broths and finished products. Anal. Chem. **21**: 659–664. 1949.
22. **Hollaender, A.** The mechanism of radiation effects and the use of radiation for the production of mutations with improved fermentation. Ann. Mo. Bot. Gard. **32**: 165–178. 1945.
23. —— **and C. W. Emmons.** Wavelength dependence of mutation production in the ultraviolet with special emphasis on fungi. Cold Sp. Harb. Symp. Quant. Biol. **9**: 179–186. 1941.
24. ——, **K. B. Raper and R. D. Coghill.** The production and characterization of ultraviolet-induced mutations in *Aspergillus terreus*. I. Production of the mutations. Amer. Jour. Bot. **32**: 160–165. 1945.
25. ——, **E. R. Sansome, E. Zimmer and M. Demerec.** Quantitative irradiation experiments on *Neurospora crassa* II. Ultraviolet irradiation. Amer. Jour. Bot. **32**: 226–235. 1945.
26. **Jarvis, F. G. and M. J. Johnson.** The role of the constituents of synthetic media for penicillin production. Jour. Am. Chem. Soc. **69**: 3010–3017. 1947.
27. ——. The mineral nutrition of *Penicillium chrysogenum* Q176. Jour. Bact. **59**: 51–60. 1950.

28. **Johnson, M. J.** Metabolism of penicillin producing molds. Ann. N. Y. Acad. Sci. **48**: 57–66. 1946.
29. ———. Recent advances in penicillin fermentation. Rendiconti Instituto Superiore Di Sanità (Roma) **16**: special fascicle: Primo Symposium Internazionale Di Chimica Microbiologica (1951), pp. 125–154. 1953.
30. **Knight, S. G. and W. C. Frazier.** The effect of corn steep liquor ash on penicillin production. Science **102**: 617–618. 1945.
31. **Koffler, H., R. L. Emerson, D. Perlman and R. H. Burris.** Chemical changes in submerged penicillin fermentations. Jour. Bact. **50**: 517–548. 1945.
32. ———, **S. G. Knight and W. C. Frazier.** The effect of certain mineral elements on the production of penicillin in shake flasks. Jour. Bact. **53**: 115–123. 1947.
33. ———, **S. G. Knight, W. C. Frazier and R. H. Burris.** Metabolic changes in submerged penicillin fermentations on synthetic media. Jour. Bact. **51**: 385–392. 1946.
34. **Perlman, D.** Production of penicillin on natural media. Bull. Torrey Bot. Club **76**: 79–88. 1949.
35. ———. Some mycological aspects of penicillin production. Bot. Rev. **16**: 449–523. 1950.
36. **Peterson, W. H.** Factors affecting the kinds and quantities of penicillin produced by molds. The Harvey Lectures **42**: 276–302. 1946–47.
37. **Pontecorvo, G. and G. Sermonti.** Recombination without sexual reproduction in *Penicillium chrysogenum*. Nature **172**: 126–127. 1953.
38. ———. Parasexual recombination in *Penicillium chrysogenum*. Jour. Gen. Microbiology **11**: 94–104. 1954.
39. **Raper, K. B.** The development of improved penicillin-producing molds. Ann. N. Y. Acad. Sci. **48**: 41–52. 1946.
40. ———. A decade of antibiotics in America. Mycologia **44**: 1–59. 1952.
41. ——— **and D. F. Alexander.** Penicillin. V. Mycological aspects of penicillin production. Jour. Elisha Mitch. Sci. Soc. **61**: 74–113. 1945.
42. ——— **and C. Thom.** A Manual of the Penicillia. 875 pp. Williams and Wilkins Co., Baltimore. 1949.
43. **Reese, E., K. Sanderson, R. Woodward and G. M. Eisenberg.** Variation and mutation in *Penicillium chrysogenum* Wis. Q176. Jour. Bact. **57**: 15–21. 1949.
44. **Roegner, F. R.** Natural and artificially-induced variation in *Penicillium chrysogenum* Thom. Ph.D. thesis, 179 pp., University of Wisconsin. 1951.
45. ———, **M. A. Stahmann and J. F. Stauffer.** Induction of variants in *Penicillium chrysogenum* by methyl-bis(β-chloroethyl)amine. Amer. Jour. Bot. **41**: 1–4. 1954.
46. **Rowley, D., J. Miller, S. Rowlands and E. Lester-Smith.** Studies with radioactive penicillin. Nature **161**: 1009–1010. 1948.
47. **Schmidt, W. H. and A. J. Moyer.** Penicillin. I. Methods of assay. Jour. Bact. **47**: 199–208. 1944.
48. **Sermonti, G.** Genetics of *Penicillium chrysogenum*. I. Heterokaryosis in *Penicillium chrysogenum*. Rendiconti Instituto Superiore Di Sanità (English Edition) **17**: 213–230. 1954.

49. **Stahmann, M. A. and J. F. Stauffer.** Induction of mutants in *P. notatum* by methyl-bis(β-chloroethyl)amine. Science **106**: 35–36. 1947.
50. **Stauffer, J. F. and M. P. Backus.** Spontaneous and induced variation in selected stocks of the *Penicillium chrysogenum* series. Ann. N. Y. Acad. Sci. **60**: 35–49. 1954.
51. ——— **and B. W. Churchill.** Stimulation of colony formation from spores of *Penicillium chrysogenum*, Wis. Q176, by ultraviolet radiation (abstr.). Amer. Jour. Bot. **36**: 816. 1949.
52. **Stefaniak, J. J., F. B. Gailey, C. S. Brown and M. J. Johnson.** Pilot plant equipment for submerged production of penicillin. Ind. Eng. Chem. **38**: 666–671. 1946.
53. ———, **F. B. Gailey, F. G. Jarvis and M. J. Johnson.** The effect of environmental conditions on penicillin fermentations with *Penicillium chrysogenum* X-1612. Jour. Bact. **52**: 119–127. 1946.
54. **Thom, C.** Mycology presents penicillin. Mycologia **37**: 460–475. 1945.
55. ——— **and K. B. Raper.** A Manual of the Aspergilli. 373 pp. Williams and Wilkins Co., Baltimore. 1945.
56. **Wakaki, S.** Mycological aspects of penicillin production, III–IV. Statistical study on the monospore isolation of *Penicillium chrysogenum* Q176, I–II. Jour. Antibiotics **4**: 545–551. 1951.
57. **Wolf, F. T.** The amino acid metabolism of *Penicillium chrysogenum* Q176. Arch. Biochem. **16**: 143–149. 1948.
58. ———. Amino acids in the biosynthesis of penicillin. Mycologia **41**: 403–410. 1949.

Use of Mutagens in the Improvement of Production Strains of Microorganisms

R. W. THOMA

The Squibb Institute for Medical Research, New Brunswick,
New Jersey, U.S.A.

Received October 17, 1970

ABSTRACT. Physical and chemical agents were employed in our laboratories to induce mutation in a variety of microorganisms used for production of antibiotics or enzymes. Improved production strains of *Penicillium chrysogenum* (penicillin-producer), *Streptomyces griseus* (streptomycin-producer), *Streptomyces nodosus* (amphotericin B-producer), *Streptomyces noursei* (nystatin-producer), *Streptomyces umbrinus* (diumycin-producer), *Streptomyces prasinus* (prasinomycin-producer), *Streptomyces roseochromogenes* (steroid-16α-hydroxylase-producer), and *Arthrobacter simplex* (steroid-1-dehydrogenase-producer) were developed by use of mutation-selection techniques.
The methods found to be most successful with each species are described. The genealogical relationships within species of a number of strains of *Penicillium chrysogenum*, *Streptomyces prasinus*, and *Streptomyces roseochromogenes* are presented.

Maintaining production strains of microorganisms without trying to improve them is futile. Periodic re-isolation and evaluation under production conditions is necessary, not only to replenish stocks of primary culture sources, but also to allow selection of strains under new fermentation conditions and with fermentation raw materials that may not be constant in composition.

Cloning procedures that do no more than allow selection of normal variants or spontaneous mutants are inefficient (Sermonti, 1969; Calam, 1970). Although re-isolation of strains on selective media without mutagenic treatment is still practiced, it is generally preferable to subject a culture to a physical or chemical mutagenic treatment before cloning, in order to increase the range of variation and the ratio of mutants to parental genotypes in the surviving population. Not only will the ratio of mutants to parental genotypes be higher after an induction treatment, but there is reason to believe that some types of mutations that can be induced would occur spontaneously only very rarely.

It is undoubtedly true that all major commercial producers of fermentation products are engaged in mutation-selection programs, at least with the strains employed in their principal processes. General reviews (Alikhanian, 1962, 1969; Bradley, 1966; Calam, 1964, 1970; Davies, 1964; Sermonti, 1969) reveal that there is rather widespread agreement as to the best strategy for a program of strain development or screening designed to improve production strains. All these authors endorse the use of mutagens. In our laboratories, mutagens have been used for more than two decades.

In contrast to the willingness of workers in the field to expose general principles, the recent literature contains few details as to the methods they employ and the results they have obtained with industrially important microorganisms. The papers of Alikhanian (1962; 1969) and co-workers are a notable exception.

It is our intention to present in this paper the results of some strain-selection programs carried out in our laboratories in the last 15 years. We have selected examples that illustrate what can be accomplished with a variety of microorganisms. We do not intend to give a complete history of the development of strains in any one of these programs, nor do we suggest that the work reported is necessarily the most recent. We shall describe a variety of mutagenic treatments that have produced, or have been associated with, positive quantitative effects.

MATERIALS AND METHODS

Ultraviolet irradiation was carried out by exposing about 15 ml of a suspension of spores, cells, or fragmented mycelium in a petri dish to a light source rich in energy at 254 nm. Irradiations and primary dilutions were carried out in subdued light to avert photoreactivation. Two different irradiation devices were employed. The older apparatus, used most often, had as a light source a 4-W,

Table 1. Comparison of the energy delivered by two irradiation devices

Distance, light source to culture (mm)	Energy delivered by	
	old device (ergs/mm²/sec)	new device (ergs/mm²/sec)
411	5	19
360	7	23
310	8	30
259	12	39
208	17	46
157	24	70

U-shaped, GE germicidal lamp, and agitation of the suspension was effected by means of a Teflon-coated magnetically rotated stirring bar. The newer apparatus had a 15-W, straight, 45-cm GE lamp with a chrome reflector and shield, and agitation of the petri dish and contents was effected by means of a serological-type shaker. A comparison of the energy delivered by the two devices is given in Table 1.

Chemical mutagens were tried initially in a range of concentrations and exposure times, and usually in several menstrua. Generally, a preliminary appreciation of toxic effect was obtained on gradient plates. An appropriate concentration and range of exposure conditions were selected for each combination of organism and agent. A list of some of the mutagens used in our laboratories is given in Table 2.

Table 3 summarizes some methods that were found to be especially effective, as measured by the productivity of one or more strains derived from application of the method.

RESULTS

Development of penicillin-producing *Penicillium chrysogenum* strains

Fig. 1 shows an abbreviated genealogical chart of *Penicillium chrysogenum* strains. The progenitor of all these shown was SC3576, an important production strain, many generations removed from Wisconsin strain 51—20. Strains P9808, P13412, P17450, P23501, at the ends of the branches shown, and several others intermediate between SC3576 and these, were used in the manufacturing plant or were regarded seriously as candidates for production use. Strains of each generation were 5 to 10% better in productivity in shaken flasks than those of the preceding generation. Other strains equivalent in productivity, but less interesting for secondary reasons, are not shown on the chart. In this program more than 25,000 strains were tested, and flask-scale titers were doubled.

Development of streptomycin-producing *Streptomyces griseus* strains

Natural selection or, more properly, cloning out followed by evaluation and preservation of a few superior strains,

Table 2. Some chemical mutagens, grouped by class

Purine analogs	2,6-diaminopurine 8-azaguanine 6-mercaptopurine
Pyrimidine analogs	5-fluorouracil 5-bromouracil 6-azauracil 5-bromodeoxyuridine 6-azauridine 5-bromodeoxycytidine 5-azacytidine
Proflavine	acriflavine
Monofunctional alkylating agents	β-propiolactone N-methyl-N′-nitro-N-nitrosoguanidine methyl ethanesulfonate ethyl methanesulfonate chlorethyl methanesulfonate n-butyl methanesulfonate ethyleneimine glycidol iodoacetate epichlorhydrin epibromhydrin dimethyl sulfate diethyl sulfate
Bifunctional alkylating agents	diepoxybutane methyl-bis-(β-chlorethyl)amine bis-(2-chlorethyl)sulfide
Trifunctional alkylating agent	triethylenemelamine

permitted us to maintain a satisfactory production culture for several years, after an initial period of intensive mutation and screening. One mutation program, conducted for a year after the period of maintainance by cloning without treatment, produced a few strains with superior qualities. Ultraviolet irradiation was the treatment that gave best results. Irradiation with a ^{60}Co source gave many variants less productive than the parent, and gave no stable positive variants. Chemical mutagens yielded no strains of lasting interest.

Development of amphotericin B-producing *Streptomyces nodosus* strains

One ultraviolet-induced strain of value emerged from a rather limited program. No success was obtained with chemical mutagens.

Development of nystatin-producing *Streptomyces noursei* strains

During a period when many mutagenic treatments were tested, only ultraviolet irradiation produced a strain accepted for production use. One stable recombinant strain, obtained by promoting

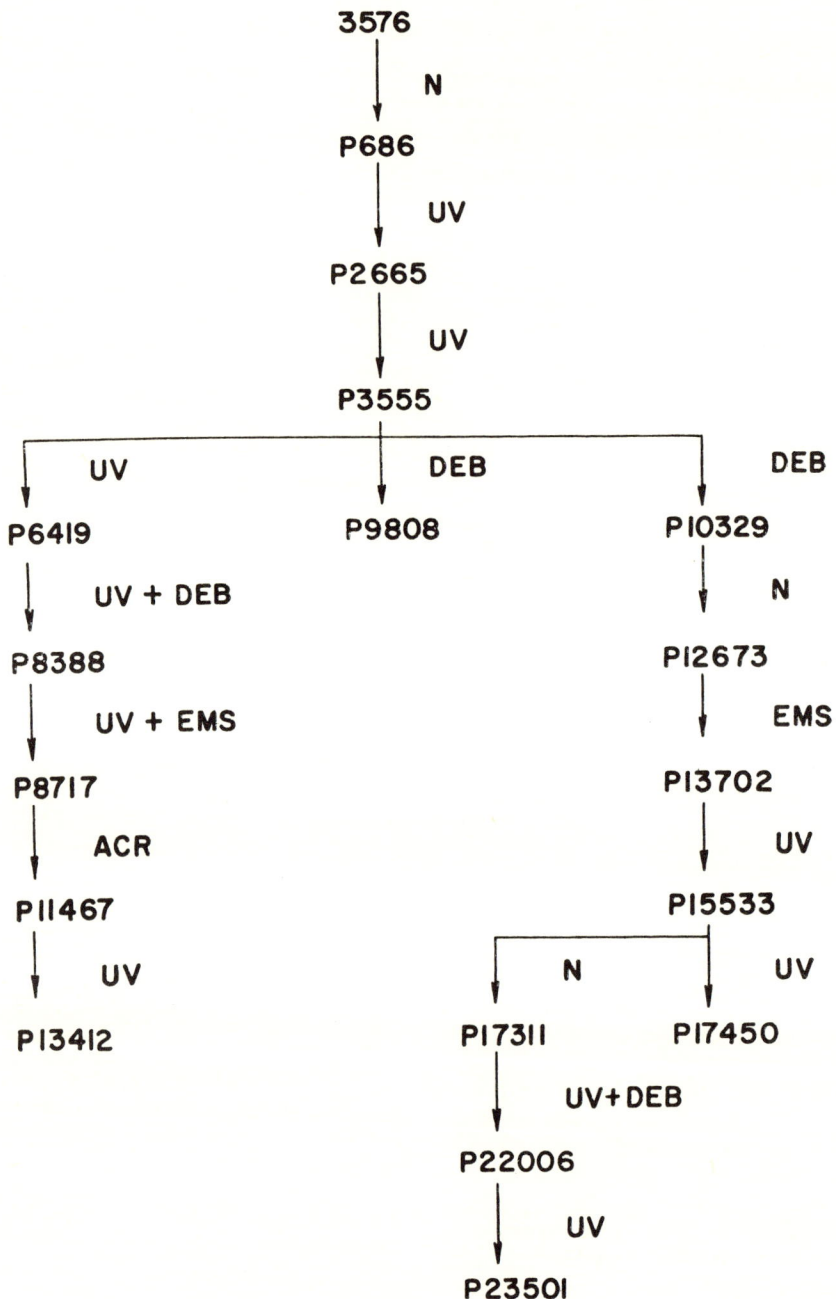

Fig. 1. Genealogy of some *Penicillium chrysogenum* strains. N — no treatment; UV — ultraviolet irradiation; DEB — diepoxybutane; EMS — ethyl methanesulfonate; ACR — acriflavine.

Table 3. Some mutagenic treatments that have produced strains with superior productivity

Species of organism (product)	Mutagenic agent and treatment conditions
Penicillium chrysogenum (penicillin)	*UV (Ultraviolet irradiation)* Surface growth from 7-day old agar slant suspended in 20 ml peptone (0.1%) solution, filtered through sterile cotton, 15 ml put into a Petri dish. Depth of suspension, 5 mm. Initial spore count, 1.1×10^7. Irradiated with UV (1,800 ergs/mm^2). Killing, 88%.
P. chrysogenum (penicillin)	*DEB (Diepoxybutane)* Surface growth from a 7-day old agar slant suspended in 18 ml solution of sodium lauryl sulfate (0.01%) and sodium chloride (0.85%) filtered through cotton. Diluted with five volumes aqueous 0.05 M phosphate (pH 7.0) with DEB (0.1%), agitated 30 min at 30 C, washed. Killing, 92%.
P. chrysogenum (penicillin)	*UV + DEB* Surface growth from a 7-day old agar slant collected with 20 ml 0.1% peptone, filtered through cotton, irradiated with UV (1800 ergs/mm^2), diluted (3 parts/50) with 0.1% peptone solution containing DEB (0.2%), agitated 40 min at 30 C, washed. Killing, 99.8%.
P. chrysogenum (penicillin)	*UV + EMS (Ethyl methanesulfonate)* Spores irradiated with UV (1440 ergs/mm^2), diluted (3 parts/50) with 0.2 M phosphate (pH 7.2) containing 0.2% EMS, agitated 30 min at 30 C, washed. Killing, 99%.
P. chrysogenum (penicillin)	*ACR (Acriflavine)* Spores from a corn culture were suspended in 2.0 M phosphate (pH 7.2), filtered through cotton, centrifuged, resuspended, diluted (3 parts/50) in fresh buffer with ACR (0.2%), shaken 30 min at 30 C, centrifuged, resuspended. Killing, 71%.
P. chrysogenum (penicillin)	*EMS* Spores from 7-day slant culture suspended in 0.2 M phosphate (pH 7.0), filtered, diluted (3 parts/50) with phosphate buffer containing EMS (1%), shaken 60 min at 30 C, washed by centrifugation. Killing, 34%.
Streptomyces griseus (streptomycin)	UV irradiation of homogenized mycelium.
S. nodosus (amphotericin)	UV irradiation.
S. noursei (nystatin)	UV irradiation of conidia from surface culture.
S. umbrinus (diumycin)	NTG (N-methyl-N'-nitro-N-nitrosoguanidine); 5-day old shaken culture, spontaneously fragmented, was filtered, centrifuged, washed, resuspended in pH 7.2 tris-maleate buffer at 30 C for 4 hours with NTG (10 µg/ml). Little if any killing effected.
S. prasinus (prasinomycin)	UV irradiation of conidia from surface culture suspended in distilled water, 2,160 ergs/mm^2. Killing, 90%.
S. prasinus (prasinomycin)	NTG + UV; 2-day shaken culture, spontaneously fragmented, filtered, centrifuged, suspended in pH 7.2 tris-maleate buffer at 30 C, for 4 hours with NTG (50 µg/ml) then irradiated with UV (1,080 ergs/mm^2). Killing, 25%.
S. roseochromogenes (steroid-16α-hydroxylase)	UV; shaken culture grown in presence of steroid; washed with 0.5 M NaCl, resuspended in 0.5 M sorbitol − 0.05 M phosphate pH 7.0, exposed to lysozyme (500 µg/ml) for 60 min at 25 C. Washed in 0.05 M NaCl − 0.05 M glycine pH 9.0, irradiated with UV (6,000 ergs/mm^2).

Species of organism (product)	Mutagenic agent and treatment conditions
Arthrobacter simplex (steroid-1-dehydrogenase)	UV irradiation of 3-day old broth culture (3,600 ergs/mm^2) after filtration and dilution.
Bacillus megaterium (penicillin acylase)	Cloning without treatment.

parasexual recombination of multiply marked strains, was chosen for use in the manufacturing plant, but was shown not to offer any advantage over the standard culture source of the ultraviolet--derived strain (Bryson et al., 1968).

Development of diumycin-producing *Streptomyces umbrinus* strains

An initial gain in productivity was obtained by ultraviolet irradiation of a strain derived from a natural source. Several third-generation strains obtained after ultraviolet treatment were superior to the second-generation parent (M124), but were surpassed in productivity by strain 6219 selected after treatment of M124 by N-methyl-N'-nitro-N-nitrosoguanidine (NTG). A total of 2300 strains was derived from 6219 by treatment with UV or UV + NTG; of these 2300, six were 20—40% superior to strain 6219.

Development of prasinomycin-producing *Streptomyces prasinus* strains

Fig. 2 shows the relationships among several prasinomycin-producing strains. The first improved strain was derived by cloning without treatment. In the next generation, one strain (6182) selected without treatment was equivalent to two others selected after ultraviolet induction. Strain 6182 was induced by treatment with NTG + UV to give 6196, the best member of this family. Strain 6196 was twice as productive as 3768. About 600 strains derived from 5426 were tested before 6196 was selected.

Development of steroid-16α-hydroxylating *Streptomyces roseochromogenes* strains

Fig. 3 shows a family of strains derived from strain 6035, the resultant of an earlier intensive program. Interesting strains were derived from 6035 by ultraviolet irradiation as well as by treatment with NTG. The strains of the second generation were only 5 to 10% better than 6035 in their ability to hydroxylate 9α-fluoro-cortisol. However, the third-generation strains, notably 6162 and 6186, were twice as active as 6035. A total of 2400 strains was tested in this program.

Development of steroid-1-dehydrogenating *Arthrobacter simplex* strains

Use of a group of chemical mutagens failed to develop any strains more active than the parent. The only superior strain was obtained by selection after ultraviolet irradiation. About 700 strains were isolated and tested.

Development of penicillin acylase-producing *Bacillus megaterium* strains

Periodic cloning without mutagenic treatment allowed a small but measurable improvement in activity to be realized over a period of several years.

DISCUSSION

We have presented some examples of successful development of strains more productive than their progenitors with respect to the synthesis of antibiotics (penicillin, streptomycin, amphotericin, nystatin, prasinomycin, and diumycin) or enzymes (penicillin acylase, steroid-1-dehydrogenase, or steroid-16α-hydroxylase). We have described the mutagenic treatments that produced a number of these useful strains. It is important to

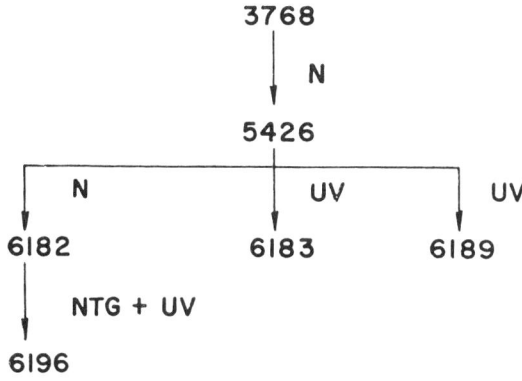

Fig. 2. Genealogy of some *Streptomyces prasinus* strains. N — no treatment; UV — ultraviolet; NTG — N-methyl-N'-nitro-N-nitrosoguanidine.

emphasize that we make no claim to having directed mutation in a desired fashion. Mutagens may differ in specificity of intragenic site of action (Freese, 1963; Kihlman, 1961, 1966; Krieg, 1963) but we know of no real example of mutagenic action that shows gene specificity (Sermonti, 1969).

Our philosophy in using mutagens is that each strain and mutagen may interact in a different manner from that in which a closely related strain and mutagen pair may interact. Therefore, in our screening programs we customarily choose several strains for treatment at any one time, and avoid the exclusive use of one type of mutagenic treatment. Occasionally, we clone without treatment, in order to allow selection of mutants whose expression may have been delayed. Our screening methods differ in few respects from those outlined by Davies, Calam, Bradley, and Sermonti.

Although we have devised some novel methods of treatment of our microorganisms in an attempt to induce mutation, usually we have selected agents reported to effect mutations in other microorganisms. Most of the chemical agents we have used in recent years have been studied by others for their effects on nucleic acids (Lawley, 1966) or bacteria-bacteriophage systems (Orgel, 1965).

The effects of ultraviolet irradiation have been studied and reviewed in

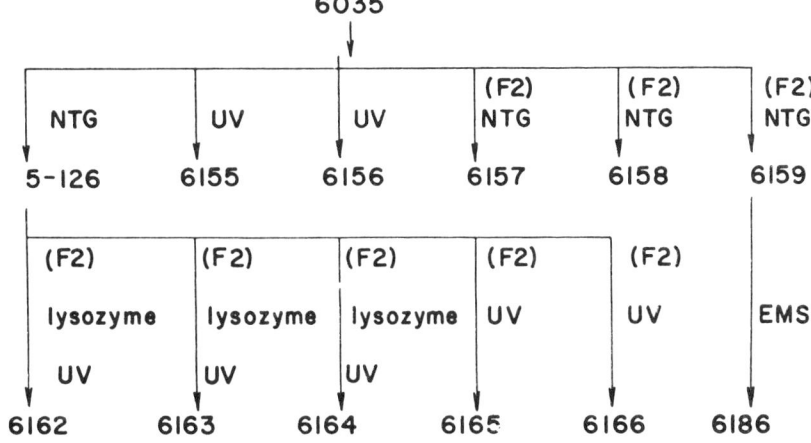

Fig. 3. Genealogy of some *Streptomyces roseochromogenes* strains. NTG — N-methyl-N'-nitro-N-nitrosoguanidine; UV — ultraviolet; (F2) — grown on 9α-fluorocortisol; lysozyme — treated before irradiation; EMS — ethyl methanesulfonate.

detail by Witkin (1969a, b), who has worked mainly with *Escherichia coli*. We are well aware that the effects of mutagenic agents on the spore, mycelial, or vegetative stage of fungi, actinomycetes or even true bacteria may be very complex, in contrast to those observed with simpler systems (Auerbach, 1961).

The use of mutagens in the improvement of production strains of microorganisms cannot be expected to effect quantitative mutations in a rational way. Rationality must be brought into a screening program by imposing selective pressures in those stages of propagation that follow induction of mutation, in order to eliminate low producers and favor the growth, or otherwise allow identification of, superior producers. Some methods of achieving this result have been suggested by Bradley (1966).

The use of mutagens in the improvement of production strains is established; the techniques and agents are available; production of mutants is never the limiting stage in a strain development program. We believe that the efficiency of screening programs will be increased by automating existing methods for selecting and evaluating strains, and by devising new methods for differentiating among strains in the first clonal stage or as soon thereafter as possible.

The work reported in this paper was done in the Department of Microbiological Development, under the direction of Dr. W. E. Brown. Investigators whose work is reported are Mrs. K. Batt, Mr. A. P. Bayan, Miss J. Baymiller, Dr. C. F. Bruno, Dr. P. F. Crosby, Dr. P. K. Lee, Dr. M. W. Nimeck, Mrs. C. Ricca, and Mr. N. Sicurella.

I am indebted to Dr. D. Frost for editorial criticism of this manuscript.

References

Alikhanian, S. I.: *Induced mutagenesis in the selection of microorganisms*. Adv. Appl. Microbiol. 4 : 1, 1962.
Alikhanian, S. I.: *Induced mutagenesis as related to variation in quantitative features (antibiotic production)*. In: Proceedings of International Symposium on Genetics and Breeding of Streptomyces. (Dubrovnik, 1968) (Sermonti, G., & Alačević, M., Eds.). Yugoslav Academy of Sciences and Arts, Zagreb, p. 69, 1969.
Auerbach, C.: *Chemicals and their effects*. In: Mutation and Plant Breeding. NAS-NRC 891 p. 120, 1961.
Bradley, S. G.: *Genetics in applied microbiology*. Adv. Appl. Microbiol. 8 : 29, 1966.
Bryson, V., Thoma, R. W., Nimeck, M. W.: *Application of microbiological genetics in industry*. In: Proc. XII Int. Congr. Genet. (Tokyo, 1968). Vol. 2, p. 250. Sci. Council Japan, Tokyo, 1968.
Calam, C. T.: *The selection, improvement, and preservation of microorganisms*. Progr. Ind. Microbiol. 5 : 1, 1964.
Calam, C. T.: *Improvement of microorganisms by mutation, hybridization, and selection*. In: Methods in Microbiology. (Norris, J. R., & Ribbons, N. W., Eds.). Vol. 3A, Chap. VII, p. 435, Acad. Press, New York, 1970.

Davies, O. L.: *Screening for improved mutants in antibiotic research*. Biometrics 20 : 576, 1964.
Freese, E.: *Molecular mechanisms of mutations*. In, Molecular Genetics. (Taylor, J. H., Ed.). Part I: p. 207, Acad. Press, New York, 1963.
Kihlman, B. A.: *Biochemical aspects of chromosome breakage*. Adv. Genet. 10 : 1, 1961.
Kihlman, B. A.: *Actions of chemicals on dividing cells*. Prentice-Hall, Englewood Cliffs, N. J., 1966.
Krieg, D. R.: *Specificity of chemical mutagenesis*. Progr. Nucleic Acid Res. Mol. Biol. 2 : 125, 1963.
Lawley, P. D.: *Effects of some chemical mutagens and carcinogens on nucleic acids*. Progr. Nucleic Acid Res. Mol. Biol. 5 : 89, 1966.
Orgel, L. E.: *The chemical basis of mutation*. Adv. Enzymol. 27 : 289, 1965.
Sermonti, G.: *Genetics of antibiotic-producing microorganisms*. Wiley-Interscience, London, 1969.
Witkin, E. M.: *Ultraviolet-induced mutation and DNA repair*. Annual Rev. Microbiol. 23 : 487, 1969a.
Witkin, E. M.: *The role of DNA repair and recombination in mutagenesis*. In: Proc. XII Int. Congr. Genet. (Tokyo, 1968). Vol. 3, p. 225. Sci. Council Japan, Tokyo, 1969b.

Parasexual Recombination in *Penicillium chrysogenum*

BY G. PONTECORVO AND G. SERMONTI

Department of Genetics, University of Glasgow

SUMMARY: Roper's (1952) technique for the isolation in filamentous fungi of strains carrying in their hyphae diploid nuclei heterozygous for known markers has been applied with minor modifications to a third species: *Penicillium chrysogenum*. One of these modifications is the use of 'dwarf' (stunted growth) mutants for growing balanced heterokaryons. Heterozygous diploid strains of *P. chrysogenum* behave like those of the other two species in yielding, vegetatively, new strains with recombined properties. This behaviour can be marshalled for the deliberate breeding of improved industrial strains.

In two species of filamentous fungi, *Aspergillus nidulans* (Roper, 1952; Pontecorvo & Roper, 1952, 1953; Pontecorvo, Tarr Gloor & Forbes, 1954) and *Aspergillus niger* (Pontecorvo, 1952; Pontecorvo, Roper & Forbes, 1953), recombination between properties of different strains may take place outside, or in the absence of, a sexual cycle. Processes leading to gene recombination otherwise than via sexual reproduction are called 'parasexual' (Pontecorvo, 1953a). The present paper reports the discovery of a parasexual cycle in a third species, *Penicillium chrysogenum*, the mould used for the industrial production of penicillin. This discovery was the outcome of the application, with minor modifications, to this species of the techniques developed for the other two species.

The differences in properties used for the present work (colours of conidia, growth rates, colony morphology and nutritional requirements) were chosen exclusively for their convenience as experimental markers. The discovery that recombination occurs in respect of these properties makes it certain that it also occurs in respect of inherited properties of other kinds. Clearly, this knowledge makes possible the deliberate breeding of improved strains for penicillin production. A preliminary report of the work in this paper has been published (Pontecorvo & Sermonti, 1953).

The production of marked strains

Unless explicitly stated, the techniques used in the present work are the same as those used as a routine for *Aspergillus nidulans* (Pontecorvo, 1953b) and applied to *A. niger* (Pontecorvo et al. 1953).

Media. Minimal medium (MM): a modified Czapek-Dox. Complete Medium (CM): a complex medium containing yeast extract, casein hydrolysate, etc. Incubation was at 26°.

Organisms. Two starting strains of *Penicillium chrysogenum* were used; one with white conidia, turning very pale green after long incubation at room temperature, and one with yellow-brown conidia. The former, kindly supplied by Dr J. C. Calam (Imperial Chemical Industries, Manchester) was a mutant obtained after nitrogen mustard treatment. It is designated here with the symbol w. The yellow strain (symbol y, code no. *1086*) was obtained previously by one of us (G.S.) after ultraviolet irradiation of strain *Q 176.47.1564*. Both were prototrophs, i.e. able to grow well on medium MM, like the green-spored strain *Q 176*, to which both ultimately traced back.

The marked strains required for the present work were strains differing from the wild type in three pairs of alternative characters and from each other in six pairs. Two pairs (white versus green conidia, yellow versus green conidia) were already available in the two starting strains mentioned above. The other four pairs were obtained by successive ultraviolet irradiations and they consisted either in differences in nutritional requirements or differences in growth rates (e.g. dwarf versus normal colony size).

Table 1. *Results of ultraviolet irradiation (4 min. at 30 cm.) of dishes of CM agar plated with about 1000 conidia each*

Strain irradiated		Viable counts (%)	Colonies isolated (No.)	Mutants among isolates				
				Nutritional			Dwarf	
Code No.	Colour and requirements			No.	Code No.	Additional requirements	No.	Code No.
w	White, no requirements	2	238	3	4w	NO_2^-	—	—
					7w	Thiosulphate		
					8w	Pyridoxine		
y	Yellow, no requirements	5	160	3	19y	Thiosulphate	—	—
					22y	Hypoxanthine		
					38y	NO_2^-		
22y	Yellow, hypoxanthine	17	325	4	22y5	NH_2 groups	1	22y dw 1
					22y13	Methionine		
					22y14	Arginine		
					22y15	Anthranilic acid		
7w	White, thiosulphate	6	150	3	7w13	Hypoxanthine	2	7w dw 4
					7w16	Adenine		7w dw 5
					7w32	Hypoxanthine		
38y	Yellow, NO_2^-	0·2	125	2	38y2	Peptone	1	38y dw 6
					38y3	Thiosulphate		
			998	15				

Suspensions of 10^4 conidia/ml. were spread in volumes of 0·1 ml./plate (1000 conidia) over the agar surface of a series of Petri dishes containing CM agar. The dishes were irradiated by exposing them to an Hanovia XI low-pressure mercury lamp for 4 min. at 30 cm. distance. Colonies which grew up (random sample) after such treatment (Table 1) were isolated and screened for mutants with additional growth factor requirements (Pontecorvo, 1953b). Furthermore, four dwarf mutants were picked by inspection from some of the irradiation series. The mutants were purified by micromanipulation of single conidia or by plating followed by single colony isolation.

Synthesis of balanced heterokaryons

Previous attempts by one of us (G.P.) to form balanced heterokaryons between pairs of strains each differing from the wild type in one additional nutritional requirement had been unsuccessful. A similar failure is mentioned by Bonner (1946) in respect of pairs of strains of *Penicillium notatum*; yet heterokaryosis in this species (Pontecorvo & Gemmell, 1944; Sansome, 1947) was known to occur. The technique used in our attempts was the one which has invariably proved effective with other species of moulds (Pontecorvo, 1947, 1953b). In its most recent version it involves the following steps: (1) inoculation into liquid CM of a thick suspension of about equal numbers of conidia of each of two strains requiring different growth factors; (2) incubation for a time sufficient to obtain a thin mesh of mycelium; (3) removal of this mesh, which is washed by dipping in MM, and thoroughly broken up by teasing out over the surface of plates of MM agar; (4) incubation of these until growth starts from a few points; (5) isolation of growing hyphal tips from these points on to MM agar. The length of incubation (under 4) required for the 'escape' of the heterokaryotic growth varies with the species and with the particular pair of strains used (4 days for most combinations in *Aspergillus nidulans* and in *A. niger*).

The growth rate on agar medium of *Penicillium chrysogenum* is about one-fifth of that of *Aspergillus nidulans*. It was therefore expected that the time of incubation necessary for the escape on MM agar of heterokaryotic growth would be much longer. In fact, in the unsuccessful attempts mentioned above, after 2 or 3 weeks the MM plates would begin to show widespread growth, but this proved to be syntrophic, *not* heterokaryotic. The conditions making it possible for balanced heterokaryotic mycelium to become established to the exclusion of, or in equilibrium with, homokaryotic mycelium have been discussed by Pontecorvo (1953b). It appeared likely that the failure just mentioned might be due to the very low growth rate of *P. chrysogenum*. A growth rate small compared to the rate of diffusion of metabolites may not confer on the heterokaryon an advantage over the two syntrophically growing homokaryons sufficient for the heterokaryon to become established.

This kind of reasoning suggested two ways out. First, the use of pairs of strains with a very much decreased growth rate not capable of being restored to normal by the supply of growth factors; the possibility of syntrophism would thus be excluded. Strains of this type constitute an appreciable proportion of those obtained after irradiation and are identified by the tiny colonies they form even on CM agar. They will be referred to as 'dwarf' strains. Balanced heterokaryons with normal growth rate, formed between pairs of dwarf strains, have been described in other species (Dodge, 1942; Pontecorvo, 1947). Dodge's work gave, in fact, the first example of balanced heterokaryosis.

Secondly, the use of pairs of strains with more than one growth factor requirement, in order to make their metabolic interdependence in the absence of the growth factors more interlocked, and therefore presumably favouring intracellular (heterokaryotic) rather than intercellular (syntrophic) co-operation. This metabolic interlocking might be particularly effective when the growth factor requirements of the two strains are metabolically related in pairs, as in the case of different blocks on the same pathway of synthesis; e.g. requirement of adenine or hypoxanthine versus adenine only; for methionine or cystine versus methionine only. Both types of strain combinations were used and both led to the successful isolation of balanced heterokaryons, as presently described.

Balanced heterokaryons between dwarf strains. The first combination attempted between two dwarf strains was that of *22y dw 1* and *7w dw 4*, i.e. of one dwarf strain with yellow conidia and requiring adenine or hypoxanthine, and another dwarf strain with white conidia and requiring cystine or thiosulphate. On MM agar, strain *7w dw 4* does not grow at all; strain *22y dw 1* grows to a barely visible size (Pl. 1, fig. 1). On CM agar *7w dw 4* forms very tiny dome-shaped colonies and *22y dw 1* irregularly shaped colonies, highly convoluted and seldom growing beyond 1 cm. diameter (Pl. 1, fig. 1).

The balanced heterokaryon was obtained by incubating for 5 days in a test-tube in liquid CM a thick mixed suspension of conidia of the two strains, removing the mycelium formed, teasing this out on the surface of MM agar plates and incubating for a further 9 days. After this time a little growth had

Table 2. *Plating on supplemented medium of conidia from balanced heterokaryons in order to recover the two component types*

Balanced heterokaryon	Medium	Colonies obtained			
		Yellow		White	
		No.	%	No.	%
22y dw 1 + 7w dw 4	CM	222	78·2	62	21·8
22y 13 + 7w 16	MM + adenine + methionine	16	8·5	171	91·5

occurred at various points, and out of some of these there grew small tufts of more vigorous mycelium with white heads. Isolation of growing tips from these tufts on to MM agar plates gave uniformly good growth with macroscopically white sporing surface (Pl. 1, fig. 1). Isolation on to CM agar gave mainly white growth of the kind just mentioned, alternating with patches obviously of *22y dw 1* type.

That most of the mycelium on MM agar, and most of the whitish regions of the mycelium on CM agar, were heterokaryotic and balanced was suggested by their growth rate and non-dwarf habit, and was proved by the following tests: (1) successive transfers both on MM agar and on CM agar, of hyphal tips from heterokaryons grown on MM or on CM, gave growth with the characters just described; (2) a proportion of single hyphae isolated from heterokaryons gave origin again to heterokaryotic growth; (3) platings of conidia from colonies derived from single hyphae gave the two types of parental strains: *22y dw 1* and *7w dw 4* (Table 2).

A second combination of dwarf strains was prepared from *7w dw 5* and *38y dw 6*. The former has white conidia and requires cystine or thiosulphate; the latter has yellow conidia and requires NH_4^+ or NO_2^-. In this case, the heterokaryons were established by isolating small bits of more vigorous mycelium which arose at the region of contact between colonies of the two types grown on CM agar. These more vigorous tufts were quite evident after

12 days from the inoculation of the plates. Evidently, even on CM agar, the heterokaryon between the two dwarf strains used had a sufficient selective advantage over either dwarf component to be able to escape from the homokaryotic mycelia.

Mass hyphal transfers on to MM agar from these more vigorous tufts grew at a rate approaching that of the wild type and developed a macroscopically yellowish sporing surface. No further work was carried out with this combination.

Following the notations used in previous papers (e.g. Pontecorvo, 1953b), heterokaryons are designated by the code numbers of the components, joined by the sign +. Thus the two heterokaryons mentioned so far will be designated: $22y\ dw1 + 7w\ dw4$ and $7w\ dw5 + 38y\ dw6$, respectively.

Balanced heterokaryons between strains each having two nutritional requirements. Balanced heterokaryons were synthesized between strains $22y13$ (yellow, requiring adenine or hypoxanthine, and methionine; inhibited by cystine) and $7w16$ (white, requiring methionine or cystine or thiosulphate, and adenine; inhibited by hypoxanthine). Neither of these two strains shows any growth on MM agar and their conidia do not even germinate on it. They both grow well on MM agar supplemented with the respective pairs of growth factors; $22y13$ grows well and $7w16$ grows poorly on CM agar.

The balanced heterokaryons $22y13 + 7w16$ were obtained in the same way as in the case of $7w\ dw5 + 38y\ dw6$, i.e. by isolation on to MM agar of tufts of vigorous mycelium arising from the region of contact between colonies of the two strains growing on CM agar. The selective advantage of the heterokaryon on CM agar may perhaps depend, in this case, on overcoming the inhibitions mentioned above.

The heterokaryon grows well on MM agar; macroscopically its sporulating surface is yellowish. The same three tests mentioned before gave proof of the heterokaryotic condition: (1) perpetuation of the heterokaryon by mass transfers of hyphae on MM agar; (2) perpetuation of the heterokaryon by single hypha isolation; (3) recovery of both component strains by plating conidia of colonies grown from single hyphae (Table 2). An additional visual test was that on addition of the appropriate growth factors to MM agar, the white, the yellow or both components would sector out.

Properties of heterokaryons. The balanced heterokaryons of *Penicillium chrysogenum*, in contrast with those of other species (*Aspergillus nidulans*, Pontecorvo, 1953b; *A. niger*, Pontecorvo et al. 1953) can be maintained by massive transfers from the sporing surface. As in those other species, the conidia of *P. chrysogenum* are uninucleate (Tonolo & Urbani, 1952). Thus in a heterokaryon/nuclei of each kind should segregate when the conidia are formed. That this is substantially so is shown by the results of platings of conidia at density of the order of 100 conidia/plate (Table 3). However, when plating massive numbers of conidia, rare heterokaryotic colonies arise within 48 hr., at the rate of a few per 10^6 plated conidia (Table 3; Pl. 1, figs. 2, 3). It is probable that these colonies originate from bits of heterokaryotic mycelium, accidentally present in the inoculum.

As mentioned before, one of the three heterokaryons formed ($22y\ dw1 + 7w\ dw4$) has macroscopically a white sporing surface; the other two have yellowish surfaces. The structure of the penicillus in *Penicillium chrysogenum*, with loose and divergent conidial chains, makes it impossible to tell whether chains of different colour arise from the same conidiophore. Furthermore, the two colours (white and yellow) are not sharply distinguishable under the microscope. Nevertheless, the fact remains that all heterokaryons have, in different

Table 3. *Heterozygous diploids obtained from conidia of balanced heterokaryons plated in minimal medium*

	Plated conidia		New heterokaryotic colonies arising*		Diploids detected as†			
					Whole colonies		Sectors from heterokaryotic colonies	
Balanced heterokaryons	Total no.	No. per dish	Total	Per 10^6 plated conidia	Total	Per 10^7 plated conidia	Total	Per 10^2 colonies
$22y\ dw1 + 7w\ dw4$	77×10^6	2.6×10^6	769	10	2	0.25	6	0.77
$22y\ 13 + 7w\ 16$	8.4×10^6	0.6×10^6	1083	138	3	3.6	16	1.5

* When plating conidia of balanced heterokaryons at densities of the order of 10^6 per plate a few tens of colonies arise within 48 hr. on MM. They turn out to be balanced heterokaryons, presumably arising from bits of mycelium or from rare bi- or multinucleate conidia carrying nuclei of the two kinds. Macroscopically their sporing surface is white in the case of heterokaryon (1), or yellowish in the case of heterokaryon (2), and their morphology normal (Pl. 1, figs. 2, 3).

† Diploids are identified either as green spored colonies of normal morphology and growth rate or as green sectors out of the newly arising heterokaryons (Pl. 1, figs. 2, 3).

proportions, both white and yellow penicilli but no green ones, while diploids (see below) have green penicilli. The colour of individual conidia seems thus to depend upon the kind of nucleus segregated into each of them. In this respect the colour markers used in *P. chrysogenum* behave more like those used in *Aspergillus nidulans* (Pontecorvo, 1953b) than like those used in *A. niger* (Pontecorvo et al. 1953).

Isolation of heterozygous diploids

Diploid from heterokaryons $22y\ dw1 + 7w\ dw4$. From heterokaryons grown for 2–3 weeks on MM agar, conidia (washed twice in water) were plated on MM agar at the rate of 2.6×10^6 conidia/plate. After 48 hr. a mean of about twenty-eight colonies/plate were visible. After a few days' further incubation all but two of these colonies developed a macroscopically visible whitish sporing surface; tests on about thirty of them showed that they were heterokaryons, as mentioned before (Table 3). Two, however, grew better than the others and developed green penicilli. Isolation, purification by single conidium micromanipulation, and further tests (see later) showed these two colonies to be the desired heterozygous diploids. Diploids are designated by interposing a fraction sign between the symbols of the strains associated in the heterokaryon which gave them origin:

$$22y\ dw1/7w\ dw4, \quad \text{or} \quad \frac{22y\ dw1}{7w\ dw4}.$$

On keeping the plates for 2 weeks or more, about 1 % of the heterokaryotic colonies produced one macroscopically visible green sector (Table 4; Pl. 1, fig. 2). Isolation and tests of strains from such sectors shows them also to be diploids. Incidentally, after such prolonged incubation the plates show a background growth of minute colonies (Pl. 1, fig. 2) of type *22y dw1*, presumably adapted; these minute colonies represent about 5 % of the conidia of this type present in the plated suspension (Table 4).

Table 4. *Properties of heterozygous diploid* 22y dw1/7w dw4 *compared with those of the heterokaryon, of the two marked haploid strains, and of the original haploid wild type* Q176 (*see also Pl. 1, fig. 1*)

Strain	Growth habit on CM	Individual conidia		Growth factor requirements of colonies arising from		Segregation through conidia and sectors
		Colour	Size	Single conidia	Hyphae	
Q176 (haploid)	Normal	Green	Normal	None	None	None
22y dw1 (haploid)	Dwarf	Yellow	Normal	Hypox.	Hypox.	None
7w dw4 (haploid)	Dwarf	White	Normal	Cystine	Cystine	None
22y dw1 + 7w dw4 (heterokaryon)	Normal	White or yellow	Normal	Hypox. or cystine	None	Parental types, except for very rare diploids (10^{-7})
22y dw1/7w dw4 (diploid)	Normal	Green	Giant	None	None	Small proportion (10^{-2}) of parental and new types

Diploid from heterokaryon 22y13+7w16. From heterokaryons grown for about 2 weeks on MM agar, conidia (washed twice in water) were plated on MM agar at the rate of 0.6×10^6 conidia/plate. The results were similar to those of the previous example with the difference that, (*a*) the newly arisen heterokaryotic colonies had macroscopically yellowish sporing surface; (*b*) relative to the number of conidia plated, the proportions of newly arisen heterokaryotic colonies and of the green diploid colonies was about ten times higher (Table 4); (*c*) the green sectors from the heterokaryotic colonies were about twice as frequent; (*d*) there was no delayed background growth of parental types (Pl. 1, fig. 3). Isolation from one of the green colonies, purification by single conidium micromanipulation and further tests led to the establishment of diploid *22y13/7w16*.

Properties of diploids. The diploids differ from the heterokaryons, from the haploids which formed the heterokaryons and from the haploid wild type in a number of ways. These have been tabulated (Table 5) for the case of diploid *22y dw1/7w dw4* and some of them are evident on Pl. 1, fig. 1. These differences also apply, *mutatis mutandis*, to the case of diploid *22y13/7w16*. The size of the conidia is not easily measurable because of considerable variations in shape. However, a visual comparison (Pl. 1, fig. 4) leaves no doubt that diploids have larger conidia, a fact suggesting that Sansome (1949) was right when she deduced that the 'gigas' strains isolated by her in the closely related species *Penicillium notatum* were (homozygous) diploid. As diploids heterozygous for nutritional requirements, colours of conidia and habit of growth, show features approaching those of the wild type (no requirement, green conidia and normal growth) all the mutant properties used as markers in this work are recessive. In the diploids segregation and recombination take place, as shown in detail in the next section, i.e. a small proportion of individual conidia or hyphae of the diploids give rise to strains different from the diploid from which they arose, in that they show one or more properties of the original marked strains associated either in the original way or in new ways.

Table 5. *First-order segregants from diploid:*

$$\frac{22y\ dw1\ (yellow,\ hypoxanthine\text{-}requir.,\ dwarf\text{-}1)}{7w\ dw4\ (white,\ cystine\ requir.,\ dwarf\text{-}4)}$$

	Prototrophs	Requiring hypoxanthine	Requiring cystine	Requiring both	Total
A. Obtained as whole colonies*					
Green					
Dwarf	5	0	0	0	5
Normal	964	1	0	0	965
Yellow					
Dwarf	2	1	0	0	3
Normal	2	0	0	0	2
White					
Dwarf	6	0	0	0	6
Normal	12	0	0	0	12
Non-sporing					
Normal	17	0	0	0	17
	1008	2	0	0	1010
B. Obtained as sectors†					
Green					
Dwarf	0	0	0	0	0
Normal	0	0	1‡	0	1
Dwarf	1	0	0	0	1
Yellow					
Normal	5	0	0	0	5
White					
Dwarf	0	0	0	0	0
Normal	14	0	0	0	14
	20	0	1	0	21

* Conidia of *22y dw1/7w dw4* were plated on CM, and 1010 colonies were classified as to colour, morphology and requirements.

† Sectors, differing in colour or growth habit from the mother colonies, were isolated from a proportion of the 964 green, normal, prototroph colonies, and classified as to colour, morphology and requirements.

‡ Though this sector had green conidia and non-drawf habit, it was isolated because of its rather thinner growth than the rest of the mother colony: it turned out to be cystine-requiring.

Segregation and recombination in diploids

In *Aspergillus nidulans* and *A. niger* (Pontecorvo & Roper, 1953; Pontecorvo et al. 1953; Pontecorvo et al. 1954) mitotic segregation and recombination occurs once about every 100 divisions in heterozygous diploids. Its detectable result is the production of diploid nuclei homozygous for one or more of the recessive markers or of haploid nuclei with one or more of the recessive markers. A patch of mycelium homokaryotic for one of these nuclei (for short, 'segregant' nuclei) will have the recessive marker property or properties. When these properties are detectable by inspection (e.g. colour of conidia) the patch of mycelium may be identified visually and a strain isolated from it. In the two other species mentioned above this visual identification is easy even in the case of a single head in a background of thousands with the wild type colour. In *Penicillium chrysogenum* individual heads differing in colour from the background are difficult to recognize, and only patches of homokaryotic segregant mycelium with a substantial number of segregant heads can be identified visually and isolated.

Thus one way of detecting mitotic segregation and recombination of colour markers is to examine a number of diploid colonies (preferably originated from single conidia) for the presence of sectors with penicilli of one of the recessive colours; isolation from these patches will permit the classification of each segregant in respect of other properties also (e.g. nutritional requirements, growth habit, etc.). Another way in which segregation, not only of colour markers but of any markers, can be recognized is by plating conidia of the diploid and classifying individually the resulting colonies. This classification can be visual as to colour of conidia and growth habit, but requires the testing of individual colonies for nutritional requirements. Segregation and recombination was analysed extensively in the case of diploid *22y dw1/7w dw4* (see later). As to diploid *22y 13/7w 16*, we only went so far as to observe that it did segregate for colour markers. An extensive analysis of this and other diploids was carried out later at the Istituto Superiore di Sanitá, Rome, by one of us (G. S.), and it will be the subject of a separate publication.

First-order segregants. First-order segregants from diploid *22y dw1/7w dw4* were obtained by plating conidia of a subculture (purified by single conidium micromanipulation) on fully supplemented medium; 1010 colonies (Table 5) from this plating were transferred to plates of minimal agar. After 2 days those colonies which showed no further growth were picked out and tested for nutritional requirements and colour. The remaining colonies were transferred back on to supplemented medium, kept until the colour appeared on the sporing surface, and then classified for colour and morphology.

A proportion of the 964 green prototroph normal colonies (out of the total of 1010) showed sectors; some were yellow, some white and some non-sporulating or with a growth habit different from that of the mother colony. Twenty-one of these sectors were isolated, each from a different colony, and further tested. Table 5 shows the classification of the forty-six segregants obtained as whole colonies after plating, and of the twenty-one segregants detected as sectors in some of the colonies arising from platings. The proportion of segregants among the conidia of the diploid is about 5% (46/1010). No recombination of parental *recessive* characters occurred among the sixty-seven first-order segregants isolated either as colonies or as sectors (Table 5).

Second-order segregants. The first-order segregant with green conidia, normal growth and hypoxanthine requirement was tested for further segregation after purification by single conidium isolation. One yellow and two white sectors from colonies of this strain gave yellow or white, hypoxanthine-requiring strains, with normal growth. A white hypoxanthine-requiring strain is an example of recombination between recessive properties of the starting strains.

CONCLUSIONS

The experiments reported show the occurrence in *Penicillium chrysogenum* of the parasexual processes already found in two other species of filamentous fungi. In Penicillium the analysis has gone only so far as to identify three out of the four steps which were studied in detail in the two other species, namely: (1) formation of heterokaryons; (2) formation of heterozygous diploid nuclei within heterokaryons; (3) recombination in the diploid nuclei. We have not yet looked for and identified the formation of haploid nuclei from diploid nuclei (Pontecorvo *et al.* 1953a; Pontecorvo, 1954). As, however, this is a consequence of accidents of mitosis of a type known to occur in all higher and lower organisms in which it has been looked for, it is a reasonable assumption that it will be found also in Penicillium.

Penicillium thus may have, like the two other species, a complete parasexual cycle (Pontecorvo, 1953a). This involves alternation of haploid and diploid stages, the possibility of storing gene variation under the cloak of dominance in both heterokaryons and heterozygotes and the possibility of gene recombination.

Because of its low growth rate and certain details of its morphology, *Penicillium chrysogenum* is a species much less suitable than *Aspergillus nidulans* or *A. niger* for the detailed analysis of the parasexual processes themselves. It is, however, a species of great economic importance. The discovery of the parasexual processes in it is relevant in two respects. First, this third species of filamentous fungus again shows the occurrence of these parasexual processes; thus they would appear to be widespread at least. Secondly, the parasexual processes in *P. chrysogenum* can be used for practical purposes: (*a*) deliberate 'cross-breeding' for the production of improved industrial strains; (*b*) the identification of different genetic blocks in the analysis of the biosynthesis of penicillin.

One of us (G.S.) is indebted to the Istituto Superiore di Sanità, Rome, for a grant enabling him to take part in this work at the Department of Genetics, University of Glasgow. This work is part of a general programme supported by the Nuffield Foundation.

REFERENCES

BONNER, D. (1946). Production of biochemical mutations in *Penicillium*. *Amer. J. Bot.* **33**, 788.
DODGE, B. O. (1942). Heterokaryotic vigor in *Neurospora*. *Bull. Torrey bot. Cl.* **69**, 75.
PONTECORVO, G. (1947). Genetic systems based on heterokaryosis. *Cold Spr. Harb. Symp. quant. Biol.* **11**, 193.
PONTECORVO, G. (1952). Non-random distribution of multiple mitotic crossing-over among nuclei of heterozygous diploid Aspergillus. *Nature, Lond.* **170**, 204.
PONTECORVO, G. (1953a). Mitotic recombination in the genetic systems of filamentous fungi. *Proc. IXth Intern. Congr. Genetics* (in the Press).
PONTECORVO, G. (1953b). The genetics of *Aspergillus nidulans*. *Advanc. Genet.* **5**, 141.

PONTECORVO, G. & GEMMEL, A. R. (1944). Genetic proof of heterokaryosis in *Penicillium notatum*. *Nature, Lond.* **154**, 514.

PONTECORVO, G. & ROPER, J. A. (1952). Genetic analysis without sexual reproduction by means of polyploidy in *Aspergillus nidulans*. *J. gen. Microbiol.* **6**, vii.

PONTECORVO, G. & ROPER, J. A. (1953). Diploids and mitotic recombination. *Advanc. Genet.* **5**, 218.

PONTECORVO, G., ROPER, J. A. & FORBES, E. (1953). Genetic recombination without sexual reproduction in *Aspergillus niger*. *J. gen. Microbiol.* **8**, 198.

PONTECORVO, G. & SERMONTI, G. (1953). Recombination without sexual reproduction in *Penicillium chrysogenum*. *Nature, Lond.* **172**, 126.

PONTECORVO, G., TARR GLOOR, E. & FORBES, E. (1954). Analysis of mitotic recombination in *Aspergillus nidulans*. *J. Genet.* **52**, 22b.

ROPER, J. A. (1952). Production of heterozygous diploids in filamentous fungi. *Experientia*, **8**, 14.

SANSOME, E. (1947). Spontaneous variation in *Penicillium notatum*. *Trans. Brit. mycol. Soc.* **31**, 66.

SANSOME, E. (1949). Spontaneous mutation in standard and 'Gigas' forms of *Penicillium notatum*. *Trans. Brit. mycol. Soc.* **32**, 305.

TONOLO, A. & URBANI, E. (1952). Observations on *Penicillium chrysogenum* in submerged culture. *Bull. World Hlth Org.* **6**, 277.

EXPLANATION OF PLATE

Fig. 1. +, the haploid green wild type; 1, the haploid dwarf, yellow, hypoxanthine-requiring $22y\ dw\ 1$; 4, the haploid dwarf, white, cystine-requiring $7w\ dw\ 4$; HK, the heterokaryon between $22y\ dw\ 1$ and $7w\ dw\ 4$; D, the diploid derived from the heterokaryon. Left: on complete medium. Right: on minimal medium. White circles indicate the positions of inoculum on MM of the two dwarf strains.

Fig. 2. Plating on MM of conidia of heterokaryon $22y\ dw\ 1 + 7w\ dw\ 4$ ($2 \cdot 6 \times 10^6$/dish): there are about twenty heterokaryotic colonies (large) per dish, and a background of minute colonies of type $22y\ dw\ 1$. About 1 % of the heterokaryotic colonies show, like the one in the centre, a green diploid sector.

Fig. 3. Plating on MM of conidia of heterokaryon $22y\ 13 + 7w\ 16$: results similar to those of Fig. 2, but no background growth. Two heterokaryotic colonies show green diploid sector.

Fig. 4. Conidia of heterokaryon $22y\ dw\ 1 + 7w\ dw\ 4$ (left) and of the diploid derived from it (right) about $\times 400$.

[Editor's Note: The Plate appears on the following page.]

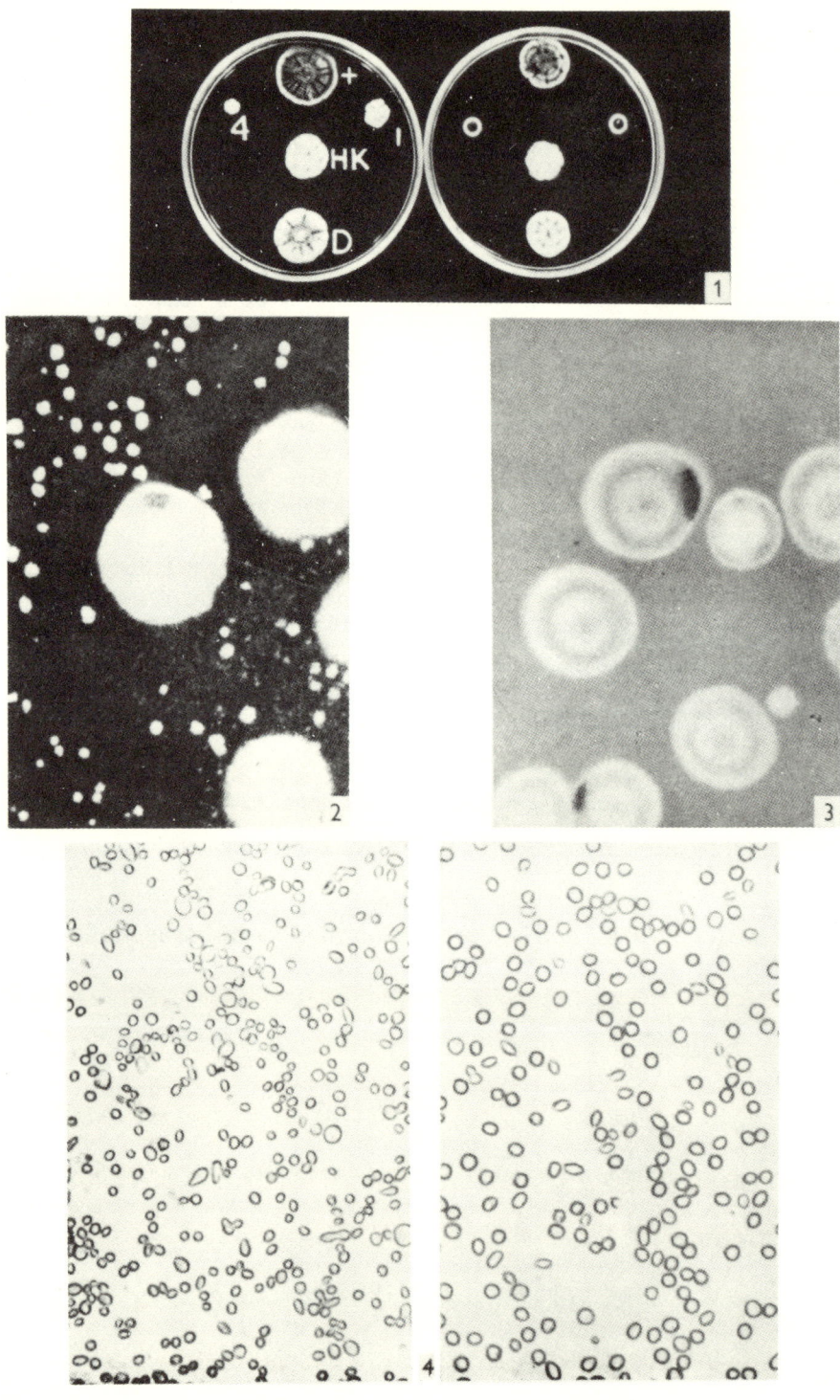

G. Pontecorvo & G. Sermonti—Parasexual recombination

Part IV

MEDIUM DEVELOPMENT

Editor's Comments
on Papers 7A Through 9

7A   MOYER and COGHILL
Penicillin. VIII. Production of Penicillin in Surface Cultures

7B   MOYER and COGHILL
Excerpt from *Penicillin. IX. The Laboratory Scale Production of Penicillin in Submerged Cultures by* Penicillium notatum Westling *(NRRL 832)*

8   RAKE and DONOVICK
Studies on the Nutritional Requirements of Streptomyces griseus *for the Formation of Streptomycin*

9   DAVEY and JOHNSON
Penicillin Production in Corn Steep Media with Continuous Carbohydrate Addition

    The papers in this group deal with the phase of fermentation process development known as *medium development* or *medium design*. Although these four papers treat only two fermentation processes—those for production of penicillin and streptomycin—they illustrate principles and contain early reports of use of materials of general and continuing importance—namely, lactose and corn steep liquor for penicillin production, and soybean meal for streptomycin production (Moyer and Coghill 1947).

    Papers 7A and B (the latter used only in part), like the others in the series on penicillin by these authors and their colleagues, contain a large body of data and thoughts that were accumulated during several years of intensive work but not published earlier because of a wartime restriction. Even though Paper 7A describes cultivation in stationary flasks only and titres of penicillin quoted are miniscule by standards set only a few years later, Moyer and Coghill's approach to medium development is embarrassingly similar to that of many present-day workers. Furthermore, the discussion reveals that their appreciation of factors affecting growth and penicillin production by *Penicillia* in stationary culture was more in harmony with present-day concepts of the kinetics of quasi-continuous fermentations than with stereotyped concepts of rate processes in batch fermentations that were enunciated in the intervening years.

The excerpt from Paper 7B shows that the transition from the stationary (but aerobic) to the shaken (more aerobic) flask-scale fermentation had been made, and that the latter would subsequently be the prevailing mode of small-scale medium development studies.

Paper 8 describes some early work with the streptomycin fermentation. The approach displayed by Rake and Donovick was highly pragmatic, but it was based on a full appreciation of what was known about the nutrition of the streptomycetes. The bases of their choice of fermentation raw materials were availability, cost, and effects not only on production but also on recoverability of the antibiotic. Like Moyer and Coghill, Rake and Donovick were not encumbered in their thinking by a rigid mechanistic view of the phases of growth and product formation in batch fermentation, although they did observe a relationship between antibiotic formation and spore formation in shaken cultures. Unlike Moyer and Coghill, in this paper they did not predict the advantages of sustaining and controlling growth and product formation by feeding of nutrients.

In Paper 9, Davey and Johnson convincingly demonstrate the practicality of using a rapidly metabolized carbohydrate such as glucose in place of a more slowly utilized source such as lactose, by employing the technique of feeding, or by supplying the nutrient only as required to sustain the desired rates of growth and antibiotic synthesis in aerated and agitated small-scale fermentors. The principle had been conceived by Moyer and Coghill (Papers 7A and B) and probably by others.

As a result of the work of the Wisconsin group in the 1940s and early 1950s, most penicillin manufacturers adopted the feeding of carbohydrates other than lactose in the 1950s. In contrast, many if not all major producers still use corn steep liquor even though the feasibility of replacing it was demonstrated in the same era (Foster *et al.* 1946). Later, the feasibility of dispensing completely with organic nitrogenous material was shown (Hosler and Johnson 1953).

REFERENCES

Foster, J. W., H. B. Woodruff, D. Perlman, L. E. McDaniel, B. L. Wilker, and D. Hendlin. 1946. Microbiological aspects of penicilllin. IX. Cottonseed meal as a substitute for corn steep liquor in penicillin production. *J. Bact.*, **51**:695-698.

Hosler, P., and M. J. Johnson. 1953. Penicillin from chemically defined media. *Ind. Eng. Chem.*, **45**:871-874.

Moyer, A. J., and R. D. Coghill. 1947. Penicillin. X. The effect of phenylacetic acid on penicillin production. *J. Bact.*, **53**:329-341.

Copyright © 1946 by the American Society for Microbiology

Reprinted from *J. Bact.*, 51, 57–78 (1946)

PENICILLIN

VIII. PRODUCTION OF PENICILLIN IN SURFACE CULTURES[1]

ANDREW J. MOYER AND ROBERT D. COGHILL

Fermentation Division, Northern Regional Research Laboratory,[2] Peoria, Illinois

Received for publication September 1, 1945

The production of penicillin by the surface culture of *Penicillium notatum* is the foundation upon which a large new industry has been built. Although cultivation of the mold in submerged culture now appears to be more economical and more practical from an industrial standpoint, surface cultures were utilized to produce the penicillin that first effected the remarkable clinical cures which indicated that the large investment of time and money made during the past three years would be justified.

Previous investigators of penicillin production have not been successful in developing a highly productive medium. Fleming (1929), the discoverer of the drug, refers only to the use of a "nutrient broth," whereas Clutterbuck, Lovell, and Raistrick (1932) used a modified Czapek-Dox medium. Although the yields of penicillin obtained by this group and by Fleming were not evaluated in terms of the Oxford unit adopted later, they were undoubtedly very low, at least when compared with present standards. Abraham *et al.* (1941), using the same modified Czapek-Dox medium with the addition of small amounts of a crude yeast extract, obtained only 2 to 6 Oxford units per ml.

This paper will report on a few of more than five hundred experiments which have led to an increase in the yield of penicillin from the range of 2 to 6 Oxford units per ml to as much as 160 to 220 Oxford units per ml. This increase in yield has been achieved primarily by the proper selection of organisms and nutrients, including the use of corn steep liquor, the use of lactose as the principal carbohydrate, and the addition of nutrients during the course of the fermentation.[3]

METHODS AND MATERIALS

All of the assays were conducted by the cylinder-plate method (Abraham *et al.*, 1941), as modified by Schmidt and Moyer (1944). The method has the advantage that it gives no response to notatin (penatin, penicillin B) and there-

---

[1] The expenses of the work described in this paper were met in part from contract OEMcmr-100 with the Office of Scientific Research and Development, recommended by the Committee on Medical Research.

[2] One of the laboratories of the Bureau of Agricultural and Industrial Chemistry, Agricultural Research Administration, U. S. Department of Agriculture.

[3] Since March, 1942, the results of this work have been distributed in accordance with government regulations by the Committee on Medical Research to penicillin producers and many research groups in this country and abroad. Owing to the strategic significance of penicillin, it was considered advisable to delay publication of these results.

fore gives a true measure of penicillin. This fact has frequently been confirmed in this laboratory by extraction into ether of the penicillin from high-yielding cultures, followed by transfer to a buffer and reassay. However, it must be borne in mind that the results obtained by the cylinder-plate method may be in error by as much as approximately 10 per cent because of the inherent limitations of the procedure.

The production cultures were grown in 200-ml pyrex Erlenmeyer flasks containing 50 ml of the nutrient medium. Inoculations were made with a generous application of dry, ungerminated spores. In each experiment a sufficient number of flasks were employed so that duplicate cultures could be harvested and assayed on several consecutive days. All production cultures were incubated at $24 C \pm 1$.

A slightly modified Czapek-Dox solution contained, as the standard salt constituents, $MgSO_4 \cdot 7H_2O$, 0.250 g; KH_2PO_4, 0.500 g; $NaNO_3$, 3.0 g; $ZnSO_4 \cdot 7H_2O$, 0.044 g; and $MnSO_4 \cdot 4H_2O$, 0.004 g per liter of final medium. This standard medium was supplemented with various carbohydrates and nitrogenous materials as desired.

Various lots of concentrated corn steep liquor,[4] typical of the commercially available product, were employed. This material is known to vary considerably in composition; however, a typical analysis is as follows: total solids, 52 per cent; total nitrogen, 4.3 per cent; ash, 7.9 per cent; free reducing sugar, 5.6 per cent; total reducing sugar calculated as glucose after acid hydrolysis, 6.8 per cent; specific gravity, 1.25; and pH, 4.0.

EXPERIMENTAL RESULTS

Organism Selections

In the first experimental work use was made of a strain of *Penicillium notatum* Westling which was supplied by Dr. H. W. Florey of Oxford University, and which was descended from the original Fleming organism. This strain was not stable with respect to sporulation or penicillin production; therefore, very early in the research program, a search was made for a superior organism.

This survey of species of *Penicillium* extended over a considerable period, during which time improvements were being made in the culture medium. About 35 strains of *P. notatum* and *Penicillium chrysogenum* were investigated, including many cultures previously obtained for the culture collection of this laboratory (largely from the Thom collection), as well as several descendants of the original Fleming strain which were kindly placed at our disposal by other

---

[4] Corn steep liquor is a by-product of the starch industry. Before the corn is ground, it is steeped for approximately 30 hours in water originally containing 0.1 to 0.3 per cent of sulfur dioxide. The water has previously washed the starch and passed through the gluten settling tank. The addition of sulfur dioxide, at the time steeping begins, inhibits fermentation. Prior to concentration, a lactic acid fermentation occurs to a variable extent. There is considerable variation in sugar and lactic acid contents. Steep liquor is sold on a basis of approximately 55 per cent solids, although different batches may vary rather widely from this value.

investigators. Some of the *P. chrysogenum* strains gave promising yields of penicillin but were not superior to some of the descendants of the original Fleming organism. The strain finally selected for further nutritional investigation was obtained through the generosity of Dr. George Harrop, of the Squibb Institute of Medical Research. This strain, now known as NRRL 1249, was a descendant

TABLE 1

Comparison of penicillin yields from several strains of penicillium

ORGANISMS*	CULTURE AGE—DAYS			
	4	5	6	7
	Penicillin yield—units per ml			
NRRL				
831................	34	45	59	69
832................	37	48	63	69
833................	20	30	36	39
828................	4	5	13	19
815................	17	29	49	58
1249—type A.......	36	55	68	76
1249.B21...........	65	91	138	147
	pH of filtrates			
831................	6.2		6.7	7.6
832................	6.6		7.2	7.5
833................	5.7		7.1	7.6
828................	5.8		6.6	6.8
815................	6.4		6.8	7.1
1249—type A.......	5.5		7.2	7.7
1249.B21...........	5.7		7.2	7.6
	Mycelium weight—g per culture			
831................	0.88		1.11	1.10
832................	0.79		1.02	1.00
833................	0.91		1.20	1.23
828................	0.68		1.03	1.34
815................	0.79		1.06	1.00
1249—type A.......	0.78		1.11	1.23
1249.B21...........	0.76		1.05	1.08

Culture medium: 10 g concentrated corn steep liquor, 4.0 g lactose per 100 ml, and standard salts.

* NRRL 831, 832, 833, and 828 are cultures of *P. notatum* obtained from the Thom collection as B-47A, B-69, B-464, and 5646II, respectively. They are not descended from the Fleming strain. NRRL 815 came from the Thom Collection as B-508 and is listed as *P. chrysogenum*. NRRL 1249 type A and 1249.B21 arose in this laboratory as variants from NRRL 1249 and are descended from the Fleming strain of *P. notatum*.

of the Fleming strain and was chosen because it was one of the best for penicillin production and produced a more luxuriant crop of spores than many of the other strains.

Data from a portion of this survey, obtained with an improved culture medium, are presented in table 1. It is evident that there was a great variation in the capacity of the various organisms to produce penicillin. Most of the strains had essentially the same growth rate. The amount of yellow pigment accumulating in the liquid medium varied from strain to strain, although there appeared to be no correlation between penicillin yield and pigment formation. This survey failed to reveal an organism which was superior to some of the descendants of the original Fleming strain.

Spore Production and Inoculation

A large number of nutrient combinations were employed in an effort to devise a medium upon which a thin, heavily sporulating mycelium would develop. Good results were obtained with the following combination, which is suitable for use as either a liquid or jellied medium:

Sporulation medium

Glycerol	7.5 g
Cane molasses (edible quality as commonly sold at retail)	7.5 g
Corn steep liquor	2.5 g
$MgSO_4 \cdot 7H_2O$	0.050 g
KH_2PO_4	0.060 g
Peptone	5.00 g
NaCl	4.00 g
Fe-tartrate	0.005 g
$CuSO_4 \cdot 5H_2O$	0.004 g
Agar	2.50 g

Distilled water to make 1 L

When this medium is inoculated with a spore suspension and shaken vigorously prior to incubation, a very uniform surface growth results. By the addition of increased quantities of agar (about 2.5 per cent), the medium can be made to jelly for use in test tube slants or petri dish cultures.

By using an alternative method, many times the number of spores can be produced per flask. Fresh whole-wheat bread, which must not contain any commercial mold inhibitor such as "mycoban," is cut into 1-cm cubes and steam-sterilized in shallow layers in Erlenmeyer flasks. It is then heavily inoculated with spores. In 4 to 5 days, at 25 to 27C, a heavy crop of spores develops inside and outside the bread. When the spores have been formed and there has has been some drying of the spore-covered pieces of bread, these cultures can be stored at 4C for at least 2 weeks without apparent loss in vitality or ability to produce penicillin. After 6 or 7 days, the bread crumbles and is easily reduced to a powdery mass, which may be used directly for inoculation or may be blended with 3 to 4 volumes of a mixture of equal parts of whole-wheat flour and finely ground oat hulls, which will float and spread rapidly over the surface of the medium. Portions of such a spore-bearing mixture can be conveniently introduced into the sterile medium by means of a spatula or atomizer. Inoculations performed in this manner result in rapid and uniform surface growth. These improvements have resulted in greater uniformity among the production cultures and a considerable decrease in the time required to make the inoculation.

Variant Strains and Culture Maintenance

Very poor sporulation was sometimes encountered in liquid cultures inoculated with spores of *P. notatum* NRRL 1249 which had been grown for several generations on the agar sporulation medium. These poorly sporulating cultures were characterized by the appearance of white aerial growth, which was either confined to small spots or generally distributed over the surface of the mycelium. Such cultures were discarded, and only the best sporulating cultures were used to inoculate production flasks.

This method of culture selection was found to be inadequate as a means of insuring uniformity in the appearance of the culture and in the yield of penicillin. When sporulating cultures so selected were used to inoculate a production medium containing lactose and corn steep liquor, the resulting mycelium occasionally showed varying numbers of small white spots; in some cases the whole mycelium exhibited practically no sporulation. Such nonsporulating cultures gave low yields of penicillin (50 to 60 units per ml), as contrasted with 100 to 120 units per ml from other production cultures having a good crop of spores. This apparent correlation between culture appearance and penicillin yield was believed to be due either to some unknown variable in the nutrient medium or to the presence of two or more strains in the stock culture. Monospore cultures were accordingly prepared from one of the stock cultures which had been grown for several generations on a good agar sporulation medium. When these monospore selections were cultivated on a lactose corn steep liquor production medium, two types of mycelia were observed: type A produced practically no spores and only 50 to 60 units of penicillin per ml, whereas type B produced a good crop of spores and penicillin yields of 100 to 120 units per ml (table 1). Type B colonies closely resembled the parent strain, NRRL 1249. One of the type B strains was again subjected to monospore selection; all mycelial growth resulting from the selection sporulated well and gave good yields of penicillin. One of these monospore selections, *P. notatum* NRRL 1249.B21 (isolated and evaluated by the senior author), has been used in most of the early investigations conducted at this laboratory and because of its desirable characteristics is today generally used throughout industry for the production of penicillin in surface cultures. The superiority of this strain has been substantiated by later surveys conducted by Raper *et al.* (1944) of this laboratory.

In order to maintain the potency of this organism and to minimize the appearance of inferior mutant strains, it is recommended that spores be maintained in lyophil or dry soil tubes and that subcultures be made from such preparations at frequent intervals.

Optimum Conditions for Penicillin Production

Medium. The culture medium recommended by Dr. H. W. Florey and Dr. N. G. Heatley (personal communication on July 16, 1941) was the Czapek-Dox medium, supplemented with 4 per cent of glucose and 5 to 10 per cent by volume of a crude yeast extract. Two to six units of penicillin per ml could be obtained on this medium in 7 to 8 days. Improvements have gradually been made in this culture medium, with the result that 150 to 200 units per ml now can be obtained in 5 to 6 days. The compositions of the original medium and of the medium considered near optimum today are as follows:

	ORIGINAL MEDIUM	IMPROVED MEDIUM
NaNO₃	3.0 g	3.0 g
MgSO₄·7H₂O	0.500 g	0.250 g
KH₂PO₄	1.00 g	0.500 g
KCl	0.50 g	None
FeSO₄·7H₂O	0.010 g	None
Glucose monohydrate	40.000 g	2.75 g
ZnSO₄·7H₂O	None	0.044 g
MnSO₄·4H₂O	None	0.004 g
Lactose monohydrate	None	44.0 g
Corn steep liquor	None	100. g
Crude yeast extract	50–100 ml	None
Initial pH	4.6	4.6
Water to make one liter		

The composition of the improved medium was determined after many experiments and will be discussed under various headings as given below.

Carbon sources. When the Czapek-Dox medium (without corn steep liquor) was used, there was no marked difference in the response to various carbon sources, as indicated by penicillin production, except that lactose was inferior because it supported only a trace of fungus growth. However, once the effectiveness of corn steep liquor was recognized and this material was added to the basic medium, a marked effect of various carbon sources on penicillin yield was observed.

The results obtained from a comparison of several common carbon sources, each of which was used in 3 per cent concentration in steep-liquor standard salt medium, are given in table 2. In the corn steep liquor there was sufficient assimilable carbohydrate, probably mainly glucose and dextrins, to support fairly good fungus growth and moderate penicillin production, even in the absence of added carbon sources. Lactose, cornstarch, and corn dextrin were equally good for penicillin production, glycerol being definitely inferior. No significant difference in penicillin production could be observed between cultures containing commercial glucose and those containing brown sugar. The pH of the broth did not rise so fast in the lactose, starch, and dextrin cultures as it did in the glucose, sucrose, and glycerol cultures.

A further comparison of these carbon sources was made at 6 per cent concentration (table 3). Again cultures containing lactose gave the highest penicillin yields. The change in the pH of the broth was slower in the cultures containing lactose than in those containing glucose, sucrose, glycerol, or sorbitol. Increasing the concentrations from 3 to 6 per cent under these conditions failed to increase the penicillin yields.

In these experiments, the cornstarch was added as a dry powder to the small culture flasks containing 50 ml of the basic medium. The starch was then

TABLE 2

Penicillin production from various carbon sources of 3 per cent concentration

CARBON SOURCE	CULTURE AGE—DAYS				
	3	4	5	6	7
	Penicillin yield—units per ml				
Control (no added carbon)	27	45	41	36	27
Glucose	18	54	91	95	66
Brown sugar	7	40	85	102	79
Lactose	36	63	112	138	146
Glycerol	14	43	80	74	52
Cornstarch	35	85	122	140	146
Corn dextrin	28	73	91	125	146
	pH of filtrates				
Control (no added carbon)	6.0	7.5	8.0	8.1	8.3
Glucose	4.6	6.1	7.3	8.1	8.2
Brown sugar	4.3	5.4	7.0	7.8	8.1
Lactose	4.7	5.7	6.7	7.4	7.8
Glycerol	4.8	6.2	7.5	8.1	8.2
Cornstarch	4.9	6.6	7.1	7.4	7.7
Corn dextrin	4.7	6.0	6.7	7.4	7.7

Culture medium: Carbon source, 3.0 g per 100 ml; corn steep liquor, 10 g per 100 ml; standard salts; and initial pH, 4.0.
Organism: *P. notatum* 1249.B21.

TABLE 3

Penicillin production from various carbon sources at 6 per cent concentration

CARBON SOURCE	CULTURE AGE—DAYS					
	6			8		
	Penicillin yield	pH of filtrates	Mycelium weight per culture	Penicillin yield	pH of filtrates	Mycelium weight per culture
	units per ml		*g*	*units per ml*		*g*
Control*	40	8.2	.51	10	8.4	.41
Lactose	105	7.2	1.14	133	7.8	1.15
Glycerol	40	8.1	1.31	14	8.4	1.13
Sorbitol	42	8.1	1.28	18	8.4	1.07
Brown sugar	45	7.5	1.48	58	8.3	1.32
Glucose	50	7.6	1.31	30	8.4	1.14

Culture medium: Carbon source, 6.0 g per 100 ml; corn steep liquor, 10 g per 100 ml; standard salts; and initial pH, 4.0.
Organism: *P. notatum* NRRL 1249.B21.
* No carbon source added.

gelatinized by heating and shaking the flasks in a water bath. Starch which was liquefied by an acid or malt treatment was easier to handle and gave good penicillin yields. A commercial 80-fluidity starch, which was readily suspended in the culture medium by heating and stirring, appeared to be slightly better for penicillin production than untreated starch.

Since cornstarch was a satisfactory carbohydrate, tests were made to determine whether crude sources of starch could be used. Whole corn and whole wheat were ground to pass a $\frac{1}{16}$-inch screen and were added to the basic medium in varying quantities. Ground corn, ground wheat, and granular wheat flour were as satisfactory as lactose for penicillin production (table 4). Although, under these conditions, ordinary starch and the crushed grains gave good penicillin yields in the small flasks, difficulty was encountered in obtaining an even distribution of the grain particles in large bottles, especially in tall bottles which had to be sterilized in an upright position and later placed horizontally.

To develop the idea of using crude carbohydrate sources further, a corn mash was prepared by mixing 150 g of ground, yellow corn with 1,500 ml of water,

TABLE 4
Penicillin production from medium containing ground corn and wheat

CARBON SOURCE		CULTURE AGE—DAYS				
Type	g/100 ml	4	5	6	7	8
		Penicillin yield—units per ml.				
Ground corn..................	4.3	46	92	95	109	140
Ground corn..................	6.0	55	80	120	150	145
Ground wheat................	4.5	46	98	110	127	140
Ground wheat................	7.4	50	85	115	140	140
Granular wheat flour.........	3.7	53	90	112	138	119
Cornstarch...................	3.0	56	68	102	127	127
Lactose......................	3.0	64	85	109	143	147

Culture medium: Corn steep liquor, 10 g per 100 ml; standard salts; and initial pH, 4.0. Organism: *P. notatum* 1249.B21.

gelatinizing the mixture in a hot-water bath, and then autoclaving it for 30 minutes at 15 lb to stop enzymatic activity. The resulting mash was fairly fluid and could be mixed readily with the basic medium. Portions of this mash were employed to provide 4 and 6 per cent starch concentrations in media for penicillin production. The penicillin yields compared favorably with those obtained with whole grain, dextrin, and unhydrolyzed starch (table 5).

Mixtures of lactose and ground corn were satisfactory for penicillin production. The results obtained when 1 per cent of lactose and 4 per cent of ground corn were employed in production media are shown in table 6. The slow utilization of the corn prevented a too rapid rise in the pH of the medium; the pH of the broth did not rise above 7.8, and there was no decrease in penicillin yield during the 9-day incubation period. Subsequent investigations showed that a slightly faster accumulation of penicillin was obtained when a mixture of 2 per cent of lactose and 2 per cent of ground corn was used.

Because of its solubility lactose is easier to handle in nutrient medium preparation than is starch or ground grain. By using the best culture conditions thus far developed, the effect of lactose concentration on penicillin production was determined (table 7). The highest penicillin yields were obtained in 5 days by using a lactose concentration of 4 per cent. The higher the lactose concentration, the slower the rise in the pH of the culture medium. The fungus growth was not significantly greater in the 5.5 and 7.0 per cent lactose cultures.

TABLE 5

Penicillin production from medium containing malted corn mash

STARCH CONTENT OF FINAL MEDIUM	CULTURE AGE—DAYS				
	4	5	6	7	8
	Penicillin yield—units per ml				
per cent					
4	41	112	122	132	110
6	26	82	117	146	129

Culture medium: Corn steep liquor, 8.75 g per 100 ml; standard salts; and initial pH, 4.2. Organism: *P. notatum* 1249.B21.

TABLE 6

Penicillin production from medium containing ground corn and lactose

CARBOHYDRATE SOURCE	CULTURE AGE—DAYS					
	4	5	6	7	8	9
	Penicillin yield—units per ml					
1% lactose....................	55	62	59	55	60	45
1% lactose + 4% corn........	44	68	111	138	157	159
	pH of filtrates					
1% lactose....................	6.6	7.4	7.8	8.0	8.15	8.25
1% lactose + 4% corn........	5.4	6.2	6.9	7.5	7.8	7.8
	Mycelium weight—g per culture					
1% lactose....................	0.54	0.60	0.62	0.58	0.54	0.48
1% lactose + 4% corn........	0.66	0.82	0.91	0.90	0.88	0.86

Culture medium: Corn steep liquor, 10 g per 100 ml; standard salts; and initial pH, 4.0. Organism: *P. notatum* 1249.B21.

Some experiments were made to determine the ability of the fungus to utilize lactic acid or salts of lactic acid which were known to be present in the corn steep liquor. Various concentrations of lactic acid were used in media low in glucose and corn steep liquor. Both *P. notatum* NRRL 1249.B21 and NRRL 832 produced more mycelial growth on media to which the lactic acid had been added (table 8). Under similar conditions calcium lactate gave increased mycelial growth. Direct determinations of the lactic acid were not made but the rise in pH and the increase in fungus growth were considered as evidence that lactic acid or its salts served as carbon sources for these organisms.

TABLE 7
Effect of lactose concentration on penicillin production

LACTOSE CONCENTRATION	CULTURE AGE—DAYS					
	3	4	5	6	7	8
	Penicillin yield—units per ml					
g per 100 ml						
1.0	42	43	34	18	3	1
2.5	53	80	80	68	39	18
4.0	53	112	192	188	168	149
5.5	71	112	180	188	184	146
7.0	66	100	160	188	178	134
	pH of filtrates					
1.0	7.2	8.05	8.2	8.4	8.7	8.8
2.5	6.9	7.55	8.0	8.3	8.4	8.7
4.0	6.5	6.9	7.5	7.8	8.2	8.4
5.5	6.4	7.0	7.4	7.8	8.2	8.35
7.0	6.2	6.9	7.2	7.5	7.8	8.2
	Mycelium weight—g per culture					
1.0	0.65	0.72	0.69	0.56	0.51	0.50
2.5	0.74	0.94	1.01	0.91	0.81	0.75
4.0	0.81	1.05	1.26	1.24	1.13	0.99
5.5	0.91	1.17	1.37	1.34	1.30	1.18
7.0	0.88	1.11	1.30	1.50	1.43	1.32

Culture medium: Corn steep liquor, 10 g per 100 ml; glucose, 0.275 g per 100 ml; standard salts; media adjusted with KOH to initial pH, 4.6.
Organism: *P. notatum* 1249.B21.

TABLE 8
Growth and pH change in media containing various concentrations of lactic acid

LACTIC ACID ADDED PER LITER	INITIAL pH	CULTURE HARVESTED AT 5 DAYS			
		P. notatum NRRL 832		*P. notatum* NRRL 1249.B21	
		pH of filtrates	Mycelium wt per culture	pH of filtrates	Mycelium wt per culture
g			g		g
0	3.9	8.0	0.60	7.8	0.54
3.6	3.3	8.15	0.64	7.8	0.59
5.4	3.2	8.0	0.71	7.9	0.60
7.2	3.1	7.95	0.74	7.75	0.64
9.0	3.0	7.8	0.75	7.6	0.66
10.8	2.95	7.6	0.78	7.3	0.67
12.6	2.9	7.6	0.83	7.2	0.71
14.4	2.85	7.3	0.84	6.9	0.76
16.2	2.8	6.85	0.83	5.2	0.77
18.0	2.75	6.1	0.81	4.55	0.65

Culture medium: Corn steep liquor, 2.5 g per 100 ml; glucose, 2.0 g per 100 ml; and standard salts.

A comparison of penicillin production, carbohydrate utilization, pH change, and growth rate on a glucose and on a lactose medium (standard salts plus 10 per cent corn steep liquor) is shown in figure 1. The maximum penicillin yields of 143 and 84 units per ml from lactose and glucose media, respectively, were attained between the fifth and sixth days, after which there was a decrease in the penicillin content of the broth. The total reducing power of the medium was determined after acid hydrolysis and was calculated as glucose. The cultures on the glucose medium grew slightly faster than those on the lactose medium, which corresponds with the observation that the glucose was consumed more

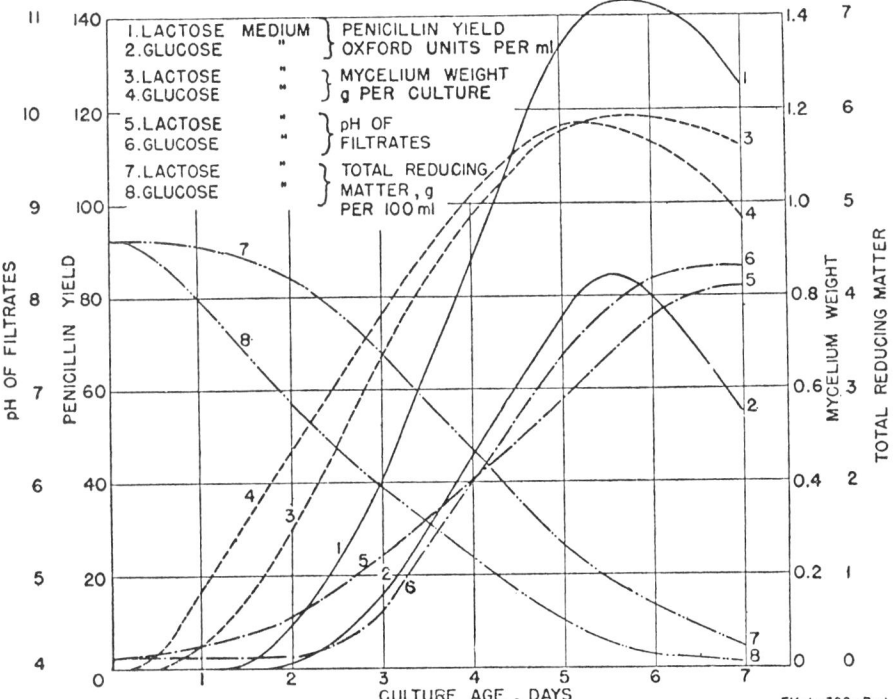

FIG. 1. PENICILLIN PRODUCTION, GROWTH, pH CHANGES, AND CARBOHYDRATE UTILIZATION IN GLUCOSE AND LACTOSE MEDIA

rapidly than the lactose. The pH of the glucose cultures during the first 3.5 days was lower than that of the lactose cultures, but after the fourth day the reverse was true.

Various lots of crude whey concentrates were found to inhibit both growth and penicillin production. Lactose crystallized from such syrups was as satisfactory as a highly purified lactose. This situation was apparently due to the salt content of the whey concentrates. Various concentrations of NaCl were added to a nutrient medium containing 8.8 per cent of corn steep liquor. A 10 per cent or greater concentration of NaCl inhibited both growth and penicillin production (table 9). A similar salt toxicity was encountered when high concentrations of corn steep liquor were neutralized with alkali, and with certain acid-hydrolyzed proteins.

The addition of a small amount of glucose to the basic medium containing lactose as the main carbohydrate source increased the penicillin yield in some cases and, furthermore, decreased the time required to attain the maximum yield. The value of adding 3 to 5 g of glucose per liter of medium seems to depend upon the glucose content of the corn steep liquor, which varies from about 2 to 12 per cent, depending upon the extent of the lactic fermentation occurring during manufacture. The advantage of adding glucose to the culture medium is most evident when highly fermented corn steep liquors (i.e., those having a low glucose content) are used.

Nitrogen sources and corn steep liquor. Various nitrogen sources, such as $NaNO_3$, KNO_3, $Ca(NO_3)_2$, $Mg(NO_3)_2$, NH_4NO_3, NH_4Cl, and urea, were employed with and without corn steep liquor. Two to six units of penicillin per ml were obtained by using $NaNO_3$, KNO_3, $Ca(NO_3)_2$, or $Mg(NO_3)_2$ in a medium without corn steep liquor. Under similar conditions, only traces of penicillin were pro-

TABLE 9
Effect of various concentrations of NaCl on growth and penicillin production

NaCl—g per 100 ml	0	1	2	3	4	5
Bulk of growth at 2 days—score	1.3	0.9	0.5	0.1	0	0
Mycelium weight at 6 days—g per culture	1.12	.61	.56	.41	.34	.26
pH of medium at 6 days	7.2	6.9	6.1	5.4	4.6	4.3
Penicillin at 6 days—units per ml	96	72	23	11	0	0

Culture medium: Lactose monohydrate, 4.4 per cent; corn steep liquor, 8.8 g per 100 ml; standard salts; and initial pH, 4.1.

duced by using NH_4Cl, NH_4NO_3, or urea as nitrogen sources. In culture media containing 8.6 per cent of corn steep liquor, in addition to these individual nitrogen sources, no significant differences in penicillin production were detected. A near-optimum concentration of the inorganic source of nitrogen was attained by the use of 3.0 g of $NaNO_3$ per liter in a medium containing 7.5 to 10 per cent of steep liquor.

The crude yeast extract used by Abraham *et al.* (1941) was employed in small quantities primarily for the purpose of accelerating the growth of *P. notatum*, which develops slowly on unsupplemented Czapek-Dox medium. Experiments were undertaken to determine whether a similar result might be achieved by using corn steep liquor, a cheap, commercially available material which has long been employed as a supplement to nutrient media for the production of yeast; for the bacterial production of butanol and acetone (Legg and Christensen, 1932), sorbose (Wells *et al.*, 1939), 2-ketogluconic acid, and 5-ketogluconic acid (Stubbs *et al.*, 1940); and for the production of gluconic acid by *Aspergillus niger* (Moyer *et al.*, 1940).

The first tests of corn steep liquor in penicillin production were made by using only 2 to 5 ml of this material per liter of Czapek-Dox medium. At these levels, corn steep liquor caused increased mold growth and gave a maximum yield of 10 units of penicillin per ml. In succeeding experiments, in which the concentration of corn steep liquor was increased, yields of 60 to 80 units of penicillin per ml were obtained in 6 to 8 days from a medium containing 3 per cent of glucose and 7.5 to 10 per cent of steep liquor.

The effect on penicillin yield, pH change, and fungus growth of various concentrations of corn steep liquor in a medium containing lactose is shown in table 10. The optimum concentration of corn steep liquor was between 7.5 and 12.5 per

TABLE 10

Effect of various concentrations of corn steep liquor on penicillin production

CORN STEEP LIQUOR CONCENTRATION	CULTURE AGE—DAYS					
	3	4	5	6	7	8
	Penicillin yield—units per ml					
g per 100 ml						
5	31	57	89	99	106	125
7.5	50	85	128	188	178	168
10	71	112	180	168	134	106
12.5	63	106	192	150	100	40
15	45	85	82	68	31	9
	pH of filtrates					
5	6.4	6.9	7.3	7.4	7.6	7.7
7.5	6.2	6.8	7.2	7.3	7.65	7.9
10	6.4	7.0	7.4	7.8	8.2	8.35
12.5	6.5	7.1	7.6	8.1	8.3	8.5
15	6.3	7.1	7.7	8.2	8.5	8.6
	Mycelium weight—g per culture					
5	0.54	0.72	0.87	0.81	0.93	0.95
7.5	0.67	0.96	1.15	1.21	1.21	1.16
10	0.91	1.17	1.37	1.34	1.30	1.18
12.5	0.99	1.33	1.46	1.40	1.24	1.04
15	1.00	1.36	1.46	1.50	1.19	1.13

Culture medium: Lactose, 5.5 g per 100 ml; glucose, 0.275 g per 100 ml; standard salts; media adjusted with KOH to initial pH 4.6.

Organism: *P. notatum* 1249.B21.

cent. The highest penicillin yield (192 Oxford units per ml) was obtained in 5 days with 12.5 per cent of steep liquor. Inhibition of penicillin accumulation was noted when 15 per cent of corn steep liquor was employed, although there was no inhibition of growth. The decrease in penicillin concentration occurring at 7 to 8 days was associated with a rise in alkalinity above pH 8.0 and with a decrease in the mycelium weight, and is attributed to a high initial concentration of corn steep liquor. When only 50 to 75 g of steep liquor per liter was employed, the alkalinity did not reach pH 8.0, and there was no significant decrease in penicillin content during the 8-day incubation period.

Other nutrient salts. In a culture medium containing 3.75 to 12.5 per cent of corn steep liquor, the concentrations of $MgSO_4 \cdot 7H_2O$ and KH_2PO_4 were increased and decreased from the amounts in the original Czapek-Dox medium without marked effect on growth or on penicillin yield; therefore, it appeared that corn steep liquor contains sufficient of these essential elements in available form. Although it was not obligatory to add $MgSO_4 \cdot 7H_2O$ and KH_2PO_4 to corn steep liquor medium, these salts were generally added at concentrations of 0.25 g and 0.50 g per liter, respectively, in order to insure adequate nutrition regardless of the quality or quantity of steep liquor employed.

Hydrogen ion concentration. Corn steep liquor usually has a pH of about 3.9 to 4.1, some variation being encountered depending on the extent of lactic acid fermentation occurring during manufacture. Because of the presence of amino acids and other protein fission products, the steep liquor is highly buffered; thus, 100 ml of one sample of steep liquor required 161 ml of 1 N NaOH to raise the pH from 4.0 to 8.5, which is approximately the pH change occurring during the 7- to 8-day growth period of the fungus on a 10 per cent corn steep liquor medium.

TABLE 11
Effect of initial pH on penicillin production

Initial pH of medium.	5.7	5.25	4.9	4.4	4.1	3.95	3.8	3.65	3.4	3.25
Penicillin yield, units per ml............	154	148	143	132	124	113	103	88	64	38
Mycelium weight, g culture............	0.98	0.99	1.00	1.03	1.13	0.95	0.95	0.95	1.04	1.02
pH of filtrates........	7.8	7.8	7.75	7.5	7.4	7.0	6.9	6.75	6.5	6.0

Culture medium: Lactose, 4.4 g per 100 ml; corn steep liquor, 7.5 g per 100 ml; and standard salts.
Organism: *P. notatum* 1249.B21.
All data reported for 6-day-old cultures.

The net result of all reactions taking place in the nutrient medium is, therefore, to render the medium alkaline.

Since there was poor penicillin accumulation in a medium more acid than pH 5.0, it seemed that there might be an advantage in changing the initial pH of the medium. The pH of a series of media was accordingly raised above 3.95 with NaOH, or lowered with lactic acid. The correlation between penicillin production and various initial pH values of the medium is shown in table 11. The amount of fungus growth was not significantly affected within the pH range, 3.25 to 5.7. Raising the initial pH increased penicillin production, the maximum yield being obtained from cultures started at pH 5.7. Decreased penicillin production was observed in cultures started at low pH levels.

In another experiment, various concentrations of corn steep liquor were employed and the initial pH was raised with NaOH. A medium containing 10 per cent corn steep liquor and inoculated at pH 5.5 resulted in poor growth, as well as in poor penicillin production. This inhibition was apparently due to an increase in the salt concentration above the tolerance level of this organism, as indicated in table 9. The level to which the initial pH of the medium can be raised is, accordingly, dependent upon the concentration of the corn steep liquor employed.

Addition of nutrients during the course of the fermentation. It is shown by data presented above that the quantity of penicillin in the broth decreases rapidly as the pH of the medium rises above 8.0 This decrease in penicillin yield was closely associated with the exhaustion of the carbohydrate, and with the beginning of mycelium autolysis (figure 1). There appeared to be a limit to the quantity of carbohydrate that could be included in the original culture medium, since an

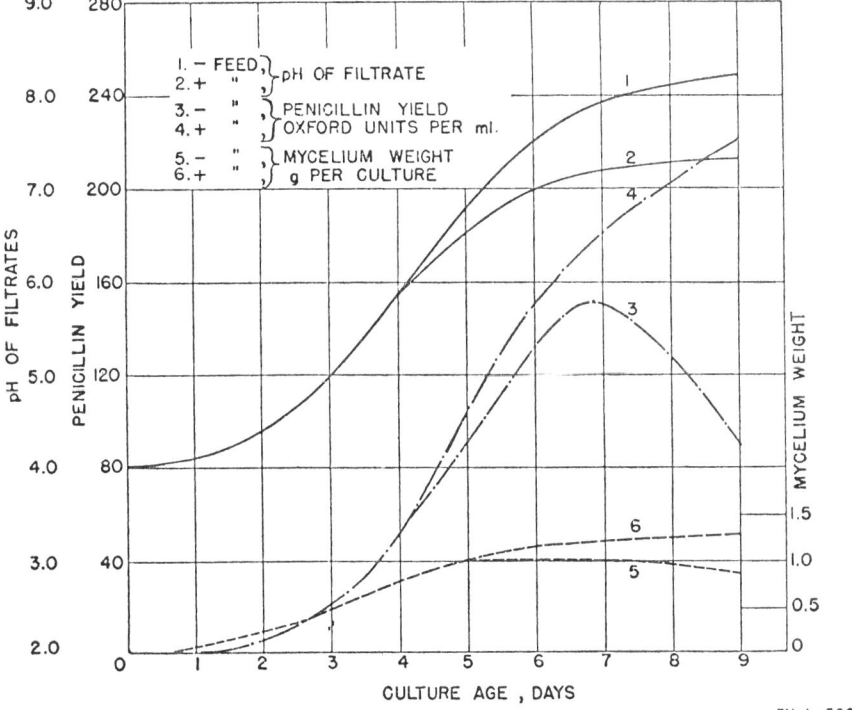

FIG. 2. EFFECT OF NUTRIENT ADDITION DURING FERMENTATION

excess of this constituent caused a delay in penicillin accumulation. The increase in the pH of the medium was most rapid when the initial supply of available carbohydrate was low. Therefore, it seemed probable that the addition of carbohydrate during the course of the fermentation might inhibit the rise of the pH to a level at which penicillin was rapidly destroyed.

To study the effect of periodic addition of nutrients during the growth period, experiments were conducted using a near-optimum medium for penicillin production. This medium contained the standard salts, 10 per cent of corn steep liquor, and 4 per cent of lactose. Cultures were allowed to grow for 4 days. Then, beginning on the fourth day and daily thereafter, nutrients were added by means of a sterile pipette. The control cultures received 1 ml of distilled water, and the fed cultures received 1 ml of a solution containing 0.20 g of lactose, 0.20 g of glucose, 0.010 g of corn starch, and 0.050 g of corn steep liquor per ml.

The effect of this feeding on penicillin yields, mycelial growth, and pH change are shown in figure 2. The addition of these nutrients caused an increase in penicillin as contrasted with the decrease in the control cultures. Penicillin yields of 220 units per ml were obtained on the ninth day, when the experiment was terminated. The pH of the fed cultures did not rise so rapidly as that of the control cultures, and there was an appreciable increase in fungus growth as a result of the added nutrients. The increase in penicillin yield was probably due both to a favorable pH and to a continued growth of the fungus.

Further experiments on the addition of nutrients showed that feeding 0.3 g of glucose per culture per day was nearly as effective as the foregoing combination of glucose, lactose, starch, and steep liquor. The addition of corn steep liquor alone was of little value in increasing penicillin yields. The addition of citric acid during the fermentation gave some pH control and an increase in penicillin production. The effect of pH control in all these cases is not clearly distinguishable from the effect of increased fungus growth. Addition of hydrochloric acid gave some pH control and prevented the decrease in penicillin yield normally occurring on the sixth or seventh day, but did not bring about appreciable increase in penicillin yield.

The optimum kind and amount of feeding, and the best time to begin it, have not been definitely established. These factors, it is believed, will vary considerably with the initial concentration of nutrients, the size of the culture flask, and the depth of the medium. Culture feeding may have some practical value in large pan cultures when mechanical stirring of the liquid is applied.

The Role of Corn Steep Liquor

The increase in penicillin yield due to the addition of steep liquor to the culture medium was so marked that the determination of the specific role of corn steep liquor became a very intriguing problem. It appeared that the steep liquor might supply any of a number of substances, such as trace elements, growth factors, amino acids, and possibly even building blocks for the penicillin molecule. Accordingly, experiments were planned to determine the effects of such substances when they were added to a medium containing no steep liquor.

Trace elements. P. notatum showed a typical, trace element requirement in a synthetic medium purified by the procedure as described by Steinburg (1935). A large number of trace elements, including zinc, iron, copper, molybdenum, manganese, and columbium were tested singly and in combination in the standard synthetic medium containing glucose but no corn steep liquor. None of these elements gave any response except zinc, which caused slight increase in growth and in penicillin production. Zinc has been included in the optimum medium containing corn steep liquor and lactose, but equally good results have been obtained when it was omitted.

P. notatum or *P. chrysogenum* gives good yields of gluconic acid when grown on the standard salt medium containing 0.5 per cent corn steep liquor and 5 to 25 per cent glucose. In such a medium, with or without $CaCO_3$, the yield of gluconic acid is greatly reduced by the addition of zinc. The accumulation of gluconic acid could not be detected in a medium containing 4 to 5 per cent glucose and 4 to 12 per cent corn steep liquor, and, under such conditions, the addition of zinc exerted little if any effect on fungus growth or penicillin accumulation.

To test further the possibility of trace elements being responsible for the action of corn steep liquor, a sample of steep liquor was ashed, and the ash, redissolved in dilute HCl, was added to the standard synthetic medium referred to above. Only a slight increase in growth and penicillin yield was observed. It was concluded that this fungus requires some trace elements for growth but that the trace elements alone are not responsible for the effect of corn steep liquor on penicillin yield.

Growth factors. To test the effect on penicillin production of certain growth factors, pimelic acid, nicotinic acid, thiamine, inositol, pyridoxine, sodium pantothenate, biotin, and *p*-aminobenzoic acid were added separately and as a composite to the standard synthetic medium. These compounds caused no increase in growth or in penicillin formation, nor did they, when added to agar plate cultures, give any improvement in growth or sporulation. In contrast, a few drops of dilute corn steep liquor, so added, greatly stimulated growth and spore production. Accordingly it appeared that the steep liquor did not function primarily as a source of these growth factors.

Amino acids. The fact that growth and penicillin yield improved with the use of increased concentration of corn steep liquor as the sole source of nitrogen suggested the possible importance of its nitrogenous constituents. Since steep liquor contains many amino acids, the effect of a number of these, including glycine, alanine, tyrosine, methionine, tryptophane, valine, proline, histidine, phenylalanine, and β-alanine, was determined by employing the standard salt medium containing 3 per cent of glucose but no steep liquor. These amino acids, when used in 0.05 M concentrations, singly or in combinations, had little effect on growth and no effect on penicillin formation.

Fractionation of corn steep liquor. Diluted corn steep liquor was dialyzed through a collodion membrane. The dialyzable fraction, after concentration, was found to be as good as the untreated corn steep liquor for growth and penicillin production.

A prolonged extraction of corn steep liquor at pH 5.0 with *n*-butanol or ethyl ether failed to remove anything which, when added to the standard medium in place of steep liquor, materially favored penicillin production.

The addition of comparatively large volumes of acetone or ethyl alcohol to steep liquor yielded copious precipitates. In general, the greater the precipitation, the less potent the filtrate became for penicillin production. It was not possible, however, to obtain any clear-cut fractionation of an active principle.

Corn steep liquor was treated with sodium hydroxide to pH 9.0, the precipitate (probably consisting largely of phytin and basic phosphates) was filtered off, and the filtrate acidified to pH 4.3 with hydrochloric acid. Steep liquor thus treated was as good as, but not better than, the untreated liquor. (In this case, the effect of high salt concentration was not apparent because of the use of only 4 per cent of steep liquor.) When the untreated steep liquor medium (containing 7.5 per cent steep liquor) was partially neutralized to pH 5.6 and steam-sterilized, a heavy precipitate formed and remained in the flasks during the fermentation. Such cultures gave better penicillin yields than those started at pH 4.0, in which there was much less precipitation. Removal of the precipitate from such cultures before inoculation did not cause a decrease in penicillin yield.

Substitutes for corn steep liquor. Since the work on corn steep liquor suggested the importance of amino acids and protein split products in penicillin production, other crude nitrogen sources were investigated. Finely ground soybean meal, fish meal, or a whole corn mash gave only slight increase in fungus growth and penicillin production. Various acid-hydrolyzed protein preparations from corn, wheat, and soybeans were all inferior to corn steep liquor when used on the basis of equal total nitrogen. It is believed that, to some extent, the ineffectiveness of these acid-hydrolyzed preparations was associated with their high salt content.

Trypsin-hydrolyzed casein, however, gave promising results when used in 1, 2, and 3 per cent concentrations (based on the dry casein), yielding in six days, 22, 20, and 19 units of penicillin per ml, respectively, as compared with 80 units per ml with 7.5 per cent corn steep liquor and 4 per cent glucose. However, concentrations greater than 3 per cent had a toxic effect and did not lead to increased penicillin yields. Subsequent to these tests, it was ascertained that British investigators had used acid-hydrolyzed casein, obtaining penicillin yields similar to those given above. Milk, whole or skimmed, has some value for penicillin formation, but probably not enough to merit commercial exploitation. By using 80 per cent of skimmed milk as the only nutrient in the medium, penicillin yields of 13, 20, 19, 26, and 31 units per ml were obtained at 4, 5, 6, 7, and 8 days, respectively.

Bacto-peptone, some commercial grades of peptone, and Difco dried yeast compound were ineffective in promoting penicillin production, although all of these products supported a good growth of the organism. The pH rise in the media was much slower than that encountered with corn steep liquor.

A malt syrup, employed in concentrations ranging up to 10 per cent by weight, produced maximum penicillin yields of 30 units per ml, which was approximately one third of the yield obtained with a corresponding weight of corn steep liquor. Very heavy precipitates were encountered upon steam sterilization. Portions of this malt syrup, when sterilized by filtration, inhibited all penicillin production without any inhibition of growth.

A barley and a soybean steep liquor, when tested over a wide range of concentration, proved worthless for penicillin production.

DISCUSSION OF RESULTS

A culture medium for good penicillin production must not only provide nutrients suitable for rapid fungus growth and penicillin formation, but must also provide conditions under which the penicillin is not too rapidly destroyed. The amount of penicillin which accumulates in the medium is, obviously, the difference between the amount produced and the amount destroyed. When low penicillin yields are obtained on a given medium, it should not necessarily be assumed that only a small amount of penicillin is formed by the fungus. In many cases it is probable that only a small amount of penicillin survives some unfavorable condition prevailing in the culture. Important factors influencing the stability of penicillin during fermentation are the temperature and pH of the medium. The instability of penicillin to numerous chemical and physical factors was recognized by early investigators, particularly in connection with isolation and purification procedures. Recent investigations by Benedict et al. (1945) on highly purified penicillin preparations have shown that penicillin inactivation is a function of temperature and pH, and that the greatest stability in solution is exhibited at pH 6.0. It is to be noted that the temperature and pH conditions which necessarily prevail in the culture medium throughout most of the growth period are not optimum for penicillin stability.

The data obtained in the course of these nutritional investigations indicate that penicillin is produced during the active growth of the fungus and is not an autolytic product released after growth has ceased. If this is correct, the the greatest quantity of penicillin should accumulate when the maximum growth occurs at or near pH 6.0, where penicillin is most stable. As shown in figure 1, penicillin formation can be detected by the time the pH has risen to 4.5. At this pH and temperature the half-life of pure penicillin is about 60 hours. In the culture liquors it is materially less stable. It is therefore probable that outside the pH limits of 5.0 to 7.5 there is substantial loss of penicillin. Near-optimum growth-pH- penicillin relationships are believed to exist when the greater part of the fungus growth occurs between pH 5.0 and 7.5. In the lactose medium (figure 1), 72 per cent of the growth occurred between pH 5.0 and 7.0, and in the glucose cultures only 31 per cent of the growth occurred in this range, 69 per cent of it occurring below pH 5.0.

There are several factors involved in the pH rise during the course of the fermentation. A typical corn steep liquor has a high buffer capacity (100 g may require as much as 160 ml of 1 N alkali to raise the pH from 4.0 to 8.5). During the fermentation the pH normally rises from 4.0 to over 8.0. This change is believed to be due both to assimilation of the lactate in the corn steep liquor (leaving the cation), and to the formation of alkaline products of metabolism such as ammonia. Both processes probably occur simultaneously.

It has been demonstrated that lactic acid or Ca-lactate can serve as a carbon source for this fungus in a medium containing corn steep liquor. Thus it would appear logical to assume that the lactate occurring in the corn steep liquor would be as readily assimilated as lactate which had been added to the culture medium. A pH rise would result from the utilization of the lactate and from the release of alkaline or alkaline earth elements originally combined as salts of lactic acid. The accumulation of sodium ions from the utilization of the nitrate ion of $NaNO_3$ will also account for part of the pH rise.

The rapidity and extent of the pH rise is associated with the amount of corn steep liquor and the amount and kind of carbon source supplied. Other investigators of fungal or bacterial metabolism have shown that deamination of protein degradation products may occur, especially when the available supply of readily assimilable carbohydrate is low, or after it has been exhausted. Kendall and his associates (1913) in their studies on bacterial metabolism, concluded that utilizable carbohydrates protect proteins from bacterial breakdown. Waksman (1917), in his investigations of the influence of available carbohydrate upon ammonia accumulation, demonstrated "the protein sparing action of the carbohydrates" in pure cultures of *Aspergillus niger* and *Citromyces glaber*. He found that ammonia accumulation was greatest when the supply of carbohydrate was low or after it had been exhausted. He concludes that, when available carbohydrates are present, the fungus utilizes all the nitrogen split off from the protein for its own metabolism; in the absence of carbohydrate, the protein molecule is attacked not only for its nitrogen but also for its carbon content. Butkewitsch (1922) showed with cultures of *Citromyces* that 75 per cent of the total nitrogen accumulates as ammonia in the culture medium when peptone is used as the sole nitrogen and carbon source. It has been shown in this paper that the pH rises more rapidly in a medium in which corn steep liquor is the sole source of carbon than in one in which glucose is used to supplement the corn steep liquor. It has also been shown that the pH rise in a steep liquor medium is less rapid when the supply of carbohydrate is increased. The addition of glucose during the fermentation also inhibits the rapidity of the pH rise. These results suggest that the accumulation of ammonia, through deamination of amino acids, can play an important part in the pH rise during the fermentation.

A difference in pH rise occurs with the various carbon sources. There is a more rapid rise in pH in the lactose cultures during the first part of the fermentation, and a slower rise later, than there is in the glucose cultures. These differences in pH change are believed to be due to differences in the availability of the carbon sources. It has been shown that lactose does not support growth of the fungus in the standard salt medium. Lactose in the presence of corn steep liquor is more slowly utilized than is glucose. A partial carbohydrate starvation may occur during the first few days of growth, before the lactose is rapidly hydrolyzed. During this period, ammonia from deamination apparently accumulates faster than it can be assimilated, resulting in a rise in pH. Starch also must undergo enzyme hydrolysis, and, in a manner similar to that of lactose, its use results in a slow rise in pH during the first few days of culture incubation.

P. notatum produces gluconic acid from low concentrations of glucose in a medium low in nitrogen, but in the standard medium with 3 to 4 per cent glucose and 6 to 8 per cent corn steep liquor, sufficient gluconic acid has not been detected to account for the slow rise in pH. So far as has been observed, *P. notatum* does not produce appreciable amounts of any organic acid from lactose under any nutrient conditions.

The use of lactose or starch as a carbon source provides a more satisfactory range of pH change during the fermentation period than can be obtained with the more readily available carbon sources such as glucose, sucrose, glycerol, or sorbitol. With lactose or starch, a higher percentage of fungus growth occurs within a pH range favorable to the accumulation of penicillin than occurs with the more readily available carbon sources. Such slowly available carbon sources as lactose and starch appear actually to prolong the productive life of the culture.

The role of corn steep liquor in the production of penicillin appears to be complex and is not clearly understood. It obviously supplies nutrients, chiefly nitrogenous compounds which favor a rapid growth of the fungus. Nitrogenous compounds such as peptone, soybean meal, fish meal, Difco yeast compound, and a whole corn mash also support a fair fungus growth but give lower penicillin yields than were obtained with corn steep liquor. Investigations of such compounds have not been complete enough to establish a critical composition difference from corn steep liquor. The role of the corn steep liquor in maintaining a proper pH change appears to be just as important as the maintenance of a vigorous fungus growth.

The value of these nitrogenous compounds may depend upon the extent to which the protein molecule is broken down. This occurs through the following stages: protein, peptones, polypeptides, and amino acids. The amino acids may be broken down by deamination or decarboxylation. Berger *et al.* (1937) made a comparison of the proteolytic activities of some common molds. The species of *Penicillium* showed low proteinase and dipeptidase activity, but they had a fairly high aminopolypeptidase activity. Kirch (1939) found that *P. luteum-purpurogenum*, when grown on soybean meal for 30 days at 23 to 26 C, showed weak proteolytic activity, and the presence of amino acids could not be detected. If *P. notatum* is correspondingly weak in its proteolytic activity, the value of an organic nitrogen supplement may be related to its content of amino acids, which this fungus can utilize.

A cheaper and more satisfactory organic nitrogen supplement than corn steep liquor has not been found for penicillin production. The use of corn steep liquor, with lactose or starch as component parts of the nutrient medium, has resulted in greatly increased penicillin yields and the establishment of commercial production.

SUMMARY AND CONCLUSIONS

Several culture methods can be utilized to produce spores for the inoculation of production cultures. Spores for laboratory inoculations can be readily grown on agar slants or on petri dish cultures. The use of dry spores mixed with a floating and spreading agent, such as whole-wheat flour, has given very satisfactory results in uniformity of surface growth and penicillin yield.

In the search for a better organism for penicillin production in surface culture, none was found superior to one of the descendants of the original Fleming strain which had been freed, insofar as possible, from degenerate, mutant strains. Continual precautions must be taken to guard against the appearance of these inferior forms. This special strain (NRRL 1249.B21) is now widely used for the industrial production of penicillin by the surface culture method.

The yield of penicillin produced in surface cultures by *Penicillium notatum* Westling, has been increased from 2 to 6 to over 200 units per ml. The addition of corn steep liquor to the culture medium greatly increases the penicillin yield.

The use of lactose or starch is found to give higher penicillin yields than can be obtained with glucose, sucrose, sorbitol, or glycerol. The use of strain NRRL 1249.B21 in a culture medium containing corn steep liquor and lactose has given penicillin yields of 190 units per ml in 5 days. The addition of nutrients during the course of the fermentation has further increased the yield up to 220 units per ml.

ACKNOWLEDGMENT

The authors wish to acknowledge the advice and assistance of Dr. Kenneth B. Raper, Dr. G. E. Ward, Mr. Wm. H. Schmidt, Mr. G. E. Nelson, and Mrs. Dorothy Alexander which were freely given during the course of these investigations.

REFERENCES

Abraham, E. P., Chain, E., Fletcher, C. M., Gardner, A. D., Heatley, N. G., Jennings, M. A., and Florey, H. W. 1941 Further observations on penicillin. Lancet, **2**, 177-189.

Benedict, R. G., Schmidt, W. H., Coghill, R. D., and Oleson, A. P. 1945 Penicillin. III. The stability of penicillin in aqueous solution. J. Bact., **49**, 85-95.

Berger, J., Johnson, M. J., and Peterson, W. H. 1937 The proteolytic enzymes of some common molds. J. Biol. Chem., **117**, 429-438.

Butkewitsch, W. 1922 Die Ausnutzung des Peptons als Kohlenstoffquelle durch Citromyces-Arten. Biochem. Z., **129**, 455-463.

Clutterbuck, P. W., Lovell, R., and Raistrick, H. 1932 The formation from glucose by members of the *Penicillium chrysogenum* series of a pigment, an alkali sol. protein and penicillin—the antibacterial substance of Fleming. Biochem. J., **26**, 1907-1918.

Fleming, A. 1929 The antibacterial action of cultures of a *Penicillium*, with special reference to their use in the isolation of *B. influenzae*. Brit. J. Exptl. Path., **10**, 226-236.

Kendall, A. I., Day, A. A., and Walker, A. W. 1913 Studies in bacterial metabolism. XIII-XXX. J. Am. Chem. Soc., **35**, 1201-1249.

Kirch, E. R. 1939 Hydrolytic cleavage and oxidation of soybean meal by *Pen. luteum-purpurogenum* and *Aspergillus niger*. Food Research, **4**, 363-370.

Legg, D. A., and Christensen, L. M. 1932 Process for the production of organic acids from cellulosic materials. U. S. patent, 1,864,746.

Moyer, A. J., Umberger, E. J., and Stubbs, J. J. 1940 Fermentation of concentrated solutions of glucose to gluconic acid. Improved process. Ind. Eng. Chem., Ind. Ed., **32**, 1379-1383.

Raper, K. B., Alexander, D. F., and Coghill, R. D. 1945 Penicillin. II. Natural variation and penicillin production in *Penicillium notatum* and allied species. J. Bact., **48**, 639-659.

Schmidt, W. H., and Moyer, A. J. 1944 Penicillin. I. Methods of Assay. J. Bact., **47**, 199-208.

Steinberg, R. A. 1935 Nutrient purification for removal of heavy metals in deficiency investigations with *Aspergillus niger*. J. Agr. Research, **51**, 413-424.

Stubbs, J. J., Lockwood, L. B., Roe, E. T., Tabenkin, B., and Ward, G. E. 1940 Ketogluconic acids from glucose-bacterial production. Ind. Eng. Chem., Ind. Ed., **32**, 1626-1631.

Waksman, S. A. 1917 Influence of available carbohydrate upon ammonia accumulation by microorganisms. J. Am. Chem. Soc., **39**, 1503-1512.

Wells, P. A., Lockwood, L. B., Stubbs, J. J., Roe, E. T., Porges, N., and Gastrock, E. A. 1939 Sorbose from sorbitol-semiplant-scale production by *Acetobacter suboxydans*. Ind. Eng. Chem., Ind. Ed., **31**, 1518-1521.

Copyright © 1946 by the American Society for Microbiology

Reprinted from pp. 79, 80, and 93 of *J. Bact.*, 51, 79–93 (1946)

PENICILLIN

IX. THE LABORATORY SCALE PRODUCTION OF PENICILLIN IN SUBMERGED CULTURES BY PENICILLIUM NOTATUM WESTLING (NRRL 832)[1]

ANDREW J. MOYER AND ROBERT D. COGHILL

Fermentation Division, Northern Regional Research Laboratory,[2] Peoria, Illinois

Received for publication September 1, 1945

Penicillin was first produced by cultivation of the fungus *Penicillium notatum* Westling on the surface of a liquid nutrient, such as Czapek-Dox glucose medium. Such a method of cultivation is very laborious, costly, and time-consuming when practiced on a large scale. Huge numbers of bottles or pans must be washed and sterilized, relatively small volumes of nutrient media must be dispensed into individual containers, and each container with its allotment of medium must be sterilized and inoculated. The incubation period for such cultures is usually 6 to 12 days, and, at the conclusion of the fermentation, considerable hand labor is required to remove the penicillin-containing liquors from the numerous fermentation vessels and from the fungus mycelium.

It was obvious that a more economical fermentation process would result from growing the mold submerged and uniformly distributed in vats or tanks such as are used in other fermentation industries. Submerged mold fermentation processes have been previously used for the production of gallic acid (Calmette, 1902), gluconic acid (Moyer *et al.*, 1940), and lactic acid (Ward *et al.*, 1938), and the adaptation of this method to the cultivation of *P. notatum* appeared to offer a means of decreasing the labor involved, of decreasing the fermentation time, and of increasing the penicillin yield.

This paper deals with the selection of a strain of *P. notatum* Westling suitable for the production of penicillin in submerged culture, and with investigations of nutrient media and culture conditions as nearly optimal as possible for this procedure.[3] (*P. notatum* NRRL 832 was originally selected and evaluated for submerged penicillin production by the senior author.)

---

[1] The expenses of the work described in this paper were met in part from Contract OEMcmr-100 with the Office of Scientific Research and Development, recommended by the Committee on Medical Research.

[2] One of the laboratories of the Bureau of Agricultural and Industrial Chemistry, Agricultural Research Administration, U. S. Department of Agriculture.

[3] The results of this work have been communicated monthly and bimonthly since March, 1942, in a series of restricted reports to Dr. A. N. Richards, chairman, Committee on Medical Research, Office of Scientific Research and Development, who, in turn, sent copies to all penicillin producers and to many research groups in this country and abroad. Owing to the strategic significance of penicillin, publication of papers covering this research has been delayed.

METHODS AND MATERIALS

The fungus was grown in the submerged condition in 300-ml Erlenmeyer flasks which were shaken continuously on Ross-Kershaw shaking machines. In this machine the flasks are secured to a flat table which is mounted excentrically and revolved at 200 cycles per minute. This movement imparts a swirling motion to the contents of the flasks and serves both to agitate and aerate the medium. Since the flasks are plugged with cotton, gas diffusion into and out of the flasks occurs readily. All cultures were maintained at 24 C. Samples were withdrawn for penicillin assay and pH determinations by means of a sterile, wide-mouthed pipette.

Assays were made by the cylinder-plate method originated by Abraham et al. (1941), and modified by Schmidt and Moyer (1944). Determinations of pH were made electrometrically.

The nutrient salts, unless otherwise specified, were $MgSO_4 \cdot 7H_2O$, 0.25 g; KH_2PO_4, 0.50 g; $NaNO_3$, 3.0 g; $ZnSO_4 \cdot 7H_2O$, 0.044 g; and $MnSO_4 \cdot 4H_2O$, 0.020 g per liter of medium. This concetration of salts will be referred to as the standard nutrient salts. The $CaCO_3$ was always sterilized dry in separate containers and added just prior to inoculation. The corn steep liquor employed contained 50 to 55 per cent total solids, 5 to 6 per cent free sugar calculated as glucose, and an acidity corresponding to pH 4.0.

Two types of inoculum were employed, namely, ungerminated spores used dry or wetted in a 0.1 per cent soap solution, and germinated spores in the form of clumps or tiny pellets. The pellet inoculum was prepared by developing mycelium from spores, in submerged culture, in a medium containing, in addition to the standard nutrient salts, 30.0 g of lactose and 55 ml of corn steep liquor per liter. One gram of sterile $CaCO_3$ and 125 ml of medium were employed for each 300-ml flask. In seeding flasks with spores, a suspension of spores from an agar slant (in a 6- by 1-inch test tube) was made in 60 ml of a 0.1 per cent soap solution; 10 ml of this suspension wree used to inoculate each culture for pellet production. When it was desired to start the fermentation with ungerminated spores, 1 ml of such a spore suspension were used for direct inoculation of production cultures. After 2 to 3 days on the Ross-Kershaw shaker, the pellets were 0.5 to 1.0 mm in diameter, and the solution had a penicillin content of 8 to 10 units per ml and a pH of 7.6 to 7.8. Usually 5.0 to 7.5 ml of this pellet preparation were employed to inoculate a production culture.

[Editor's Note: Material has been omitted at this point.]

REFERENCES

ABRAHAM, E. P., CHAIN, E., FLETCHER, C. M., GARDNER, A. D., HEATLEY, N. G., JENNINGS, M. A., AND FLOREY, H. W. 1941 Further observations on penicillin. Lancet, II, 177–189.

CALMETTE, A. 1902 Verfahren zur Umwandlung von Tannin in Gallussaure. German patent, 129,164.

MOYER, A. J., AND COGHILL, R. D. 1945 Penicillin production in surface cultures. *In preparation*.

MOYER, A. J., UMBERGER, E. J., AND STUBBS, J. J. 1940 Fermentation of concentrated solutions of glucose to gluconic acid. Improved process. Ind. Eng. Chem., Ind. Ed., **32**, 1379–1383.

RAPER, K. B., ALEXANDER, D. F., AND COGHILL, R. D. 1945 Penicillin. II. Natural variation and penicillin production in *Penicillium notatum* and allied species. J. Bact., **48**, 639–659.

SCHMIDT, W. H., AND MOYER, A. J. 1944 Penicillin. I. Methods of assay. J. Bact., **47**, 199–208.

WARD, G. E., LOCKWOOD, L. B., TABENKIN, B., AND WELLS, P. A. 1938 Rapid fermentation process for dextrolactic acid. Ind. Eng. Chem., Ind. Ed., **30**, 1233–1235.

8

Copyright © 1946 by the American Society for Microbiology

Reprinted from *J. Bact.*, 52, 223–226 (1946)

STUDIES ON THE NUTRITIONAL REQUIREMENTS OF STREPTOMYCES GRISEUS FOR THE FORMATION OF STREPTOMYCIN

GEOFFREY RAKE AND RICHARD DONOVICK

Division of Microbiology, The Squibb Institute for Medical Research, New Brunswick, New Jersey

Received for publication May 3, 1946

In view of the marked activity of streptomycin *in vitro* against a number of species of organisms relatively resistant to penicillin, such as *Escherichia coli* (Schatz, Bugie, and Waksman, 1944), certain of the *Salmonella* (Robinson, Smith, and Graessle, 1944), *Klebsiella pneumoniae* (Donovick, Hamre, Kavanagh, and Rake, 1945), and *Mycobacterium tuberculosis* (Schatz and Waksman, 1944), as well as its *in vivo* therapeutic behavior against such infecting agents (Jones, Metzger, Schatz, and Waksman, 1944; Robinson, Smith, and Graessle, 1944), investigations of this antibiotic as well as of the characteristics and growth requirements of the causative organism, *Streptomyces griseus*, are now being vigorously investigated in many laboratories. Chemical studies on streptomycin have already resulted in the preparation of crystalline derivatives (Fried and Wintersteiner, 1945; Kuehl, Peck, Walti, and Folkers, 1945).

Schatz, Bugie, and Waksman (1944) state that the production of streptomycin by *Streptomyces griseus* requires in the culture medium the presence of a specific growth-promoting substance supplied by beef extract or corn steep liquor. The medium recommended contained peptone, beef extract, glucose, and sodium chloride. We have studied various other materials as sources of nutrition for *Streptomyces griseus* and have found that it is possible to devise media including neither beef extract nor corn steep liquor that yield as much as 250 units[1] of streptomycin per ml, and from which streptomycin is recovered more readily in a purified state. The latter advantage arises because certain of the basic constituents of beef extract are concentrated in a manner similar to that of streptomycin and are found as impurities in the end product. The present paper deals chiefly with media employing soybean meal as the source of nitrogen.

The experiments described below were all conducted in a uniform fashion. A dilute suspension of *S. griseus* spores was prepared by suspending in distilled water the surface growth of this organism grown on Krainsky's asparagine glucose agar. Since the spores wet with great difficulty, the suspension was shaken with glass beads for half an hour. The resultant even suspension was stored at 4 C and was used for some months to inoculate the various media tested.

The media to be tested were dispensed in 200-ml amounts in the earlier experiments, and later in 100-ml amounts in 500-ml Erlenmeyer flasks. After being autoclaved, each flask was inoculated with 0.5 ml of spore suspension. The

---

[1] This is an average peak figure for medium no. 8. Individual shake flasks have on occasion been found to contain more than 350 units per ml of broth.

flasks were then incubated at 24 C on a shaking apparatus oscillating approximately 100 strokes per minute.

From the third through the sixth or seventh days of incubation, samples were taken daily for pH determination and streptomycin assay. The samples were clarified by centrifugation; the supernatant fluids were removed and heated in a boiling water bath for two minutes and assayed by the 2-ml broth dilution method (Donovick, Hamre, Kavanagh, and Rake, 1945). The constituents of the media employed and the streptomycin concentrations obtained in the various broths

TABLE 1
Streptomycin production in various media

MEDIUM NO.	CONSTITUENTS*				NO. OF REPLICATE FLASKS	VOL. OF MEDIUM PER FLASK		DAYS OF INCUBATION				
	Soybean meal†	Glucose	Beef extract	Sodium chloride				3	4	5	6	7
	%	%	%	%		ml						
1	1.5	1.0			2	100	pH	6.8	7.0	6.9	7.5	7.8
							u/ml‡	1.6		5.7	5.7	9.8
2§	1.5	1.0	0.5	0.5	4	200	pH	7.0	7.1	7.4	7.2	7.4
							u/ml	6.1	10.7	27.2	37.4	73.2
3	1.5	1.0	0.1		3	100	pH	7.3	7.8	8.2	8.3	8.5
							u/ml	37.1	44.1	96.2	121.0	114.0
4	1.5	1.0	0.1	0.5	6	100	pH	7.1	7.7	8.2	8.4	
							u/ml	123.5	147.0	146.0	158.0	
5	1.5	1.0	0.2	0.5	5	100	pH	6.8	7.7	8.2		
							u/ml	69.2	156.0	160.0		
6§	1.5	1.0	0.5	0.5	7	100	pH	7.2	7.9	8.3	8.6	
							u/ml	41.5	100.0	168.0	181.0	
7	1.5	1.0		0.5	16	100	pH	7.1	7.6	8.1	8.3	..
							u/ml	120.5	170.0	187.5	212.0	
8	1.0	1.0		0.5	6	100	pH	7.0	7.1	7.4	7.9	8.3
							u/ml	129.0	148.0	201.0	236.0	237.0

* Made up in distilled water.
† The soybean meal employed contained from 41 to 44 per cent protein.
‡ Units streptomycin per ml broth.
§ Media "2" and "6" were the same except that "2" was dispensed in 200 ml per flask while "6" was in 100-ml amounts.

are shown in table 1 and figure 1. These data represent a summary of the results of many replicate experiments.

Under the conditions employed in these studies the volume of medium per flask was extremely important. For example, in medium no. 2 the average flask yielded in the broth only 73.2 units of streptomycin per ml in 7 days, rising to 140 units per ml at a later date. When only 100 ml of the same medium was employed per flask (see no. 6), the average flask contained 181 units per ml of broth on the seventh day. Consequently, in most of the experiments only 100 ml of medium per 500-ml Erlenmeyer flasks was employed.

If the media are considered in the ascending order of activity obtained in the broth, it will be noted that a medium consisting of only soybean meal and glucose (no. 1) is distinctly deficient, yielding less than 10 u per ml in 7 days. The addition of 0.1 per cent beef extract but no sodium chloride (no. 3) improves the medium considerably but is only approximately half as effective as adding sodium chloride and no beef extract to a soybean glucose preparation (nos. 7 and 8). It would appear from these results that beef extract may supply certain necessary salts but is otherwise not required when soybean meal is employed. In fact, when media 4, 5, and 6 are compared with 7 and 8 it appears that the addition of beef

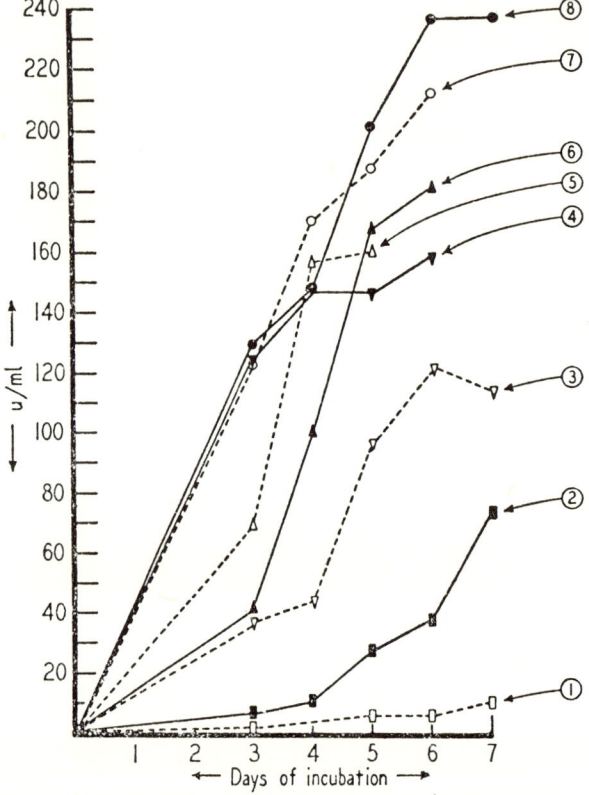

FIG. 1. STREPTOMYCIN PRODUCTION IN VARIOUS MEDIA

extract to a soybean meal, glucose, sodium chloride mixture is somewhat detrimental to the rise in streptomycin content in the broth. We have routinely observed that the rise in activity in a culture of *S. griseus* is greatest during or shortly after sporulation has begun. Thus, whereas mycelial growth is excellent in the presence of beef extract, sporulation is delayed. This may account for the lower streptomycin yields obtained in broths nos. 4, 5, and 6.

Preliminary studies have shown that substitution of the sulfate ion for the chloride ion in the salt added to soybean medium causes no appreciable change in the streptomycin concentration obtained in the broth. On the other hand, substitution of magnesium for the sodium ion gave lower streptomycin concentrations.

SUMMARY

The volume of medium in shake-flask cultures of *Streptomyces griseus* plays an important role in the concentration of streptomycin obtained in the broth.

Beef extract is not required for streptomycin production in a medium containing soybean meal, glucose, and sodium chloride. In fact, the addition of beef extract to such a medium in shake flasks delays somewhat the production of this antibiotic.

It is necessary to add an inorganic salt, e.g., sodium chloride, to soybean meal media for streptomycin production.

Preliminary studies indicate that sodium sulfate may be substituted for sodium chloride, but the substitution of magnesium chloride for either of these two electrolytes gives lower streptomycin yields.

Beef extract may supply a certain amount of the salt required in a soybean meal medium.

REFERENCES

DONOVICK, R., HAMRE, D., KAVANAGH, F., AND RAKE, G. 1945 A broth dilution method of assaying streptothricin and streptomycin. J. Bact., 50, 623–628.

FRIED, J., AND WINTERSTEINER, O. 1945 Crystalline reineckates of streptothricin and streptomycin. Science, 101, 613–615.

JONES, D., METZGER, H. J., SCHATZ, A., AND WAKSMAN, S. A. 1944 Control of gram-negative bacteria in experimental animals by streptomycin. Science, 100, 103–105.

KUEHL, F. A., JR., PECK, R. L., WALTI, A., AND FOLKERS, K. 1945 *Streptomyces* antibiotics. I. Crystalline salts of streptomycin and streptothricin. Science, 102, 34–35.

ROBINSON, H. J., SMITH, D. G., AND GRAESSLE, O. E. 1944 Chemotherapeutic properties of streptomycin. Proc. Soc. Exptl. Biol. Med., 57, 226–231.

SCHATZ, A., BUGIE, E., AND WAKSMAN, S. 1944 Streptomycin, a substance exhibiting antibiotic activity against gram-positive and gram-negative bacteria. Proc. Soc. Exptl. Biol. Med., 55, 66–69.

SCHATZ, A., AND WAKSMAN, S. A. 1944 Effect of streptomycin and other antibiotic substances upon *Mycobacterium tuberculosis* and related organisms. Proc. Soc. Exptl. Biol. Med., 57, 244–248.

Copyright © 1953 by the American Society for Microbiology

Reprinted from *Appl. Microbiol.*, **1**, 208–211 (1953)

Penicillin Production in Corn Steep Media with Continuous Carbohydrate Addition[1]

V. F. DAVEY[2] AND MARVIN J. JOHNSON

Department of Biochemistry, College of Agriculture, University of Wisconsin, Madison, Wisconsin

Received for publication April 13, 1953

The work reported here presents the results of an investigation into the replacement of lactose by glucose or by sucrose in a corn steep liquor medium for the production of penicillin. Previous work has established (Johnson, 1946) that glucose and sucrose are metabolized by *Penicillium chrysogenum* at a more rapid rate than is lactose. Also, it has been shown that a maximum rate of penicillin production is obtained under fermentation conditions which support a slow rate of growth, such as those of the conventional lactose fermentation. In order to approximate those conditions when glucose or sucrose was used as a carbohydrate source, these sugars were fed at rates chosen so as to restrict their utilization. The optimum feed rate of each sugar for penicillin production was then determined. This work was done exclusively in 30-liter stirred and aerated stainless steel fermentors. Work of a similar nature employing a synthetic medium has been reported elsewhere (Hosler and Johnson, 1953).

EXPERIMENTAL METHODS

Fermentation Techniques

Penicillium chrysogenum W49-133 was used throughout these experiments. Previous studies on penicillin production by this strain, and its derivation from the parent strain *Penicillium chrysogenum* Q176, have been reported elsewhere (Anderson et al., 1953).

A vegetative inoculum for the 30-liter fermentors was prepared through several stages. Firstly, in a 500-ml Erlenmeyer flask a mixture of 5 g bran and 5 g crushed corn, moistened with tap water and sterilized by autoclaving at 15 pounds pressure for 45 minutes, was inoculated with a small quantity of a soil stock and incubated at 25 C for 7 days, by which time sporulation was abundant. Several flasks were prepared in this way at one time, and stored at 5 C until required. Secondly, to the sporulated corn-bran flask was added 100 ml of a solution containing 6 g dextrin and 2 g corn steep solids, previously sterilized by autoclaving. The spore suspension thus obtained was incubated at 25 C on a rotary shaker operating at 250 rpm and describing a 2-inch circle. After 24 hours the contents of the flask were added aseptically to a 30-liter seed fermentor, similar in design to the fermentors described by Brown and Peterson (1950). This seed fermentor contained 16 liters of a medium comprising (in grams per liter) corn steep solids 25, lactose 30, glucose 2.0, KH_2PO_4 0.25, $MgSO_4 \cdot 7H_2O$ 0.1, and $CaCO_3$ 1.0. To control foaming, 1.0 g per liter of a 6 per cent solution of Alkaterge C[3] in lard oil was added, and the pH of the medium was adjusted to 6.2 with NaOH before sterilization.

The seed fermentor was incubated in a water bath at 25 C, agitated at 380 rpm and aerated at the rate of 0.5 volumes of air per volume of medium per minute. After 24 hours' growth the required volume of inoculum for the 30-liter fermentors was withdrawn into sterile 1-liter Erlenmeyer flasks. Each fermentor received 5 per cent by volume of this inoculum.

The medium used in the 30-liter fermentors contained (in grams per liter) corn steep solids 30, glucose 5, $CaCO_3$ 5, Na_2SO_4 1.0, and Alkaterge C in lard oil 1.0, made up to a volume of 15 liters. The sugar to be added during the run was dissolved separately in tap water, the quantity of potassium phenylacetate (precursor) solution equivalent to a concentration in the fermentor of 3 g phenylacetic acid per liter, recommended by Brown and Peterson (1950), was added to it, and the volume made up to 3 liters. For the control runs, in which all of the sugar was added initially, the potassium phenylacetate was made up in 3 liters of solution and fed during the fermentation as for a sugar solution.

An antifoam reservoir filled with the solution of Alkaterge C in lard oil was fixed to the fermentor head, and a metering valve attached after sterilization, to permit the addition of antifoam to the fermentor on demand. The construction of this antifoam metering valve been described by Anderson et al. (1953) and a general description of the fermentors given by Brown and Peterson (1950).

Each fermentor was incubated at 25 C in the water bath, agitated at 500 rpm and aerated at the rate of 1.0 volumes of air per volume of medium per minute. No provision for pH control during the runs was found

---

[1] Published with the approval of the Director of the Wisconsin Agricultural Experiment Station. Supported in part by grants from Parke, Davis and Company, Detroit, Michigan, and the Heyden Chemical Corporation, New York City, New York.

[2] Biochemist, Commonwealth Serum Laboratories, Parkville, Victoria, Australia.

[3] Commercial Solvents Corp., Terre Haute, Indiana.

140

necessary. After sterilization, the pH of the medium was in the range 5.6 to 5.8, and 20 to 24 hours after inoculation, when sugar feeding was begun, the pH had risen to about 7.0 in almost every run.

The sugar and precursor solution was fed to the fermentors by means of a small 1-rpm motor-operated positive displacement pump shown in figure 1. The rubber tubing connecting the sugar reservoir to the fermentor passed along the semi-circular channel of the pump so that the tubing was always compressed by at least one of the rollers which travelled along the channel. The pump was caused to operate intermittently by means of a timer adjusted so that the sugar solution was fed at the desired rate.

Samples were withdrawn from the fermentors at intervals for penicillin assays and chemical analyses, and were handled in the manner described by Gailey et al., (1946).

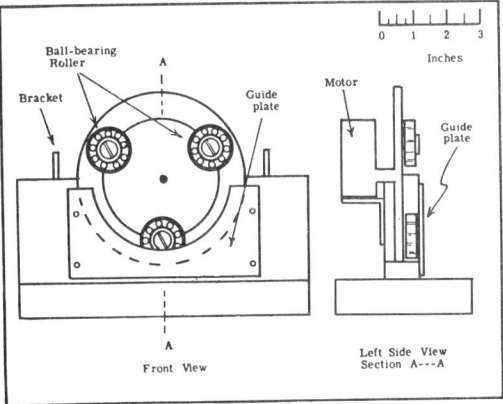

FIG. 1. Rotary positive displacement pump

Analytical Procedures

The pH of each sample was determined immediately after its withdrawal by means of a glass electrode.

The total penicillin titer was measured by a modification of the cylinder-plate method of Schmidt and Moyer (1944). The test organism used was *Micrococcus pyogenes* var. *aureus*.

All sugars were determined by the Shaffer and Somogyi (1933) method, with reagent 50 containing 5 g KI, and titrations were referred to standard curves prepared for each sugar. Lactose was hydrolyzed in 0.5 N HCl in an autoclave at 15 pounds pressure for 20 minutes. Sucrose was similarly hydrolyzed, except that the heating time was reduced to 5 minutes.

Soluble Kjeldahl nitrogen was determined by the micro-method of Johnson (1941) and the corresponding mycelial nitrogen content for any sample calculated by difference from the soluble nitrogen value of the sample taken at the time of inoculation.

For the determination of ammonia, a small aliquot of the undiluted sample filtrate was made alkaline, the solution then gently aerated, and the ammonia evolved absorbed in an aliquot of standard sulfuric acid.

RESULTS AND DISCUSSION

It was found that, with a concentration of 0.5 per cent glucose present initially in the fermentor medium, growth from the inoculum was rapid and the growth phase was virtually complete in 24 hours. The mold appeared to be indifferent to the presence of another sugar during this phase. Continuous sugar feeding was begun at 20 to 24 hours after inoculation and continued for 72 hours; extension of the feeding time beyond this period appeared to have no advantage, as a decrease in penicillin titer was found to occur at about 100 hours in most runs even when the sugar supply was not limiting.

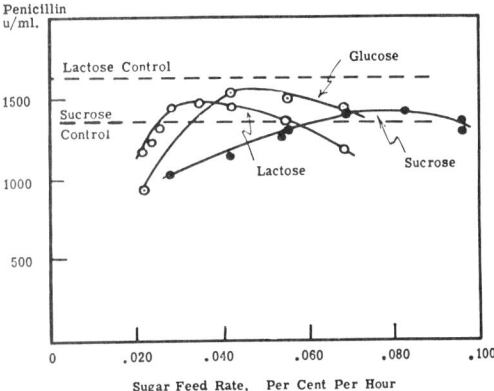

FIG. 2. Effect of sugar feed rate on penicillin production

For purposes of comparison of the value of the feeding technique for the three sugars tested, the conventional fermentation with 3 per cent lactose present initially was taken as a standard. None of the fed runs reached a penicillin titer as high as that in the lactose standard, which was 1610 units per ml (average of 3 runs). The penicillin yields from fermentations fed glucose, sucrose or lactose are shown in figure 2.

For glucose, the best feed rate tested was found to be 0.042 per cent per hour, equivalent to 3 per cent overall, giving a maximum penicillin titer of 1520 units per ml at 83 hours.

In the series of sucrose-fed runs it was found that the total concentration of sucrose required to give penicillin yields comparable with those from glucose or lactose-fed media was considerably higher than for those sugars.

The penicillin titer of control runs, in which 5 per cent sucrose was present initially, was 1360 units per

ence in the rates of sugar utilization. This is shown in figure 3. Apparently the sucrose is assimilated so rapidly by the mold that the condition of semi-starvation typical of the lactose fermentation, and which appears to be necessary to a high penicillin yield, does not occur. The 5 per cent sucrose was used almost entirely by the mold in 48 hours. However, when the same total concentration of sucrose was fed slowly over a period of 72 hours the 5 per cent level was not reached before 90 hours and the penicillin yield was higher, 1410 units per ml, although still not as high as that in the lactose-fed fermentations. The chemical changes occurring in the sucrose-fed run are shown in figure 4.

For the lactose-fed series, the optimum feed rate was found to be 2.4 per cent overall, or 0.034 per cent per hour, giving a penicillin titer of 1470 units per ml. The chemical changes in the fermentation to which 2.4 per cent lactose was fed revealed no differences from the changes in the control runs sufficient to explain the lower yield in the fed fermentation.

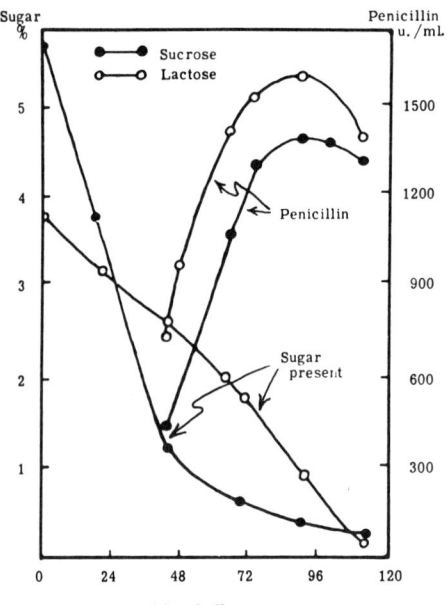

FIG. 3. Comparison of rates of sugar utilization and penicillin production for lactose and sucrose control fermentations

TABLE 1. *Effect of time of addition of lactose on penicillin yield*

LACTOSE CONCENTRATION	TIME OF ADDITION TO MEDIUM	MAXIMUM PENICILLIN YIELD
%		units per ml
3.0	Before sterilization	1610
2.4	Before sterilization	1590
2.4	With inoculum	1630
2.4	After 16 hours growth	1542
2.4	After 22 hours growth	1490
2.4	Fed from 16th hour	1560
2.4	Fed from 22nd hour	1470

Some further experiments with lactose-fed runs were conducted in an attempt to clarify the anomaly apparent in figure 2; namely, that lactose when slowly fed would not give as high a penicillin yield as when present initially in the medium. The results of these experiments, which are shown in table 1, suggest that the lactose solution carried some factor which was effective during the growth phase of the fermentation in stimulating a higher yield of penicillin during the later phase. The effect was not due to the earlier addition of the precursor, as this was present initially in the media of all these experiments. Samples of the commercial grade of lactose used in this work assayed 95 to 97 per cent lactose against a standard USP lactose monohydrate.

SUMMARY

With a slow feeding technique, more rapidly assimilated sugars, such as glucose or sucrose, have been employed successfully in place of lactose as the carbohydrate source in a corn steep medium for the production of penicillin. The work was done with 30-liter fermentors, and the penicillin-producing organism used was *Penicillium chrysogenum* W49-133. For the three sugars studied, the optimum feed rates found were:

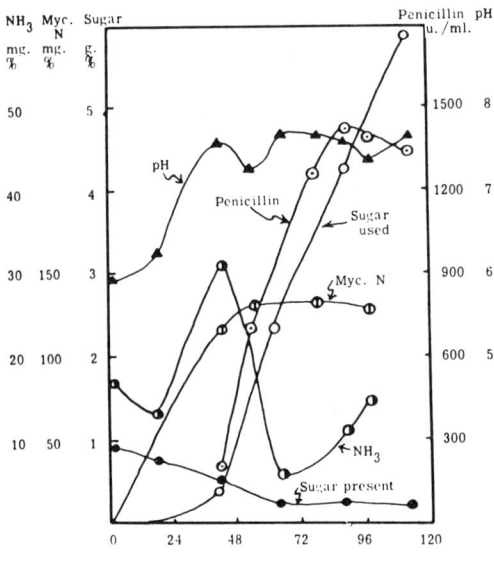

FIG. 4. Chemical changes in a fermentation fed 5 per cent sucrose

ml (average of three). While there was close agreement in the chemical changes occurring in media containing lactose or sucrose respectively, as regards ammonia and mycelial nitrogen levels, there was a striking differ-

for glucose, 0.042 per cent per hour (3.0 per cent total); for lactose, 0.034 per cent per hour (2.4 per cent total); and for sucrose, 0.070 per cent per hour (5.0 per cent total). The optimum yields of penicillin obtained, corresponding to these feed rates, were: for glucose, 1520 units per ml; for lactose, 1470 units per ml; and for sucrose, 1410 units per ml.

Penicillin yields in media containing commercial lactose as the carbohydrate source were found to be improved when the lactose was present early in the growth phase of the fermentation.

REFERENCES

ANDERSON, R. F., WHITMORE, L. M., JR., BROWN, W. E., PETERSON, W. H., CHURCHILL, B. W., ROEGNER, F. R., CAMPBELL, T. H., BACKUS, M. P., AND STAUFFER, J. F. 1953 Penicillin production by pigment-free molds. Ind. Eng. Chem., **45,** 768–773.

BROWN, W. E., AND PETERSON, W. H. 1950 Factors affecting production of penicillin in semi-pilot plant equipment. Ind. Eng. Chem., **42,** 1769–1774.

GAILEY, F. B., STEFANIAK, J. J., OLSON, B. H., AND JOHNSON, M. J. 1946 A comparison of penicillin-producing strains of *Penicillium notatum-chrysogenum*. J. Bact., **52,** 129–142.

HOSLER, P., AND JOHNSON, M. J. 1953 Penicillin from chemically defined media. Ind. Eng. Chem., **45,** 871–874.

JOHNSON, M. J. 1941 Isolation and properties of a pure yeast polypeptidase. J. Biol. Chem., **137,** 575–586.

JOHNSON, M. J. 1946 Metabolism of penicillin-producing molds. Ann. N. Y. Acad. Sci., **48,** 57–66.

SCHMIDT, W. H., AND MOYER, A. J. 1944 Penicillin: I Methods of assay. J. Bact., **47,** 199–208.

SHAFFER, P. A., AND SOMOGYI, M. 1933 Copper-iodometric reagents for sugar determination. J. Biol. Chem., **100,** 695–713.

Part V
APPARATUS AND EQUIPMENT

Editor's Comments
on Papers 10 Through 13

10   SUMINO, AKIYAMA, and FUKUDA
Performance of the Shaking Flask (I) Power Consumption

11   KROLL et al.
Equipment for Small Scale Fermentations

12   WEST et al.
An Improved pH Electrode Assembly for Pilot Plant and Plant Fermentors

13   JOHNSON, BORKOWSKI, and ENGBLOM
Steam Sterilizable Probes for Dissolved Oxygen Measurement

    This group of papers deals with design of fermentors for small-scale fermentation studies, and with design of ancillary apparatus for measurement *in situ* of pH and dissolved oxygen. These papers were written by authors who were familiar with fermentation technology from investigational as well as production aspects, and who were sensitive to the need to devise and improve small-scale fermentors and ancillary equipment for experimentation that would yield data meaningful in relation to conduction of processes in a manufacturing plant.

    Paper 10 is a rather recent report of an attempt to describe power consumption in the shaken flask fermentor. Reports of such studies are rare, although previous papers (cited in Paper 10) dealt with the problems of expressing in quantitative terms oxgen transfer and mixing rates in shaken flask fermentations. (For an excellent review of the impact of flask-shaking machines on fermentation technology see Freedman 1969.)

    Paper 11 is a brief but excellent account of the design of a 40-liter stainless steel stirred, aerated fermentor and auxiliary equipment. Although the fermentor described is inadequate for present-day highly sophisticated fermentation processes, it served a need for a number of years for scale-up studies with processes of simple batch or controlled batch design. Other reports that appeared simultaneously with Paper 11 described generally similar systems (Nelson, Maxon, and Elferdink 1956; Friedland, Peterson, and Sylvester 1956).

Although for many years the *in situ* measurement of pH and dissolved oxygen had interested those concerned with investigational as well as manufacturing aspects of fermentation processes, the construction of completely practical steam-sterilizable probes had presented many difficulties. Paper 12 is one of the first reports of design, fabrication, and use of a truly versatile pH electrode assembly. Similarly, Paper 13 is one of the first reports of a practical steam-sterilizable probe for measuring dissolved oxygen. (See Steel and Maxon 1966, which is an early, often-cited account of the use of dissolved oxygen measurement *in situ* together with analysis of effluent gas for oxygen content, to characterize the respiratory activity of a fermentation.)

REFERENCES

Freedman, D. 1969. The shaker in bioengineering. *Process Biochem.*, 4:35–40.

Friedland, W. C., M. H. Peterson, and J. C. Sylvester. 1956. Fermenter design for small scale submerged fermentation. *Ind. Eng. Chem.*, 48:2180–2182.

Nelson, H. A., W. D. Maxon, and T. H. Elferdink. 1956. Equipment for detailed fermentation studies. *Ind. Eng. Chem.*, 48:2138–2189.

Steel, R., and W. D. Maxon. 1966. Dissolved oxygen measurements in pilot- and production-scale novobiocin fermentations. *Biotech. Bioeng.*, 8:97–108.

Copyright © 1972 by the Society of Fermentation Technology, Japan

Reprinted from *J. Ferment. Technol.*, **50**(3), 203–208 (1972)

Performance of the Shaking Flask

(I) Power Consumption

Yasuhiro Sumino, Shun-ichi Akiyama, and Hideo Fukuda

(Research and Development Division, Takeda Chemical Industries, Ltd., Higashiyodogawa-ku, Osaka)

Abstract

In order to clarify the agitation intensity in a shaking flask fermentation, the power consumption per unit volume in the flask was measured by a method based on the heat transfer rate of the shaking flask containing liquid. The degree of power consumption per unit volume in a creased flask was about 5 times that in a conical smooth flask at 204 rpm on a rotary shaker with an eccentricity of 2.5 cm. An increase of the liquid volume in the flask gave a decrease of the power consumption per unit volume. The relationship between the power consumption and the shaking speed agreed with the well-known fact that in agitated tanks the power consumption is proportional to the cube of the impeller speed. From the results of a comparison of power efficiencies, it was found the agitation intensity in shaking flasks was not always lower than that in agitated tanks.

Introduction

The oxygen transfer in shaking flask fermentation has been investigated in detail from the viewpoint of scale-up for a long time. However, in the case of a novel fermentation using hydrocarbon as one of the raw materials, it seems that not only the oxygen transfer but also the mixing or dispersing of hydrocarbon should be important. There have been only a few reports which dealt with these additional factors in shaking flask fermentation; Rhodes and Gaden (1957)[1] studied the liquid-solid mass transfer and mixing time in shaking flasks and compared them with those in an agitated tank. Brandle et al. (1966)[2] compared the dissolution rate of methyl red in a shaking flask and in a fermenter. Hara (1969)[3] made an attempt to express the flow characteristics in shaking flasks in quantitative terms. Tanaka et al. (1969)[4] and Tanaka et al. (1970)[5] made a study on the effect of agitation intensity on the destruction of fungus mycelia.

On the other hand, the power input per unit volume is generally considered to have a close relationship to these characteristics, including the oxygen transfer, the mixing of the liquid, and so on. Therefore, many investigators engaged in scale-up studies have, up to now, dealt with the power consumption in agitated tanks, but not in shaking flasks.

This report was presented at the Annual Meetings of the Society of Fermentation Technology of Japan, held at Osaka on November 19, 1970.

In this paper, power consumption in shaking flasks under certain experimental conditions is discussed.

Materials and Methods

1. **Apparatus**   Conical and creased flasks of 200 ml capacity and a 2 *l* Sakaguchi flask were used for shaking flask experiments and a 2,000 *l* fermenter for the agitated tank experiments. The shapes and dimensions of these vessels are shown in Fig. 1.

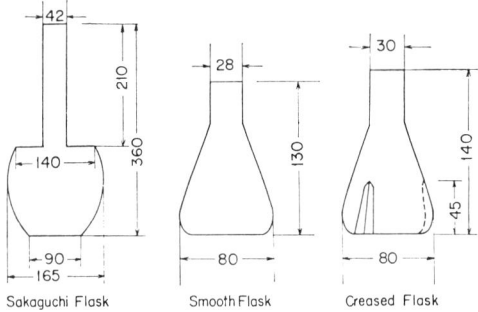

Fig. 1. Shapes and dimensions of flasks (mm).

A rotary shaker with an eccentricity of 2.5 cm and a reciprocating shaker with an amplitude of 10 cm were employed at 28°C.

2. **Measurement of the oxygen transfer coefficient**   The volumetric oxygen transfer coefficient was measured by the sulfite oxidation method.[6]

3. **Measurement of power consumption**   In the case of the tank experiments, the power consumption was calculated from the net value of the electric power consumption of the agitator. In the case of the shaking flask experiments, the following method was employed:

Liquid, heated previously to about 70°C, was poured into each flask which was covered with cotton and polyethylene film as shown in Fig. 2. The flask was

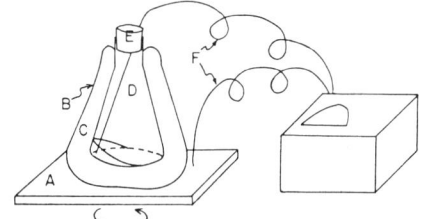

Fig. 2. Schematic diagram of measurement.
A: Shaker
B: Polyethylene film
C: Cotton
D: Flask
E: Rubber stopper
F: Thermistor

plugged with a rubber stopper and shaken on a shaker. The temperatures of the liquid and the air outside the flask were continuously measured with a thermistor (Takara Thermistor SPD-02, Takara Instruments Co., Osaka) corrected previously by a standard thermometer.

Now, the heat balance equation of the liquid in the flask may be expressed as follows:

$$\frac{dQ}{d\theta} = \frac{dQ_0}{d\theta} - \frac{dQ_T}{d\theta} - \frac{dQ_V}{d\theta} \quad\quad\quad (1)$$

In this equation, the term of the evaporation heat is able to be neglected under these specific conditions and $dQ/d\theta$ and $dQ_T/d\theta$ are expressed as the cooling rate, $(-c\,dt_F/d\theta)$, and the over-all heat transfer rate, $UA(t_F-t_0)$, respectively.

Therefore,

$$\frac{dQ_V}{d\theta}=0, \quad \frac{dQ}{d\theta}=c\frac{dt_F}{d\theta}, \quad \frac{dQ_T}{d\theta}=UA(t_F-t_0) \quad \cdots\cdots\cdots\cdots (2)$$

Substituting these values into equation (1), the following equation is obtained:

$$-c\frac{dt_F}{d\theta}=UA(t_F-t_0)-\frac{dQ_0}{d\theta} \quad \cdots\cdots\cdots\cdots (3)$$

A graphic explanation of this equation is shown in Fig. 3. When regarding

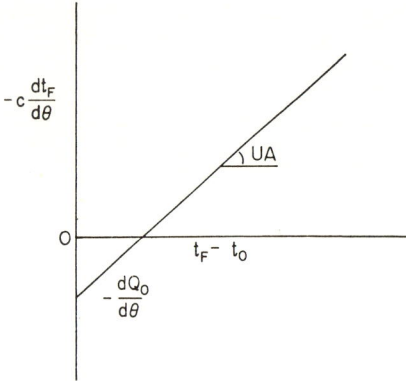

Fig. 3. Principle of power estimation.

the temperature difference, (t_F-t_0), as the abscissa and the cooling rate, $(-c\,dt_F/d\theta)$, as the ordinate, a straight line should be obtained. The intercept on the ordinate represents the heat of agitation from which the power consumption is calculated, and that on the abscissa represents the equilibrium value of the temperature difference between the liquid and the surroundings.

Results and Discussion

1. Example of the power measurement in a shaking flask Figure 4 shows an example of the time-temperature profile in the course of cooling the liquid in a creased flask containing 40 ml of the liquid and shaken at 204 rpm on a rotary

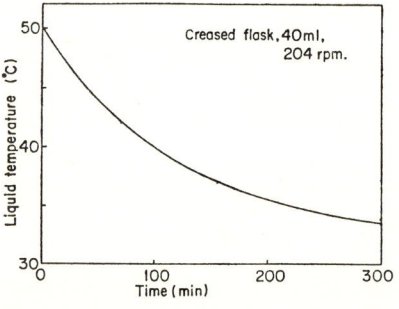

Fig. 4. t_F vs. θ.

shaker. The temperature difference and the cooling rate were calculated from this figure and plotted as shown in Fig. 5. A linear relationship was obtained as was expected from Fig. 3. The intercept on the ordinate was determined to be -0.0197 by the least squares method, and the power consumption under these conditions was calculated to be 1.38 kw/kl, wherein, $c = 1.0$ cal/ml/°C.

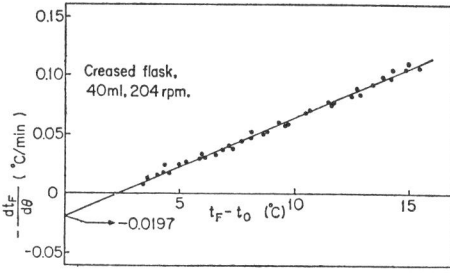

Fig. 5. $-(dt_F/d\theta)$ vs. (t_w-t_0).

In the case of experiments where the flask and its cover were not sufficiently warmed in advance, a linear relationship between the temperature difference and the cooling rate was not obtained, as is shown in Fig. 6, because of excessive consumption of heat in warming up the system.

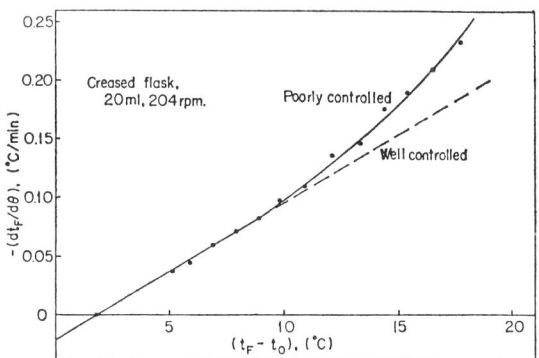

Fig. 6. $-(dt_F/d\theta)$ vs. (t_F-t_0).

2. Effects of the liquid volume and the shape of the flask

Figure 7 gives

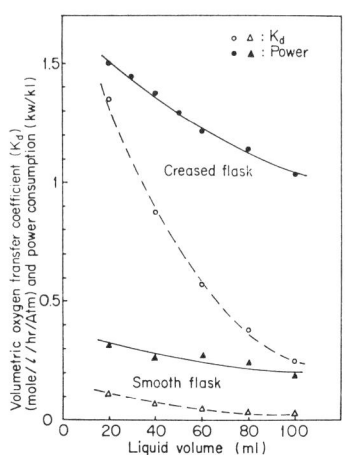

Fig. 7. Effect of liquid volume on power consumption and volumetric oxygen transfer coefficient.

the power consumption and the oxygen transfer coefficient in smooth and creased flasks containing different volumes of liquid. As is clear from this figure, the creased flask brought about an increase in the power consumption per unit volume of about 5 times and in the oxygen transfer coefficient of about 15 times that of the smooth flask, regardless of the liquid volumes in these flasks. An increase in the liquid volume gives an exponential decrease in the power consumption and the oxygen transfer coefficient in both flasks.

3. Effects of shaking speed and type of shaker   The effect of the shaking speed on the power consumption is shown in Fig. 8. The experiments were carried

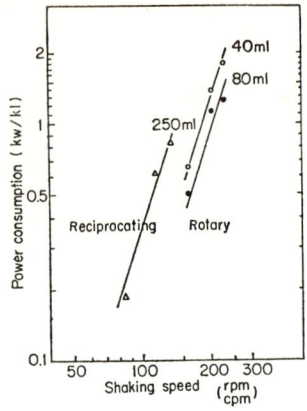

Fig. 8. Effect of shaking speed on power consumption.

out using a Sakaguchi flask containing 250 ml of liquid on a reciprocating shaker and creased flasks containing 40 ml and 80 ml of the liquid on a rotary shaker.

The power consumption in these flasks was approximately proportional to the cube of the shaking speed, independent of the type of shaking. These results agreed with those well-known[7] for agitated tanks. But, in order to confirm this fact more clearly, further studies may be required due to the comparatively narrow range of experimental conditions employed here.

4. Comparison of the agitation intensities of flasks and a fermenter   In the comparison of the agitation intensities of the flasks and a fermenter, the power efficiency for the oxygen transfer was employed as shown in Fig. 9.

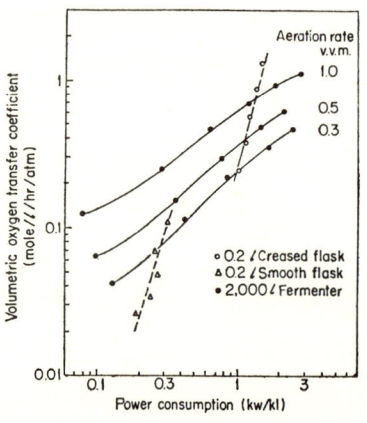

Fig. 9. Comparison of power efficiency between shaking flasks and an agitated fermenter.

This figure may be used to obtain the same experimental conditions for the power consumption and the oxygen transfer coefficient in the fermenter as in the flasks. For example, the shaking conditions of the creased and the smooth flasks containing 40 ml each of liquid at 204 rpm on the rotary shaker correspond to the operating conditions of the fermenter at an impeller speed of 210 rpm with an aeration rate of 1 V/V/M and 100 rpm with 0.3 V/V/M, respectively.

From these results, it would seem that the agitation intensity in shaking flasks is not so poor as had been thought, although it is dependent on the shape of the flask, the type of shaker, the shaking speed, and the liquid volume.

Acknowledgement

The authors are indebted to Professor Hisaharu Taguchi of Osaka University for the discussion of the results; we also wish to thank Dr. Rokuro Takeda and Dr. Yasunari Ishida for their helpful advice and Mr. Kenjiro Okazaki for his technical assistance.

Nomenclature

Q	:	Heat evolution	cal/ml
Q_F	:	Heat of fermentation	cal/ml
Q_0	:	Heat of agitation	cal/ml
Q_V	:	Heat of evaporation	cal/ml
Q_T	:	Heat loss to the surroundings	cal/ml
U	:	Over-all heat transfer coefficient	cal/cm²/°C
A	:	Heat transfer area	cm²/ml
t_F	:	Liquid temperature	°C
t_0	:	Room temperature	°C
θ	:	Time	min
c	:	Heat capacity	cal/ml/°C.

References

1) Rhodes, R. P., Gaden Jr., E. L.: *Ind. Eng. Chem.*, **49**, 1233 (1957).
2) Brandle, E., Schmid, A., Steiner, H.: *Biotech. Bioeng.*, **8**, 297 (1966).
3) Hara, M., *J. Ferment. Technol.*: **43**, 590, 597 (1965).
4) Tanaka, H., Takebe, H., Takahashi, J., Ueda, K.: (*Abstr.*) *Proc. Annual Meeting of Ferment Technol, Japan*, 28 (1969).
5) Tanaka, H., Takahashi, J., Ueda, K.: (*Abstr.*) *Proc. Annual Meeting of Ferment. Technol., Japan*, 124 (1970).
6) Cooper, C. M., Fernstrom, G. A., Miller, S. A.: *Ind. Eng. Chem.*, **36**, 504 (1944).
7) Rushton, J. H., Costich, E. W., Everett, J. H.: *Chem. Eng. Prog.*, **46**, 395, 467 (1950).

(Received July 3, 1971)

11

Copyright © 1956 by the American Chemical Society

Reprinted from *Ind. Eng. Chem.*, **48**, 2190–2193 (1956)

Equipment for Small Scale Fermentations

C. L. KROLL, STANLEY FORMANEK, A. S. COVERT[1], L. A. CUTTER[2], J. M. WEST, and W. E. BROWN
The Squibb Institute for Medical Research, New Brunswick, N. J.

▶ The small fermentor, normally used for initial scale-up of processes from the flask stage and for detailed study of the effect of major variables, is one of the most important tools available to the antibiotic industry for the development of fermentation processes. The 50-liter stainless steel fermentor and auxiliary equipment described were built to fill those needs. Although the goals of maximum versatility and minimum handling and maintenance requirements can be met in other ways, this equipment has given superior performance.

IN FERMENTATIONS, as in many other chemical processes, small scale equipment offers a number of advantages. The quantity of information that can be obtained is roughly proportional to the number of experimental units in use, with tank size having little influence; thus the same information can be obtained from a 25-gallon fermentation as from a 1000-gallon fermentation. On the other hand, most of the items that contribute to the cost per fermentation increase with increasing tank size—e.g., labor, original price of equipment, raw materials, maintenance, and physical space required. Consequently, for developmental work, where information

[1] Present address, American Cyanamid Co., New York, N. Y.
[2] Present address, Columbia University, New York, N. Y.

is the most important product, small equipment is highly desirable; for the same dollar outlay more experimental data can be obtained. On the other hand, small fermentors have the drawback that the percentage of the batch lost by sampling and evaporation can be appreciable, particularly when aerated vigorously over long periods of time. In addition, special problems related to contamination-free operation are often encountered.

The two basic types of small fermentors in common use today are the stationary type and the battery jar or portable type. The stationary type has several distinct advantages. The first is the reduced need for human labor. This is in distinct contrast to the battery jar fermentor, which must, after each fermentation, be lifted out of the water bath, cleaned, batched, carried to the autoclave for sterilization, and returned to the water bath. The second advantage is associated closely with the first: A fermentor which is not manually lifted has much less stringent size limitation. Consequently, it is possible to operate a fermentor which has sufficient capacity so that (1) operational conditions of aeration and agitation are not drastically changed during the course of the fermentation because of loss of liquid by sampling or evaporation; and (2) sufficient material can be provided for small scale recovery and purification studies from one fermentor or from the pooled contents of several. Finally, the design of the stationary fermentor permits sterilization of the medium by steam injection, with close control over the heating and cooling cycles. This procedure allows duplication of production practice and minimizes this process scale-up problem. In contrast, the medium in the battery jar fermentor is usually sterilized in the static state in an autoclave. The heating and cooling cycles are prolonged and there is a nonuniform heat treatment of material in the various positions in the jar, because circulation is effected only by convection.

Disadvantages of the stationary fermentor are the greater problems of contamination because of the additional piping—e.g., bottom drop valve—and the relatively greater expense. Finally, it is common practice to disassemble the stationary-type fermentor by lifting the motor, drive, fermentor top, and agitation shaft by means of a hoist. Excessive handling of this type often leads to shaft misalignment and leaking seals or stuffing boxes. This problem, to a more limited extent, is also encountered with the battery jar type.

After consideration of the factors discussed above, a fermentor was designed and built to satisfy as many as possible of the features thought to be desirable in small scale equipment. This vessel, diagrammatically shown in Figure 1, is extremely versatile, yet requires a minimum of physical handling and upkeep. In the sketch the fermentor has been greatly simplified to stress the salient features of its design. Stainless steel construction, in general, satisfies the requirement for long life, chemical inertness, absence of toxic effects on the fermentation, and ease in cleaning. While the total capacity is about 50 liters, the fermentor is normally operated at 30 liters, because head room is required to accommodate increased liquid volume and foam head which might develop with aeration.

Outstanding Characteristics of Design

The motor, drive, shaft, and top of the fermentor are permanently in place. Thus, as these parts are rigidly mounted to the supporting frame, shafts once aligned tend to remain in alignment. As a result, the maintenance required is considerably less than in a portable installation, where shaft couplings must be dis-

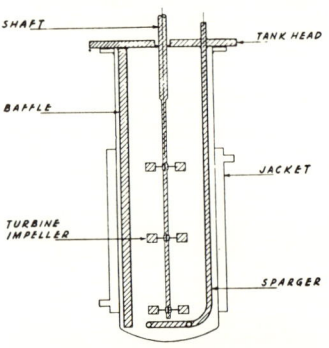

Figure 1. Fermentor detail

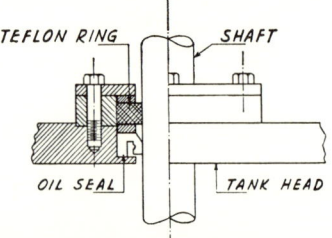

Figure 2. Shaft seal detail

connected in order to prepare for the next run. All inside attachments, including baffles, sparger, and sampling tube, are suspended from the top. To disassemble the fermentor for cleaning, it is only necessary to remove the bolts that fasten the shell to the head, uncouple the inlet and outlet lines to the cooling jacket, and lower the shell by means of a pneumatic lift.

The standard stuffing box, used to maintain the necessary close fit between tank head and shaft, has been replaced by the neoprene oil seal assembly shown in Figure 2. Although the assembly was designed for two seals, it was necessary to replace the upper one with a Teflon ring which acts as a steady bearing. With this modification the life of the neoprene seal is now about 6 months. The performance, from the standpoint of contamination-free operation, has been completely satisfactory.

The removal of the bottom valve the small fermentor is often overlooked.

Each fermentor is provided with a $1/2$-hp. motor and variable-speed drive, so that shaft speeds of 89 to 890 r.p.m. are obtainable. Mounted on the shaft are three turbine impellers, each with six flat blades, for providing adequate mixing and air dispersion. Aeration is achieved in the conventional manner through a ring sparger, with holes on the upper side, situated just below the bottom turbine. Compressed air of controlled humidity is filtered through sterile glass wool just prior to its introduction into the fermentor, and the flow rate, measured with a rotameter, is controlled manually.

The fermentors are mounted with the bottoms 4 feet off the floor to permit removal of the shells for cleaning. They are arranged in groups of 16, two rows of eight placed back to back, with a surrounding platform from which all operations can be carried out. Service lines

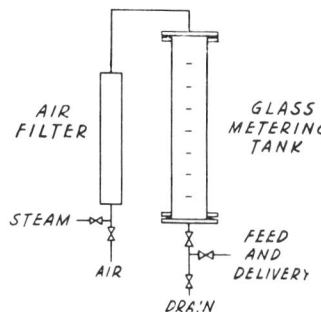

Figure 4. Sterile metering system

must also be considered. Some of this equipment may also prove of value in large scale design, while other equipment may be necessary only to solve a problem imposed by the size of the equipment. In either case the versatility of the installation can be greatly enhanced by careful selection or design of the necessary items.

Humidification of Process Air. Large volumes of air must be passed through the medium in the fermentor when aerobic fermentations are carried out. Loss of batch volume due to evaporation during aeration is a problem which, although not critical in operations on a production scale, is of major importance in small fermentors. With removal of water and the resulting lower batch volumes, the agitation-aeration pattern, power input per unit volume, and concentrations of both raw materials and products change markedly.

This dependence on fermentor size is caused by the fact that small fermentors require larger volumes of air per volume of medium than do large ones, if scale-up is based on equivalent superficial linear air velocities. This air, unless previously saturated, will pick up large quantities of moisture from the broth; losses as high

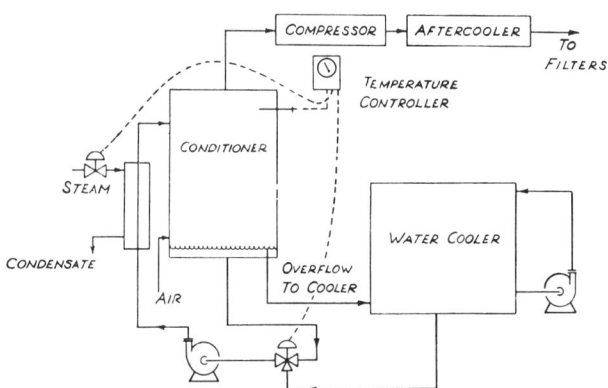

Figure 3. Air humidification system

eliminates a major source of contamination and a maintenance problem. Samples are removed through the sampling line, a simple dip tube with its lower end at the same level as the top of the sparger, or are taken directly from the shell, when lowered at the end of the fermentation. The sample line, which holds 80 cc. of material, is normally flushed by discarding 100 cc. of broth. In this way only 0.33% of the total volume is discarded at each sampling. (Such a percentage loss is not considered excessive.) The shell is emptied, after lowering, by removing broth with a suction hose. Cleaning is readily accomplished, as all internal parts are completely accessible.

Medium is sterilized in this unit by injection of steam through the sampling line and the air sparger, the common practice in most manufacturing units. This is not a technique unusual to fermentor operation, but it is emphasized because consideration of this feature in for steam, air, water, and electricity run between the two banks with branch lines extending to each fermentor.

The batch temperature, measured with a 0° to 150° C. dial thermometer, is controlled by the flow of water through the jacket. A common reservoir, in which the water volume and temperature are regulated, supplies eight fermentors in parallel through a manifold system. Desired minor variations in the individual fermentor temperatures are obtained by throttling the flow of water into their jackets. By interconnecting the two reservoir manifold systems required for a group of 16 fermentors, either system can be used to supply all 16 if the other has to be shut down.

Auxiliary Equipment

In order to carry out pure culture fermentations, tank design must be the first consideration, but auxiliary equipment

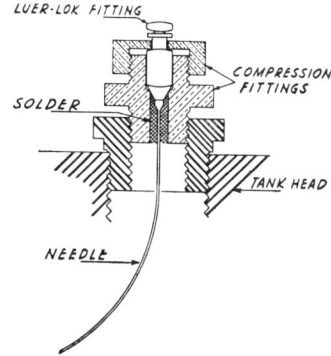

Figure 5. Feed inlet detail

Figure 6. Syringe Assembly

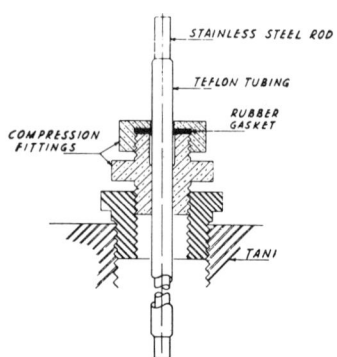

Figure 8. Antifoam probe detail

regulates both the steam valve and a three-port valve. The latter operates as follows: When the temperature is below the control point, water is simply recirculated back to the conditioner and heated; when the temperature of the spray water is too high, it is reduced to the proper level by adding cold water to the recirculating stream.

In theory at least, it should be possible, by using this system, to adjust the humidity of the air before compression to the same level that it would normally have on leaving the fermentor. Thus, if the air is prevented from picking up or losing moisture other than in the conditioner, no evaporation from the fermentation batch should occur. In practice, the most serious problem which was encountered in accomplishing this goal was the condensation of moisture from the air after compression. Even when the aftercooler was not utilized, the temperature of the compressed air in the lines dropped below the dew point and moisture condensed. By insulating the air lines it has been possible to maintain the process air temperature sufficiently high to eliminate condensation.

Addition of Materials. The maintenance of contamination-free operations when materials are added to a fermentor is more difficult as the size of the equipment decreases. This is true particularly in the case of slurries, as the small clearances that must be maintained in small scale equipment approach the size of the particles being passed. When large tanks are involved, standard diaphragm pumps have sufficiently low rates to be satisfactory for metering sterile liquids and single additions of nutrient solution or inoculum can be made by transferring the entire contents of a vessel. However, where small quantities are involved, specialized equipment must be devised in order to maintain sufficiently low flow rates or to divide the contents of a single vessel among a number of fermentors.

as 10% per day have been experienced in the 50-liter tanks described.

In the system selected for correcting this condition, process air is drawn from the atmosphere through a saturator to the compressors. After compression the air is cooled, filtered, and distributed to the fermentors. Humidification is accomplished, as shown in Figure 3, by bringing the air in contact with a spray of water in the conditioner. The degree of humidification is regulated by controlling the temperature of the outlet air, based on the assumption that it is 95 to 100% saturated. The temperature of the outlet air is dependent on the temperature of the spray water, which is controlled by the addition of cold water from the water cooler or the introduction of steam into the shell side of the heat exchanger. The temperature controller

A metering tank, such as shown in Figure 4, has been employed for adding relatively large quantities of liquid to small fermentors. The construction is simple; the reservoir is merely a section of borosilicate glass pipe. In practice the entire unit, mounted permanently in place, is first sterilized with steam and then the piping is filled with the material to be transferred. The glass pipe is filled to a given level and the material is forced by air pressure to the receiving vessel. The advantages of this type of unit are that standard items are used for construction with the exception of the top and bottom flanges, the material being transferred can be seen, and a continuous range of quantities can be metered if the pipe is calibrated.

Other systems have been devised for intermittent or continuous addition of smaller quantities of solutions. Material previously sterilized in a storage bottle is transferred through flexible rubber tubing and standard syringe fittings, mounted on each fermentor as shown in Figure 5. A long syringe needle is inserted through standard compression fittings and held in place with soft solder. The female Luer-Lok (Becton, Dickinson & Co., Rutherford, N. J.) connection at the top permits rapid attachment of the delivery equipment. Cornwall syringes (Becton, Dickinson & Co.), used for intermittent feed, are shown in Figure 6. Two check valves ensure that the syringe barrel is filled with liquid from the reservoir on the suction stroke and the contents are delivered to the fermentor on the pressure stroke. Peristaltic action pumps, which press the liquid through the tubing, have been successfully employed for continuous feed of liquids where accurate control of flow rate is unimportant.

Automatic Foam Control. Foaming is a serious problem because of the individual attention required by each fermentor to prevent loss of material through the exhaust line and possible contamination of the batch. Thus an effective means of automatically controlling the foam is desirable. The

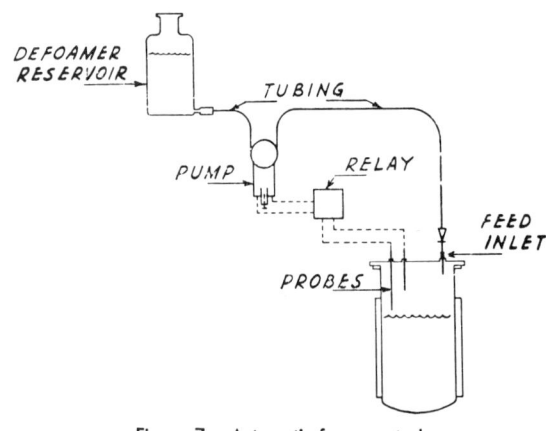

Figure 7. Automatic foam control

system shown in Figure 7 utilizes an electrical relay design for pump-down operation. When foam reaches the upper probe, an electrical connection is made between the probe and the wall of the tank, causing the relay to energize a peristaltic action pump which forces defoamer from a sterile reservoir through flexible silicone tubing into the fermentor. When the foam head has been broken, and the foam falls below the level of the lower probe, the electrical circuit is opened and the pump is shut off. A check valve situated in the line just before the inlet fitting prevents the flow of exhaust gas from the fermentor during periods when no defoamer is added.

The probes are constructed and installed as shown in Figure 8. A stainless steel rod is covered with a Teflon tube and the combined unit is slipped into the fermentor through a packing gland constructed of standard compression fittings. The complete insulation of the probe as described here has eliminated shorting of the electrical circuit at the tank head, the major problem encountered in previous unsuccessful trials.

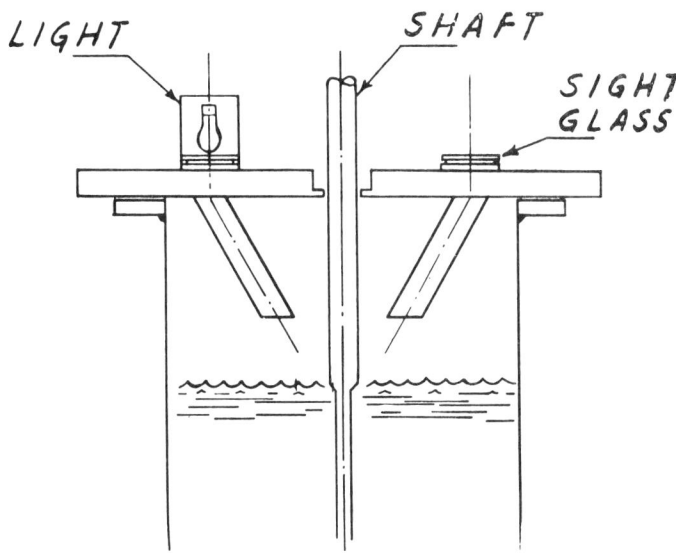

Figure 10. Sight glass protector

Continuous Fermentation

The fermentor described has been modified slightly to permit its use for study of continuous fermentation techniques. As the flow rates involved with such small volumes are below the range of available positive-displacement pumps, intermittent hourly feed of fresh nutrient medium and withdrawal of broth were employed, with the arrangement pictured in Figure 9. Two storage tanks are necessary to ensure a constant supply of sterile medium, because several hours are required to clean, batch, and sterilize tanks. The metering tank is operated in the same way as the one shown in Figure 4 and is steamed constantly when not in use.

With the equipment shown, and with proper maintenance of steam locks on all valves adjacent to the fermentor, antibiotic fermentations have been carried out without contamination for as long as 2 months. Up to the present time no tests have been made with fermentations in which no antibiotic was present, so that the sterility record may be due, in part at least, to the nature of the fermentation.

It is customary to provide fermentors with two sight glasses—one to allow entrance of light, the other to allow observation. In continuous fermentations, particularly, the clouding of these glasses becomes a serious problem. Foam, or even occasional splashing continuing over a long period, soon results in sufficient coating of the inner surface of both glasses so that the batch cannot be seen. The problem was resolved by the installation of tubes cut on a 15° angle, so that light and vision were directed to the same spot, the liquid surface at the center of the tank, as shown in Figure 10. The angle is sufficient to protect the glasses from splashing liquid; however, satisfactory foam control procedures are required to prevent foam from rising into the tubes. The batch can be observed over the entire fermentation period, unless for some reason the foam becomes unmanageable. A small weep hole in the upper end of each tube allows steam passage during sterilization and thus eliminates the possibility of a dead space which might not be adequately sterilized.

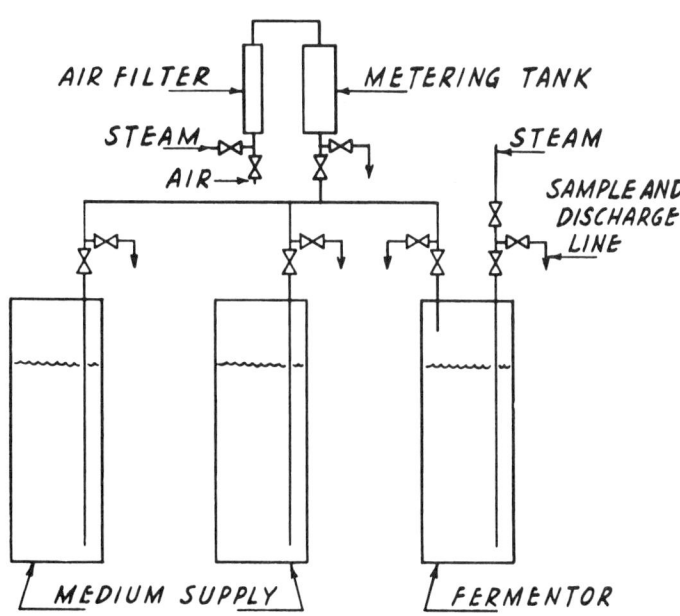

Figure 9. Continuous fermentation system

RECEIVED for review April 13, 1956
ACCEPTED August 8, 1956

12

Copyright © 1961 by John Wiley and Sons, Inc.

Reprinted from *J. Biochem. Microbiol. Technol. Eng.*, 3(2), 125–137 (1961)

An Improved pH Electrode Assembly for Pilot Plant and Plant Fermentors*

J. M. West,[†] G. P. Stickle, K. D. Walter[‡] and W. E. Brown, *The Squibb Institute for Medical Research, New Brunswick, New Jersey*

Summary. A modified pH electrode assembly is described which is capable of steam sterilization and which can be installed in steel fermentors of all sizes in a variety of ways. Its design is such that electrode life is prolonged and contamination hazards are reduced.

Introduction

Reviewing the past dozen years, the main deterrent to continuous pH measurement and control in submerged pure culture fermentations has been the unavailability of electrode assemblies which are capable of withstanding steam sterilization. The problem has been circumvented in small fermentors by sterilizing the electrodes separately with an agent such as formaldehyde, ethylene oxide, or ultraviolet light.[1,4,5,7] Use of one of these agents, however, involves introducing one or both of the sterile electrodes into the fermentor without contaminating the system. Obviously, this approach is not a desirable one for small-scale equipment and is completely impractical for plant-scale fermentors.

Although failure of both glass and reference electrodes has been observed when they are sterilized with steam under pressure, the more serious problems lie with the glass electrode. Two main types of failure have been noted when glass electrodes have been exposed to sterilization conditions, viz. failure due to breakage of the glass and failure due to moisture entering the electrode at the point of connection with the lead wire. The former problem has

---

* Presented 138th National Meeting, American Chemical Society, New York, N.Y., September, 1960.
† Present Address: Nopco Chemical Co., Newark, N.J.
‡ Present Address: Griffith Laboratories, Chicago, Ill.

been resolved in recent years by the use of heat-resistant glass. The latter problem has, unfortunately, not been resolved completely as both Nelson et al.[2] and Denison et al.[6] have noted.

With the reference electrode the main cause of failure has undoubtedly been mechanical breakage. Of particular concern to all in the fermentation field, however, is the possibility of a contaminant entering the system from the electrolyte solution used in the salt bridge. This problem appears to have been resolved for all practical purposes by the addition of a suitable bactericidal agent such as phenol.

Two types of electrode assembly which can be steam-sterilized have been reported. The first of these, the immersion type, described by Nelson et al.[2] and Huang[8], has the advantages that sterilization of the entire system is ensured and that the pressure on the electrolyte reservoir is always equal to the fermentor pressure.

Studies made with an immersion type of assembly in our laboratory demonstrated that the design was not a practical solution to our needs because of the short life of the glass electrode as a result of failures associated with both steam-sterilization and mechanical breakage. Furthermore, the lack of versatility of the immersion assembly was a major disadvantage. The assembly was too large for convenient installation in 10-gal fermentors and too frequently broken in plant-scale equipment because of the turbulence and vibration encountered.

The second system (and in our opinion the more practical of the two) is that described by Denison et al.[6] He and his co-workers installed a standard pH flow assembly in an external leg on a 50-gal fermentor. They reported that the assembly was capable of withstanding steam-sterilization and was successfully used for both measurement and control of pH. The disadvantages of this system, including problems with contamination and mechanical breakage as well as lack of versatility of installation, are all associated with the basic design of the electrode glands.

In this paper a modified pH electrode assembly is described which is capable of steam-sterilization and which can be installed in steel fermentors of all sizes in a variety of ways. Its design is such that electrode life is prolonged and contamination hazards are reduced.

Design of Assembly

When experimentation with pH electrode systems was started some years ago it was felt that the most reasonable approach involved the adaptation of commercially available equipment to fermentation use. Studies with an immersion type of assembly revealed its basic disadvantages and attention was then focused on the electrode glands manufactured by Beckman Instruments, Inc., illustrated in Fig. 1. The problem of ensuring a positive flow of

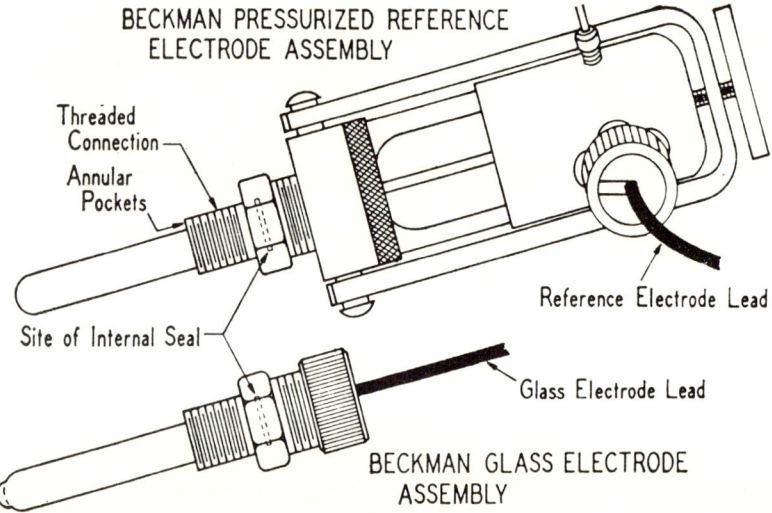

Fig. 1. Beckman pH electrode glands

electrolyte solution from the reference electrode against the pressure encountered in a fermentor was solved by the manufacturer by the application of external pressure to the electrolyte solution. No such provision was necessary for the glass electrode since the pressure problem does not exist. It was believed that the Beckman glands, if properly installed, would ensure longer life of both the glass and the reference electrodes since only a portion of the electrodes would need to be exposed to sterilization temperatures.

On the other hand, there are two basic disadvantages to the Beckman design. First, the annular pockets around the electrodes and the threaded connection tend to collect fermentation material and thus are potential sources of contamination. Secondly, the excessive protrusion of the electrodes through the gland assembly makes them susceptible to mechanical breakage. As a consequence, the glands were modified as illustrated in Fig. 2 to overcome these disadvantages.

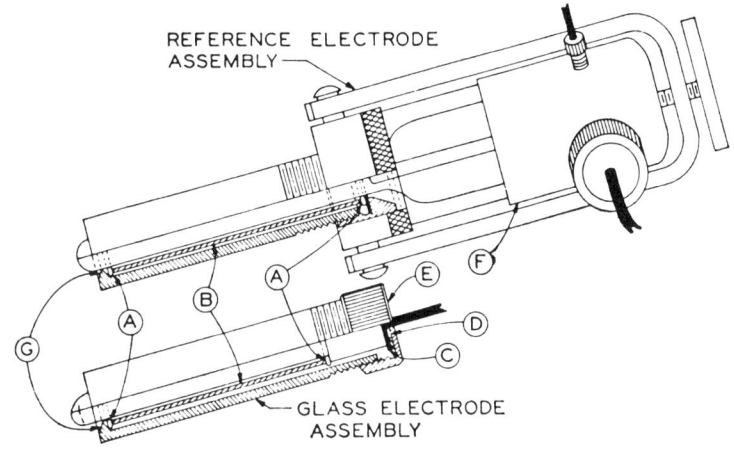

Fig. 2. Squibb modified glands (quarter-sectional view)
A, 'Quad' ring seals; B, gland sleeve; C, rubber washer; D, metal washer; E, cap; F, dome pressure assembly; G, chamfered edge

The basic differences between the modified and the original Beckman glands are seen when the two types are compared. The gland was lengthened so that only $\frac{1}{4}$ in.–$\frac{1}{2}$ in. instead of $2\frac{1}{2}$ in. of the sensitive tip of the electrode was in contact with the broth. The gland housing was modified so that it could be welded in place, eliminating the threaded connection. Additional 'quad' rings were inserted so that both assemblies were sealed at two points. In the glass electrode assembly one 'quad' ring was located at the pH-sensitive end and another was located at the top, below the plastic cap. The former compensated for variability in electrode dimensions, and the latter contained the broth should the electrode tip be shattered. Two 'quad' rings were also placed in the reference electrode for the same reasons, the upper one adjacent to the gland sleeve and the lower one near the tip of the electrode. The electrodes used for the modified gland assemblies were standard Beckman industrial electrodes (manufacturer's code 19101 and 19700).

The relationship the various parts of the gland assemblies bear to each other is demonstrated in Fig. 2. Note that the electrode

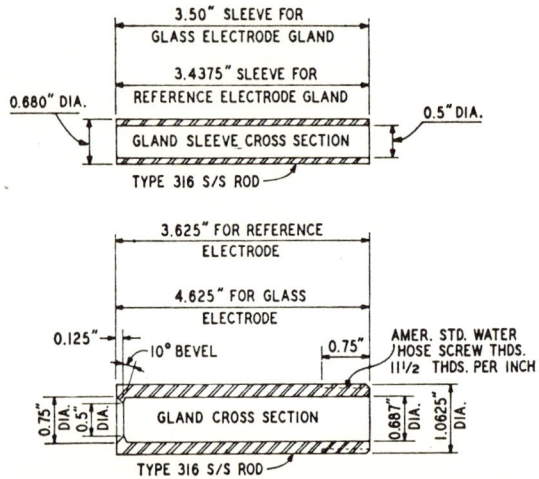

Fig. 3. Details of the Squibb glands and gland sleeves

is inside the gland sleeve and the gland sleeve inside the main body of the gland. The positions of the 'quad' rings, rubber washer, metal washer, cap, and dome assembly, are all shown. Note also that the inside edge of the gland next to the tip of the electrode is chamfered. The dimensions of the gland and the gland sleeve differ depending on which electrode they will be protecting. These details are given in Fig. 3. The inside diameter of the gland is set at 0·687 in. to permit use of a standard reamer to correct the distortion caused by welding during installation.

Method of Installation and Operation

The glands described are extremely versatile. They can be installed either directly in the side wall of a steel fermentor of any size or in an external circulating system. Although the operational advantages of the external system were recognized at the time these studies were initiated, it was felt that in small equipment the complications introduced by so equipping fermentors were not justified by the operating advantages gained. Consequently, developmental studies at the pilot and plant scale were concerned at the beginning with installation directly in the fermentor wall (direct-entry). Subsequently, an external circulating system was developed for use in production-size fermentors and this too was provided with the modified pH electrode assembly. Both methods of installation, directly in the wall of the fermentor and in an external circulating system, are described.

Installation in the Fermentor Wall

Our experience with the use of the glands in the direct-entry method of installation has been associated with fermentors of 10-gal, 100-gal, 200-gal, 1,300-gal and 17,000-gal total capacity. In all sizes of equipment two glands, one to hold the glass electrode and one to hold the reference electrode, are welded into the wall of the vessel as indicated in Fig. 4. Since fermentations are normally conducted at constant temperature, a third gland for a thermocompensator is unnecessary. The glands are installed at a minimum angle of 15° with the horizontal to ensure that the electrolyte maintains contact with the internal silver–silver chloride portion of the electrode. Although experience has shown it to be unnecessary to protect the electrode tips inside the smaller pilot plant equipment, the installation has been modified slightly in the larger production fermentors to minimize contact of the electrodes with solids in the fermentor. To accomplish this purpose, a 2-in. length of 6-in. pipe is welded to the inside wall to provide a protection ring, as shown in Fig. 4.

Fig. 5 shows the gland assembly in a pilot plant fermentor of 10-gal total capacity. Note that the water jacket has been cut away to allow installation directly through the inner wall of the

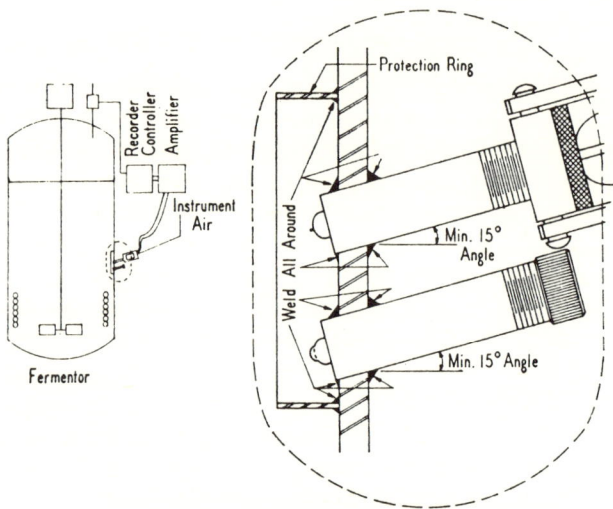

Fig. 4. Direct-entry installation

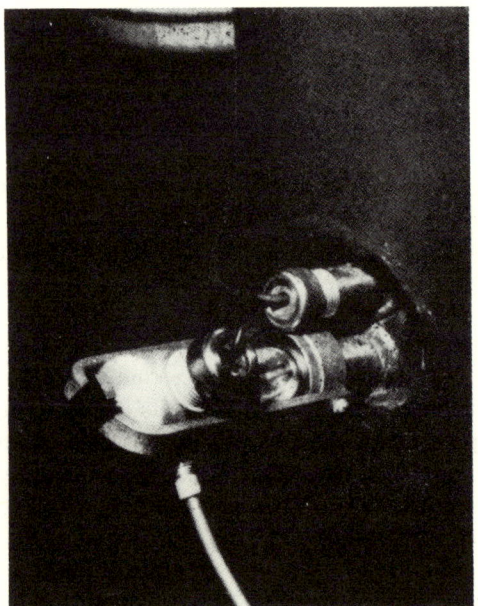

Fig. 5. View of pH gland installation in a 10-gal fermentor

fermentor. Fig. 6 shows a similar installation in a tank of 1,300-gal capacity. Here a third gland has been inserted to enable the use of a thermocompensator, a redox electrode or a second glass electrode.

The physical conditions to which the electrodes are exposed depend upon the size of the fermentor. In small fermentors, i.e. of 10-gal to 1,300-gal capacity, 20–25 p.s.i.g. air pressure is

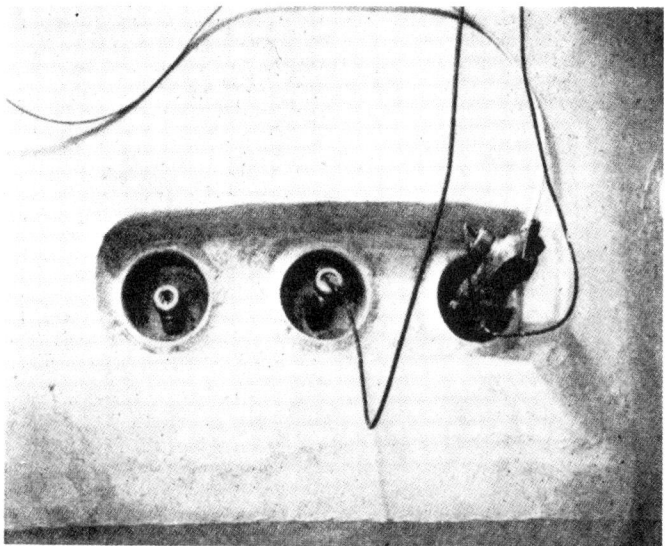

Fig. 6. View of pH gland installation in a 1,300-gal fermentor

maintained continuously on the reference electrode. It has been found necessary to replenish the electrolyte solution every 2–3 weeks. The electrodes are subjected to steam, hot water, and hot medium at 121 °C and 15 p.s.i.g. for 2–6 h each week. Once installed, the electrodes are not removed unless failure is suspected. Under these conditions of operation the average life of both glass and reference electrodes has been 6 months or more. The long life is due in part to the fact that the temperature of the electrode external to the fermentor is well below 121 °C during batch sterilization. Measurement of the temperature of the electrolyte solution in the reference electrode has indicated that it does not rise above 85°C when the fermentor is heated to 130°C for 2 h.

In the 17,000-gal fermentors, the electrodes are placed 2·5 ft from the bottom of the vessel, and it is necessary to maintain 35 p.s.i.g. air pressure on the reference electrode to compensate for the fermentation-liquid height and the fermentor back-pressure to ensure a positive flow of electrolyte solution during fermentor operation. Location of the electrodes higher up the fermentor nearer the surface of the liquid would reduce the pressure requirement, but in our particular case would complicate the servicing problem. When the fermentor is not in operation, the pressure on

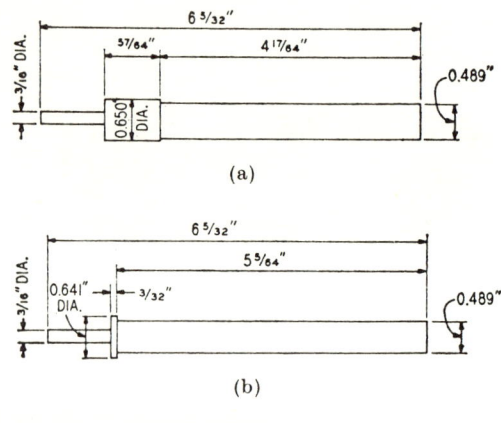

Fig. 7. Standard plugs for: (a) Glass electrode glands, and (b) reference electrode glands

the reference electrode is vented to atmosphere. Because of the severe vibration encountered during batch sterilization of the medium, it has been found necessary to fill the electrolyte reservoir to only one-half of its capacity in order to prevent electrolyte solution from splashing out and shorting the reference electrode. Under these conditions the reservoir has required filling once every 5–6 days. As in the pilot plant, the electrodes are subjected to steam, hot water, and hot medium at 121°C and 15 p.s.i.g. for 2–6 h each week. The electrodes are removed frequently during routine cleaning of the fermentors, and, because of this, mechanical breakage is greater in the plant than in the pilot plant. When the electrodes are removed for any length of time, dummy electrodes made of stainless steel as illustrated in Fig. 7 are inserted to prevent debris of any type entering the glands.

Electrode life in the plant fermentors has been surprisingly good and, as one might expect, has improved as plant personnel have gained experience. For example, during the operation of the first 100 fermentation batches using this system, 12 glass electrodes and 15 reference electrodes were replaced. Electrode failures now range from 3 to 5 per 100 batches for both electrode types and even now the great majority of failures are due to accidental breakage. Our experience has demonstrated adequately that electrode life is sufficiently long in both the pilot plant and plant to justify installation directly in the wall of the fermentor.

Installation in an External Circulating System

Although electrode failure using the direct-entry technique has been minimal, the occasional failures do require unscheduled manual operations to maintain pH control. Installation of electrodes in an external circulating system has the advantage, particularly attractive to manufacturing operations, of enabling one to isolate the pH sensing system during the course of a fermentation. Consequently, an experimental circulating system was installed in a 17,000-gal fermentor. This system, which is illustrated in Fig. 8, consists of a pump, capable of being steam-sterilized and of running contamination-free, a pH gland flow assembly, and appropriate piping. The gland flow assembly, as illustrated in Fig. 8, consists of the two electrode glands welded to a short section of 1-in., schedule 40, type 316 stainless-steel pipe. This pipe in turn is welded at each end to a type 316 stainless-steel slip-on flange. Installed, the flanged connections are joined using a Teflon gasket. As in the direct-entry method of installation, the flow assembly must be installed at a minimum angle of 15° with the horizontal to ensure that the electrolyte maintains contact with the internal silver–silver chloride portion of the electrode.

This method of electrode installation has been demonstrated to have the following advantages:

(1) The electrodes can be removed, checked, and, if necessary, replaced without interrupting the fermentation.

(2) The location of the electrode assembly is not restricted and can be located for most convenient operation and service.

(3) The life of the electrodes is extended, since failure due to operating pressure, turbulence and vibration effects is minimized.

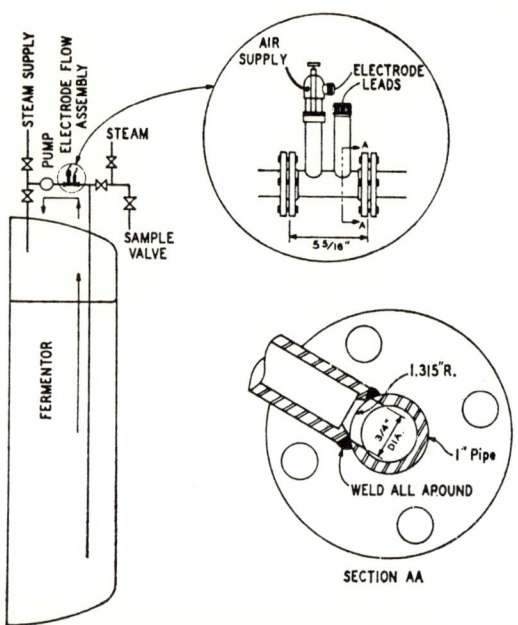

Fig. 8. External circulation installation

Use of Electrode Assembly for Monitoring pH

The electrode assembly as described has been used in a variety of pH monitoring systems. The simplest and most common application in the pilot plant has been its use with a standard pH meter, thus enabling continuous measurement of pH without the need for frequent sampling. More sophisticated systems have been built for use in both the pilot plant and plant for continuous reading and recording and for continuous recording and control.

In all of these systems the electrodes have given completely satisfactory service for periods up to 6 months or more without replacement.

The electrode assembly as described has also been used in monitoring fermentation oxidation–reduction potential. In this application, the O/R electrode (manufacturing code 19001) is used in the glass electrode gland in conjunction with a reference electrode assembly.

Pure Culture Considerations

The ability to maintain a pure culture is, of course, a foremost consideration in any pH monitoring system in which electrodes contact directly the fermentation broth. Serious consideration was given to the design and installation of the assembly to prevent its proving to be a source of contamination. Obviously, the electrolyte solution in the reference electrode had also to be considered as a potential source of contamination. In order to reduce the hazard from this source, a saturated potassium chloride solution containing 0·2 per cent phenol was used. Maximum precautions were also taken to keep the electrolyte solution free of micro-organisms during its preparation and addition to the electrode. Two other factors aid contamination-free operation: the electrolyte solution reaches pasteurization temperatures, i.e. 80°–85°C, during sterilization of the medium in the fermentor; and in addition, it is quite possible that the palladium junction acts as a bacteriological filter. This proposal is based on the fact that the annular opening around the junction of the Beckman industrial reference electrode, although an estimated 4–5 μ in size, is very much like a filter, owing to the sandblasting the palladium wire receives and to the fact that the rough surface becomes imaged on the glass as the glass cools in the manufacture of the electrode.[3]

The experience in our pilot plant and plant in a variety of fermentation processes and with a variety of micro-organisms has been that equipping fermentors with pH glands has not increased the incidence of contamination, but has provided valuable information and has made possible much better process control.

References

[1] Watson, R. W., Clement, M. T. and Muirhead, D. R. *Canad. J. Res.*, **C28**, 183 (1950)

[2] Nelson, H. A., Maxon, W. D. and Elferdink, T. H. *Industr. Engng Chem. (Anal.)*, **48**, 2183 (1956)

[3] Kehoe, T. J., Beckman Instruments, Inc., Fullerton, Calif., private communication

[4] Deindorfer, F. H. and Wilker, B. L. *Industr. Engng Chem. (Anal.)*, **49**, 1223 (1957)

[5] Dworshack, R. G. *Proc. nat. Conf. Instrum. Methods Anal.*, Chicago (1957)

[6] Denison, Jr., F. W., West, J. C., Peterson, M. H. and Sylvester, J. C. *Industr. Engng Chem. (Anal.)*, **50**, 1260 (1958)

[7] Sher, H. N. *Chem. & Ind. (Rev.)*, No. 16, 425 (1960)

[8] Huang, H. T. *Proceedings, First International Fermentation Symposium*, Rome (May 1960)

13

Copyright © 1964 by John Wiley and Sons, Inc.

Reprinted from *Biotechnol. Bioeng.*, **6**, 457–468 (1964)

Steam Sterilizable Probes for Dissolved Oxygen Measurement*

MARVIN J. JOHNSON, JOHN BORKOWSKI, and CURT ENGBLOM, *Department of Biochemistry, University of Wisconsin, Madison, Wisconsin*

Summary

Steam-sterilizable membrane probes for monitoring the dissolved oxygen level in fermentors, or the oxygen content of gas streams, are described. The probes have a silver cathode, a lead anode, and an acetate buffer as an electrolyte. The membrane is Teflon. The current output of the probes in the absence of oxygen is negligible.

INTRODUCTION

The importance of dissolved oxygen determinations in aerobic fermentors needs no emphasis. Amperometric oxygen determinations in biological systems are difficult because of rapid polarization and poisoning of electrodes. The membrane probe introduced by Clark[1] was a great improvement over previous amperometric methods. In a membrane probe, the electrochemical cell is separated from the solution being analyzed by a membrane permeable to oxygen but impermeable to water and electrolytes. Inside the membrane, the oxygen content is kept very low by the electrochemical reaction. The rate-limiting step for electrochemical oxygen reduction becomes the rate of oxygen diffusion through the membrane. This rate is proportional to the oxygen partial pressure (tension) outside the membrane. The operating principle of membrane probes has been discussed by Phillips and Johnson[2] and by Kinsey and Bottomley.[3] Various electrode compositions, electrolytes, and membrane mate-

* Supported in part by grant E-2967 from the National Institute of Allergy and Infectious Diseases, National Institutes of Health, Public Health Service.

rials have been used in membrane probes.[1–11] Membrane materials have included polyethylene, polypropylene, polystyrene, Teflon, silicone rubber, and gum rubber. Electrolysis cells have been of two general types, those in which a potential is imposed on the cell, and those in which the cell itself provides the potential for oxygen reduction (so-called "self-generating" cells). Various cells of this type have been described by Tödt.[12] In membrane probes, self-generating silver–lead[5,10] and gold–cadmium[6] cells have been used.

Only two membrane probes designed to withstand steam sterilization have been described.[2,3] In both designs, the electrolytic cell is sealed. In our laboratory, sealed cells have, on repeated sterilization, accumulated gas bubbles and been damaged by pressure. Pittman[7] has described a membrane probe that is not sealed. We have found that with proper construction, unsealed probes can be used very successfully in fermentors.

The probe described in this paper uses a lead anode, a silver cathode, and an acetate buffer as electrolyte. Alkaline electrolytes have been used[5,10] with lead–silver cells. We have observed that with lead–silver cells, alkaline electrolytes are not suitable for use in the presence of high concentrations of carbon dioxide, such as are often encountered in fermentations. When such cells are exposed to carbon dioxide, their response characteristics are altered. Teske[13] has pointed out the advantages of using acetate buffer in a lead–silver cell. The lead–silver cell was selected for the probe to be described because of its very low residual current (the current in the absence of oxygen is about 0.1% of the current in the presence of 0.2 atmospheres of oxygen), and because silver is cheaper than the noble metals often used as cathodes. Teflon was selected as a membrane material because of its resistance to heat and its chemical inertness. An unsealed mechanical design was adopted to circumvent difficulties encountered with sealed probes.

CONSTRUCTION OF PROBES

A piece of 6 mm. glass tubing, 15 cm. or more in length, having one end evenly and squarely cut, is prepared. The evenly cut end is fire-polished just enough to round the sharp edges; the end of the tube should still have a flat face after polishing. A piece of 22 gage (0.644 mm. diameter) silver wire, for construction of the cathode, is cleaned with fine emery cloth. It is then tightly wound into a

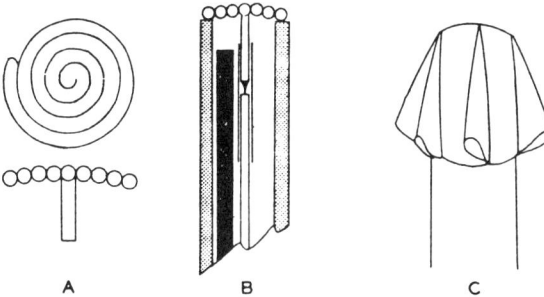

Fig. 1. A: spiral cathode; B: assembled probe (not to scale) showing silver cathode and lead anode in position; C: method of attaching Teflon membrane.

pancake spiral whose diameter is equal to the outside diameter of the glass tube (Fig. 1A). Care should be taken to avoid leaving an opening at the center of the spiral. The spiral is made slightly convex, as indicated in the figure. The cut end of the wire at the outer end of the spiral is rounded off to prevent cutting of the Teflon membrane described below. The other end of the wire, at the center of the spiral, is made to extend from the center of the spiral in a direction normal to the plane of the spiral. It is made about 1 cm. long. Surgical splinter forceps or very small needle-nose pliers are useful tools in making spirals.

The end of the silver wire is soldered to a piece of small diameter insulated wire (one conductor from Belden No. 8430 phono pickup cable). The soldered joint should not have a diameter greater than that of the silver wire. The stem of the silver spiral, the soldered joint, and the first 4 or 5 cm. of the insulated wire is covered with a thin layer of contact cement. The cement is prepared by adding to some Formica brand contact cement 40% of its volume of methyl ethyl ketone, and some crystal violet to make the cement visible. Of four brands of contact cement tested, Formica brand was found most satisfactory for use with sterilizable probes. The cement is allowed to dry for 15 min. or more. A piece of silicone rubber tubing, about 4 cm. long, 0.5 mm. bore, 0.25 mm. wall (Esco Rubber Ltd., Seal St., London E. 8) is swelled in chloroform for an hour or more, and then slipped over the insulated wire and the stem of the spiral. As the chloroform evaporates, a tight seal is formed. The silver cathode is now ready for insertion in the glass tube.

For the anode, a piece of lead foil, $1/32$ in. (0.8 mm.) thick, about 10 cm. long, and about 4 mm. wide, is cut. At one end, the width is cut to about 1 mm., and the end section rounded off, so that the anode has at one end a section of lead wire, 1 mm. in diameter, and about 8 mm. long. This end is soldered to a piece of insulated wire and the soldered joint sealed in the manner described above for the cathode. Care must be taken that the silicone rubber tubing does not slip off from the lead wire before the chloroform has evaporated. The lead anode is cleaned in dilute (1:1) nitric acid and polished with a clean cloth. The cathode and the anode are now assembled in the glass tube as shown in Figure 1B. The silver spiral should lie flat on the flat end of the glass tube. The two insulated wires are brought out through the other end of the glass tube. The tube is mounted in a suitable clamp, spiral side up. The glass tube is now coated with the diluted contact cement described above. It is coated on the outside for a distance of about 15 mm., beginning at the end bearing the spiral. Care is taken not to coat the spiral. A circular piece of Teflon (type C, Dilectrix Corp., Farmingdale, L. I., New York) 2 mils (50 μ) thick and 28 mm. in diameter, is coated on one side with cement, except for a circular area, 6 mm. in diameter, at its center. After drying at room temperature 10 min. or more, the coated Teflon circle and the coated glass tube are placed in a 100°C. oven for 5 min. to insure complete evaporation of solvent. After removal from the oven, the Teflon circle is placed on the end of the glass tube, coated side down, with the uncoated center section of the circle covering the spiral. The coated side of the Teflon circle is now brought into contact with the coated glass tube. An attempt is made to make eight folds (or flutes) in the Teflon (see Fig. 1C), although fewer than eight are satisfactory. The folds are pressed tight and wound loosely around the tube. The whole operation is performed in a way such that the Teflon fits snugly over the spiral. There should be no space between the Teflon membrane and the spiral.

A Nalgene nylon tube connector, for connecting $1/4$ in. tubing to $1/8$ in. pipe, is bored out with a $1/4$ in. drill. The fitting is then pushed on to the glass tube from the open end, and the nut is pushed on to the glass tube from the Teflon covered end. It may be necessary to enlarge the opening in the compression sleeve inside the nut. This may be done by forcing the tip of a 10 ml. Mohr pipette into it as far as possible and letting it stand for a few minutes. The tube is pushed through the nut until the Teflon covered end of the tube is flush with the surface of the nut. The fitting is then screwed tightly on to the nut. Some nylon nuts are so constructed that a Teflon circle larger than 28 mm. must be used in order for the compression sleeve to grip the tube in a position such as to press the Teflon against the glass.

The tube is now filled with electrolyte to a depth of 12 cm. or more. The electrolyte surface should be at least a centimeter or two above the upper end of the anode. The electrolyte is an aqueous solution containing $0.1 M$ sodium acetate and $1 M$ acetic acid. A convenient way of removing air bubbles and dissolved oxygen from the completed probe is to short-circuit it and steam sterilize it at 120°C. If the probe is sterilized in a beaker of water and cooled, the electrical resistance between the cathode or anode and the water outside the probe should be greater than 20 megohms, as measured with an ohmmeter.

The probe may be fastened to a length of pipe by the pipe thread on the nylon tube fitting, or it may be used in small vessels with no protection for the glass stem. A probe that is to be mounted in a larger fermentor, and steam sterilized with the fermentor, must be constructed so that the pressure inside and outside the probe are always equal. The probe is mounted in a piece of pipe long enough to extend from the probe position to the top of the fermentor. To the upper part of this pipe is fastened a bushing by means of which the pipe may be screwed into a threaded opening in the fermentor top. To the part of the pipe extending above the bushing (outside the fermentor) is fitted a nylon tube fitting similar to the one used in making the probe. The glass stem of the probe is made long enough so that it extends almost to this fitting. Into the nut of the fitting is pushed a rubber cylinder (cut from a rubber stopper with a cork borer) $1/4$ in. (6.3 mm.) in diameter and about 1 cm. long. The cylinder carries two small holes to accommodate the insulated wires from the probe. When the nut is tightened, a good seal is obtained. For pressure equalization, a small hole is drilled in the side of the pipe, just below the bushing (inside the fermentor). Because the hole is below the top of the glass stem of the probe, small amounts of foam that might enter the hole will not enter the probe.

Probes may be constructed without the use of tube fittings. A probe is constructed as outlined above, except that the Teflon circle used is 50 mm. in diameter. After the Teflon circle has been folded into place on the probe, the outside of the folded Teflon is coated with cement. The coating is made to extend beyond the Teflon, so that the probe is coated for a length of about 5 cm. A piece of Tygon (polyvinyl chloride) tubing ($1/8$ in. bore, $1/16$ in. wall) about 3 cm. long is swelled overnight in chloroform and slipped over the end of the probe to a position such that it extends from a point very close to the end of the glass tube to a point well beyond the end of the Teflon covered portion of the tube. The position of the Tygon tube may be adjusted during the period the chloroform is evaporating. Probes sealed in this manner have a smaller outside diameter than those made with tube fittings. They withstand repeated sterillization well, but are mechanically not as well protected.

In use, the anode and cathode leads from the probe are connected through a load resistor of 500–1000 ohms. The voltage drop through this resistor, produced by the electrolysis current, is measured with a 0–10 mv. strip chart recorder. The load resistor should be such that the voltage drop through it is never more than 10 millivolts. When the probe is used to measure oxygen tensions in an aqueous solution, the solution must be sufficiently well agitated that there is no appreciable liquid–film resistance at the Teflon membrane. The level of agitation normally used in fermentors is more than adequate. The probes have a temperature coefficient of about 2% per degree C., which must be taken into account if the probe is used in a fermentor without temperature control. Devices for temperature compensation of membrane probes have been described.[4,8]

PERFORMANCE OF PROBES

Speed of Response: Residual Current

The solid curve in Figure 2 is reproduction of a strip-chart recorder tracing obtained when a probe was placed alternately in an agitated, air saturated water bath at 30°, and in a sodium sulfite solution. It will be seen that the output in the absence of oxygen is very low, and that when oxygen is admitted or removed, 90% response is obtained in about 1 min., and 99% response is obtained in about 3 min. The dotted curve of the figure is a reproduction of the curve drawn by the strip-chart recorder when a probe, in a current of dry gas, was changed from air to a nitrogen–air mixture (containing 13.6%

PROBES FOR DISSOLVED OXYGEN MEASUREMENT 463

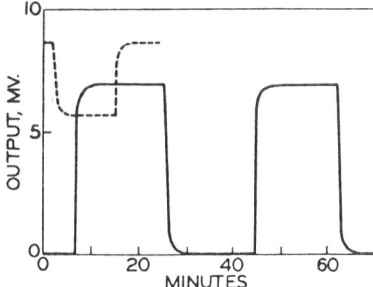

Fig. 2. Solid curve: response of a probe alternately placed in an air-saturated 30°C. water bath and in a sodium sulfite solution. The load resistance was 667 ohms. Dotted curve: reponse of a probe in a gas stream. The composition of the gas was changed from 20.9% O_2 to 13.6% O_2, then back to 20.9% O_2.

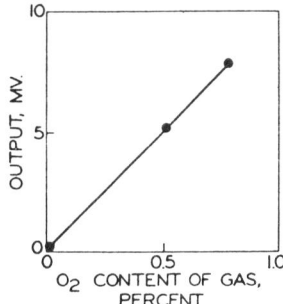

Fig. 3. Response of a probe to gases of low oxygen content. One gas was cylinder nitrogen with a known oxygen content of 0.01%, the other two were nitrogen–air mixtures, analyzed with a Beckman paramagnetic oxygen analyzer operated at a 0–1% range. The load resistance was 7500 ohms. The probe had a 1 mil Teflon membrane.

O_2, as analyzed with a Beckman paramagnetic oxygen analyzer), then back to air. It will be noted that the response speed is about the same as that obtained in aqueous solutions. Probes made with 1 mil Teflon rather than 2 mil Teflon have a higher output and a somewhat faster response. The behavior of probes made with 0.5 mil Teflon is like that of those with 1 mil Teflon, probably because the liquid film inside the membrane limits the oxygen diffusion rate.

The magnitude of the residual current is best studied with a higher load resistance. Figure 3 shows results obtained with a probe that

netic oxygen analyzer and by probes. Both the probes and the analyzer were calibrated with dry air (20.9% O_2). It will be seen from the table that probes 1, 2, and 3 gave reliable analyses, but probe 4 gave a nonlinear response. The likelihood of nonlinear behavior is greatly reduced if more concentrated electrolyte ($5M$ acetic acid; $0.5M$ sodium acetate) is used, and if the distance between the silver and lead electrodes is not more than 5 mm.

TABLE I
Analytical Performance of Probes

Probe no.	Probe output in air, µA	Oxygen content of gas sample	
		Paramagnetic O_2 analyzer, %	Membrane probe, %
1	11.9	15.5	15.7
1		12.4	12.6
1		3.8	4.0
2	10.8	12.1	12.2
2		9.1	9.2
3	11.2	17.3	17.3
3		15.2	15.2
3		7.9	8.0
4	7.54	13.0	13.7
4		8.5	9.8
4		2.1	2.5

Temperature Coefficient

Probe output varies with temperature. Temperature coefficients for three probes were measured, and were found to be identical. Results are shown in Table II. Probe 5 had a 1 mil Teflon membrane, the other two probes had 2 mil membranes.

USE OF PROBES

Probes usually exhibit stable behavior immediately after their initial steam sterilization. Sometimes, however, 24 hr. storage in air-saturated water on closed circuit is necessary as a "break-in" procedure. Probes may be stored for long periods on open circuit.

was left overnight in nitrogen, then, with a 7500 ohm load resistor, was subjected to gas mixtures containing low concentrations of oxygen. It will be seen from the figure that the output in the absence of oxygen is very low, and that analysis of gases of low oxygen content is feasible. In the experiment of Figure 3, the probe responded rapidly. However, when this probe was exposed to air for a few hours, about 3 hr. were required to bring the residual current down to the value shown in the figure. Residual current is minimized by the use of a long anode. Oxygen diffusing into the probe from the top is consumed by reaction with the lead.

Linearity

If a probe is to give an accurate analysis for oxygen, its output should be directly proportional to the oxygen partial pressure. Sev-

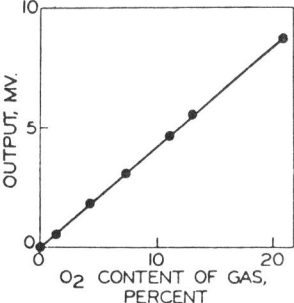

Fig. 4. Probe output at various oxygen concentrations. The probe was in a gas stream, whose oxygen content was varied. The oxygen content of the gas stream was monitored with a paramagnetic oxygen analyzer. The load resistance was 780 ohms.

eral experiments have been carried out to check this point. Various air–nitrogen mixtures were analyzed by the Beckman paramagnetic oxygen analyzer and simultaneously by a probe. Typical results are shown in Figure 4. It will be noted that the output is closely proportional to the oxygen tension. The data of Figure 3 show that linearity is also obtained at very low oxygen tensions.

Nonlinear behavior has been observed with some probes of low output. Table I gives data obtained in an experiment in which gas mixtures (air–nitrogen) were analyzed by the Beckman paramag-

TABLE II
Temperature Coefficients of Probes in Air-saturated Water

Probe no.	Output at 30°C., μA	Output at various temperatures, as percentage of output at 30°C.			
		21°	25°	30°	35°
2	11.25	82.1	90.1	100	109.8
5	21.2	82.1	90.0	100	110.5
6	10.5	82.0	90.0	100	110.0

However, since such storage results in accumulation of dissolved oxygen inside the probe, the probe must be steam sterilized or left on closed circuit for a few hours before use. When a probe is mounted in a fermentor, it is short-circuited during sterilization. After cooling, the load resistor and recorder are connected. Aeration and agitation are begun before inoculation. The output of the probe under these conditions is used as a calibration. Since the probe measures total oxygen tension, the pressure in the fermentor, if higher than atmospheric, must be taken into account in the calibration. Because probes have essentially zero output in the absence of oxygen, zero point calibration is not necessary, except when a thorough check of probe performance is desired.

Probes have been successfully used in fermentors ranging in volume from one liter to 500 gallons. A probe mounted in an intermittently used fermentor has a useful life of a month or more. The life or a probe in continuous service apparently depends on the type of fermentation. A probe in use in a continuous fermentation with a mixed culture became inoperative (slow response, low output) after 12 days. Other probes have been used for two weeks without damage to the probe. A probe stored in air on closed circuit and checked periodically changed in output only 1% during a 15 day period.

USE OF PROBES WITHOUT A POTENTIOMETRIC RECORDER

With a simple amplifier, probes may be used with a Rustrak recording microammeter (0–100 μA), which costs less than $100 (Rustrak Instrument Co., Manchester, N. H.). A suitable circuit is shown in Figure 5. Any 100 μA meter having a resistance of less than 5000 ohms may be substituted for the Rustrak recording meter. The amplifier is adjusted as follows: The probe is connected as indicated (no external load resistor is used). S_1, the zero check switch, is placed in the "down" position. R_1 is adjusted until the meter reading is just above the zero point. S_1 is thrown to the "up" position, and R_2, the output adjusting control, is adjusted so that, with the

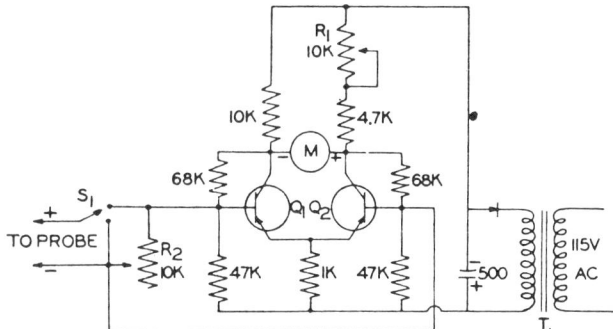

Fig. 5. Probe amplifier. Q_1 and Q_2 are audio pnp transistors, 2N217, RCA SK-3004, or the like. To minimize zero drift due to temperature changes, the transistors are mounted in snugly-fitting holes in the same small copper block. M is a 0-100 µA Rustrak recording microammeter. R_1 is a wire-wound linear potentiometer; R_2 has a clockwise log taper. Both are connected so that clockwise rotation increases resistance. T_1 is a Triad F-13X filament transformer, 6.3 volts. The 500 µF filter capacitor has a 15 volt rating. Resistors are ±5%, 0.5 watt.

probe in air-saturated water, the meter reading is almost full scale. S_1 is thrown to the "down" position, and the zero readjusted, if necessary, with R_1. During use of the amplifier, the zero should be checked from time to time, and readjusted if necessary.

References

1. Clark, L. C., Jr., *Trans. Am. Soc. Artificial Internal Organs*, **2**, 41 (1956).
2. Phillips, D. H., and M. J. Johnson, *J. Biochem. Microbiol. Technol. Eng.*, **3**, 261 (1961).
3. Kinsey, D. W., and R. A. Bottomley, *J. Inst. Brewing*, **69**, 164 (1963).
4. Carrit, D. E., and J. M. Kanwisher, *Anal. Chem.*, **31**, 5 (1959).
5. Mancy, K. H., D. A. Okun, and C. N. Reilley, *J. Electroanal. Chem.*, **4**, 65 (1962).
6. Neville, J. R., *Rev. Sci. Instr.*, **33**, 51 (1962).
7. Pittman, R. W., *Nature*, **195**, 449 (1962).
8. Willey, C. R., and C. B. Tanner, *Soil Sci. Soc. Am. Proc.*, **27**, 511 (1963).
9. Fatt, I., *J. Appl. Physiol.*, **19**, 326 (1964).
10. Mackereth, F. J. H., *J. Sci. Instr.*, **41**, 38 (1964).

11. Hospodka, J., and Z. Časlavsky, K. Beran, and F. Stros, in *Proc. 2nd Symposium on Continuous Cultivation of Microorganisms, Prague, 1962*. Czechoslovak Academy of Sciences, Prague, 1964.

12. Tödt, F., ed., *Electrochemische Sauerstoffmessungen*, de Gruyter, Berlin, 1958.

13. Teske, G., in Tödt, F., ed., *Electrochemische Sauerstoffmessungen*, de Gruyter Berlin, 1958.

14. Briggs, R., and M. Viney, *J. Sci. Instr.*, **41,** 78 (1964).

Part VI
STERILIZATION

Editor's Comments
on Papers 14 and 15

14 **DEINDOERFER and HUMPHREY**
    Analytical Method for Calculating Heat Sterilization Times

15 **HUMPHREY and GADEN**
    Air Sterilization by Fibrous Media

Pure-culture fermentation operations require that the liquid medium be sterile prior to inoculation with the seed, and that all liquid and gaseous feed streams likewise be sterile. Sterilization (or virtual sterilization) of liquid media is usually achieved by heat; sterilization of gas streams is usually achieved by filtration. (For a general discussion of sterilization theory and practice in fermentation operations, see Aiba, Humphrey, and Millis 1965).

Deindoerfer and Humphrey (in Paper 14 and in an earlier paper, 1957) were the first to develop equations for expressing the kinetics of heating and cooling cycles to which liquid media should be subjected in various types of batch and continuous sterilization procedures. They developed their equation on the assumption that thermal destruction of bacterial endospores appears to follow first-order reaction kinetics, and that the relationship between the specific reaction rate of destruction and the absolute temperature follows the Arrhenius equation.

Although aeration of fermentation processes for production of yeast cells was practiced prior to 1920, the introduction of new aerobic fermentations requiring more rigorous asepsis, in the period from 1920 to 1940, necessitated the development of methods of sparging large quantities of quasi-sterile gas. Fibrous filters were introduced prior to 1940 and were available as part of the technology on which the rapid development, in 1940 to 1950, of processes for production of antibiotics and other fine chemicals was based. However, no rigorous theoretical analysis or exposition of principles and experimental methods for design of fibrous filters for air sterilization appeared until the paper by Humphrey and Gaden (Paper 15) was published in 1955.

REFERENCES

Aiba, S., A. E. Humphrey, and N. F. Millis. 1965. Media sterilization, and air sterilization. Pages 186–216 and 217–244 in *Biochemical Engineering*. Academic Press, New York.

Deindoerfer, F. H., and A. E. Humphrey. 1957. Calculation of heat sterilization times for fermentation media. *Appl. Microbiol.*, 5:221–228.

14

Copyright © 1959 by the American Society for Microbiology

Reprinted from *Appl. Microbiol.*, 7(4), 256–264 (1959)

Analytical Method for Calculating Heat Sterilization Times

F. H. DEINDOERFER AND A. E. HUMPHREY

The School of Chemical Engineering, University of Pennsylvania, Philadelphia, Pennsylvania

Received for publication December 1, 1958

The determination of conditions necessary to adequately heat sterilize various media is a problem of considerable importance in the fermentation and food industries, or other such areas where a lack of sterile techniques preclude satisfactory performance. Until recently, however, practical ways to handle this problem were either wholly empirical ones, or theoretical ones which depended on some simplifying assumptions regarding the shape of the heating curve. Along the latter line is the development of design procedures for the sterilization of canned foods by Ball (1923) and Levine (1956), among others.

A theoretical, yet practical and completely general, method for calculating heat sterilization times for fermentation media (as well as other liquid media), based on the thermal-death kinetics of bacterial spores and the actual heating curve, was proposed by Deindoerfer (1957). For batch sterilizations, the method involved a graphical integration of the specific reaction rate for thermal spore destruction along the heating curve. All that was required for this integration was a knowledge of the temperature dependence of the specific reaction rate. For isothermal portions of the heating curve a simple analytical integration was possible, as the specific reaction rate remained constant. In certain types of continuous sterilizers this simple integration was sufficient to determine the time-temperature relationship for any desired degree of sterilization.

Humphrey (1957) was successful in analytically integrating the specific reaction rate over certain nonisothermal periods of a batch sterilization. Also, the integrated solutions for several temperature-time profiles of an element of medium flowing through continuous sterilizers were presented by Deindoerfer and Humphrey (1958) in a discussion of the principles entering into the design of these units.

This paper reviews briefly the underlying theory of heat sterilization and develops, using this theory and common expressions for heat transfer in various equipment, the analytical integrations mentioned above. The practical solution of many sterilization problems can be carried out directly by employing the equations developed. Values of difficult to compute exponential integral functions occasionally needed for particular solutions of these equations are also included in this work.

THEORY

Thermal destruction of bacterial spores may be correlated by apparent first-order reaction kinetics. The rate of destruction at a particular temperature is mathematically represented by equation 1.

$$\frac{dN}{dt} = -kN \qquad (1)$$

See the nomenclature at the end of this report for an explanation of any unfamiliar symbols.

Sterilizations are designed to reduce the viable spore population from its initial value to some predetermined level adequate for the degree of sterilization desired. An expression of this objective results from the integration of equation 1 over a particular time interval.

$$\nabla = \ln \frac{N_0}{N} = \int_0^t k \, dt \qquad (2)$$

In this expression, ∇, the design criterion is a measure of the size of the job to be accomplished. The solution of the integral on the right hand side of equation 2 leads to useful expressions for evaluating and predicting the performance of a sterilizing unit. A confronting problem exists, however, in expressing k as an explicit function of t. This problem will be handled for several types of sterilizers in subsequent developments through the kinetic relationship of k with absolute temperature, and the relationship of temperature, in turn, with time, through heat transfer rate and heat balance equations.

It is assumed that the relationship between the specific reaction rate for spore destruction and absolute temperature follows the familiar Arrhenius equation.[1]

$$k = Be^{-\mu/RT} \qquad (3)$$

The same type of relationship can be obtained by

[1] This appears a realistic assumption based on isothermal experiments. Whether it is applicable to the dynamic case where temperature changes rapidly over a time interval can be questioned. Research is currently under way to establish the validity of this assumption.

applying the theory of absolute reaction rates to the thermal destruction process.

$$k = gTe^{\Delta S^*/R}e^{-\Delta H^*/RT} \quad (4)$$

Equation 4 sheds more light on the significance of the constant B in equation 3 since, for all practical purposes, equation 4 can be reduced to an expression identical to equation 3.

Application of absolute reaction rate theory to the thermal denaturation of biological substances is discussed by Johnson et al. (1954). They, and Pollard (1953), visualize thermal inactivation resulting from the intramolecular breaking of bonds in activated molecules, thus rearranging the molecular structure and resulting in a change in biological activity. More recently Charm (1958) suggested that activated water molecules striking sensitive areas of the spore cause its inactivation. Based on this model he derives an equation similar to equation 3.

Substitution of the Arrhenius expression for k (equation 3) into equation 2 yields the integral which will be evaluated for a number of sterilization cases.

$$\nabla = B \int_0^t e^{-\mu/RT} \, dt \quad (5)$$

The use of equation 5 and its analytical integration requires a tacit assumption to be made. The Arrhenius equation must represent the data over the temperature range of the sterilization. In other words, μ must remain constant. As a matter of convenience in determining a value for ∇, it is also assumed that the entire bacterial population consists of spores of the design species. Although this latter assumption is incorrect, it offers a logical basis for design with an appropriate safety factor if the most heat resistant spores in the medium are used as the design species.

STERILIZATION CASES

Various common methods of sterilization that will be considered here are listed below.[2] Applications are illustrated in figures 1 and 2.
  A. Batch sterilization
    1. Constant rate of heat flow
      a. Constant rate of addition to medium mass (figure 1a)
      b. No change in medium mass (figure 1b)
    2. Changing rate of heat flow
      a. Isothermal heat source or sink (figure 1c)
      b. Nonisothermal heat source or sink (figure 1d)

[2] Another common method of sterilizing media in fermentors of pilot plant size is to simultaneously sparge steam into the vessel and into the vessel jacket or coils. This leads to a heating curve which does not permit analytical integrations of the type that will be illustrated. The sterilization time for this case can be calculated best by the graphical procedure described by Deindoerfer (1957).

  B. Continuous sterilization
    1. Constant rate of heat flow
      a. Constant energy loss (figure 2a)
      b. Nonisothermal heat source or sink with equal and countercurrent mass flow (figure 2b)
    2. Changing rate of heat flow
      a. Isothermal heat source or sink (figure 2c, d)

Situations which can be treated as constant rate of heat flow cases include the batch heating of media by

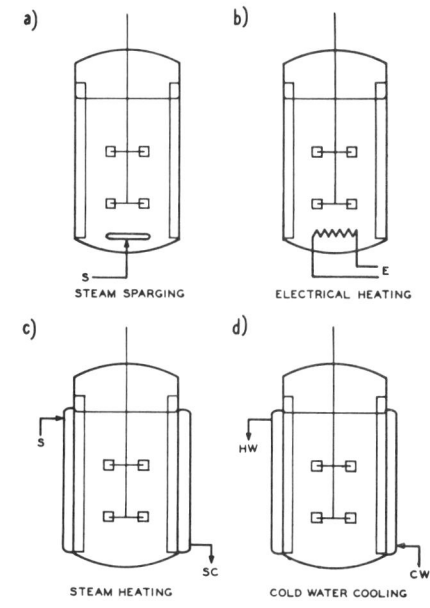

Figure 1. Methods of heating and cooling during batch sterilizations.

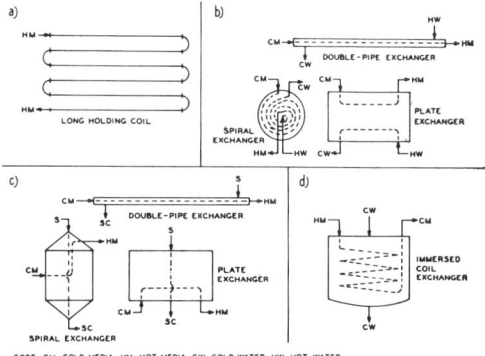

Figure 2. Methods of heating and cooling during continuous sterilizations.

direct steam sparging, or by electrical heaters, and the continuous cooling of media due to energy losses in a flow-type sterilizer. Cases where media are heated by a chamber or coils containing constant pressure condensing steam, or cooled by a constant temperature water bath, fall in the category of changing rate of heat flow from or to isothermal sources or sinks. Media which are heated or cooled by sources or sinks, which themselves cool or heat as heat is transferred, constitute the changing rate of heat flow, nonisothermal source or sink category.

The complexity of the general countercurrent flow-type sterilizer case does not permit an analytical solution by the procedures outlined in this paper, except for the special condition of a countercurrent source or sink which is equal in mass flow rate and heat capacity to the medium being heated or cooled. This condition, however, reduces the situation to a case of constant rate of heat flow.

TYPES OF TEMPERATURE-TIME PROFILES

For batch sterilizations the temperature-time profile is simply the heating curve of the medium in the particular vessel. The entire contents of the sterilizer is always at a common, but usually varying, temperature. For continuous sterilizers, corresponding curves can be constructed by visualizing the temperature change in an element of medium as it flows through the sterilizer. Equations describing the temperature change with time for the sterilization cases under consideration can be derived from heat balance and heat transfer rate equations.[3] Several equations are worked out by Kern (1950); others can be derived in a similar manner. The derived equations for the cases under consideration illustrated in figures 1 and 2 are tabulated in table 1. Notice that all these cases reduce to one of three equations. They are either linear, and of the form

$$T = \Im(1 + Kt) \tag{6}$$

or exponential of the form

$$T = \Im(1 + be^{-Kt}) \tag{7}$$

[3] Another assumption made is that the over-all heat transfer coefficient remains constant. Although it varies with temperature slightly, the variation is small enough so that an average coefficient is sufficient in these equations.

TABLE 1
Temperature-time profiles of various portions of sterilization cycles

Portion and Type of Cycle (Sterilizer)	Temperature-Time Profile		b	K	$\Im$
Batch heating with constant rate of steam addition into medium (figure 1a)	Hyperbolic	$T = T_0 \left(1 + \dfrac{hs/McT_0}{1 + (s/M)t} t\right)$	*	$\dfrac{hs}{McT_0}$	T_0
Batch heating with constant rate of heat flow (no change in medium mass) (figure 1b)	Linear	$T = T_0 \left(1 + \dfrac{q}{McT_0} t\right)$		$\dfrac{q}{McT_0}$	T_0
Batch heating using isothermal heat source (figure 1c)	Exponential	$T = T_H \left(1 + \dfrac{T_0 - T_H}{T_H} e^{(-UA/Mc)t}\right)$	$\dfrac{T_0 - T_H}{T_H}$	$\dfrac{UA}{Mc}$	T_H
Batch cooling using continuous nonisothermal heat sink (figure 1d)	Exponential	$T = T_{c0} \left\{1 + \dfrac{T_0 - T_{c0}}{T_{c0}} e^{(-wc/Mc)(1 - e^{-UA/wc})t}\right\}$	$\dfrac{T_0 - T_{c0}}{T_{c0}}$	$\dfrac{wc}{Mc}(1 - e^{-UA/wc})$	T_{c0}
Cooling due to constant rate of energy loss (figure 2a)	Linear	$T = T_0 \left(1 + \dfrac{-q}{WcT_0} t\right)$		$\dfrac{-q}{WcT_0}$	T_0
Flow heating using isothermal heat source (figure 2c)	Exponential	$T = T_H \left(1 + \dfrac{T_0 - T_H}{T_H} e^{(-UA/Wc)t}\right)$	$\dfrac{T_0 - T_H}{T_H}$	$\dfrac{UA}{Wc}$	T_H
Flow cooling using isothermal heat sink (figure 2d)	Exponential	$T = T_c \left(1 + \dfrac{T_0 - T_c}{T_c} e^{(-UA/Wc)t}\right)$	$\dfrac{T_0 - T_c}{T_c}$	$\dfrac{UA}{Wc}$	T_c
Flow heating using countercurrent heat source of equal flow rate and heat capacity (figure 2b)	Linear	$T = T_{c0} \left(1 + \dfrac{\Delta T}{T_{c0}} \dfrac{UA}{Wc} t\right)$		$\dfrac{\Delta T}{T_{c0}} \dfrac{UA}{Wc}$	T_{c0}
Flow cooling using countercurrent heat sink of equal flow rate and heat capacity	Linear	$T = T_{H0} \left(1 + \dfrac{\Delta T}{T_{H0}} \dfrac{UA}{Wc} t\right)$		$-\dfrac{\Delta T}{T_{H0}} \dfrac{UA}{Wc}$	T_{H0}

* Let $r = \dfrac{s}{M}$

or hyperbolic of the form

$$T = 3\left(1 + \frac{Kt}{1 + rt}\right) \quad (8)$$

where 3, b, K, and r are general parameters characteristic of each individual case as described in table 1. These temperature-time profiles (equations 6 to 8) are illustrated in figure 3 for media being heated and cooled.

ANALYTICAL INTEGRATION

Substitution of each of the general temperature-time profiles (equations 6, 7, and 8) into equation 5 results in integrals which can be solved analytically. The integrations are carried out as follows:

In the following integrations, let $a = \dfrac{\mu}{R3}$

Linear temperature-time profile:

$$\triangledown = B \int_0^t e^{-a/(1+Kt)} \, dt \quad (9)$$

Change variable by letting $x = \dfrac{a}{1 + Kt}$, then

$$dx = -\frac{aK \, dt}{(1 + Kt)^2} \quad \text{and} \quad dt = -\frac{a \, dx}{Kx^2}$$

When $t = 0$, $x = a$. Substitute the new variable x for t, then

$$\triangledown = -\frac{Ba}{K} \int_a^{a/(1+Kt)} \frac{e^{-x}}{x^2} \, dx \quad (10)$$

Equation 10 can be written

$$\triangledown = -\frac{Ba}{K} \int_a^{\infty} \frac{e^{-x}}{x^2} \, dx + \frac{Ba}{K} \int_{a/(1+Kt)}^{\infty} \frac{e^{-x}}{x^2} \, dx \quad (11)$$

The integrals in equation 11 are second-order exponential integrals which have been numerically evaluated for various values of their lower limits, or arguments. Rewritten in common mathematical notation

$$\triangledown = \frac{Ba}{K}\left[E_2\left(\frac{a}{1 + Kt}\right) - E_2(a)\right] \quad (12)$$

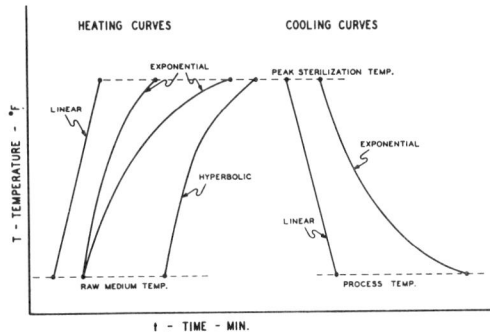

Figure 3. Temperature-time profiles for heating and cooling portions of sterilization cycles.

where, by definition,

$$E_2(z) = \int_z^{\infty} \frac{e^{-x}}{x^2} \, dx \quad (13)$$

Exponential temperature-time profile:

$$\triangledown = B \int_0^t e^{-a/(1+be^{-Kt})} \, dt \quad (14)$$

Change variables by letting $x = \dfrac{a}{1 + be^{-Kt}}$, then

$$dx = \frac{abKe^{-Kt}}{(1 + be^{-Kt})^2} \quad \text{and} \quad dt = \frac{a \, dx}{Kx(a - x)}$$

When $t = 0$, $x = \dfrac{a}{1 + b}$. Substitute the new variable x for t, then

$$\triangledown = \frac{Ba}{K} \int_{a/(1+b)}^{a/(1+be^{-Kt})} \frac{e^{-x}}{x(a - x)} \, dx \quad (15)$$

Using partial fractions, equation 15 can be written

$$\triangledown = \frac{B}{K} \int_{a/(1+b)}^{a/(1+be^{-Kt})} \frac{e^{-x}}{x} \, dx + \frac{B}{K} \int_{a/(1+b)}^{a/(1+be^{-Kt})} \frac{e^{-x}}{a - x} \, dx$$

Change variables in the second integral by letting $y = x - a$, then

$$x = y + a \quad \text{and} \quad dx = dy$$

Substitute the new variable y for x, then

$$\triangledown = \frac{B}{K} \int_{a/(1+b)}^{a/(1+be^{-Kt})} \frac{e^{-x}}{x} \, dx \\ - \frac{Be^{-a}}{K} \int_{a/(1+b)-a}^{a/(1+be^{-Kt})-a} \frac{e^{-y}}{y} \, dy \quad (16)$$

Equation 16 can be written

$$\triangledown = \frac{B}{K} \int_{a/(1+b)}^{\infty} \frac{e^{-x}}{x} \, dx - \frac{B}{K} \int_{a/(1+be^{-Kt})}^{\infty} \frac{e^{-x}}{x} \, dx \\ - \frac{Be^{-a}}{K} \int_{a/(1+b)-a}^{\infty} \frac{e^{-y}}{y} \, dy \\ + \frac{Be^{-a}}{K} \int_{a/(1+be^{-Kt})-a}^{\infty} \frac{e^{-y}}{y} \, dy \quad (17)$$

All the integrals in equation 17 are first-order exponential integrals which have been numerically evaluated for various values of their lower limits, or arguments. Thus

$$\triangledown = \frac{B}{K}\left[E_1\left(\frac{a}{1+b}\right) - E_1\left(\frac{a}{1 + be^{-Kt}}\right)\right] \\ - \frac{Be^{-a}}{K}\left[E_1\left(\frac{a}{1+b} - a\right) \\ - E_1\left(\frac{a}{1 + be^{-Kt}} - a\right)\right] \quad (18)$$

where, by definition,

$$E_1(z) = \int_z^\infty \frac{e^{-x}}{x} dx \quad (19)$$

Hyperbolic temperature-time profile:
For this case, let $p = K + r$. Then equation 8 can be written

$$T = 3\left(\frac{1 + pt}{1 + rt}\right)$$

Also,

$$\nabla = B \int_0^t e^{-a(1+rt)/(1+pt)} dt \quad (20)$$

Change variables by $x = \dfrac{1 + rt}{1 + pt}$, then

$$t = \frac{x - 1}{r - px} \quad \text{and} \quad dt = \frac{(r - p)}{(r - px)^2} dx$$

When $t = 0$, $x = 1$. Substitute the new variable x for t, then

$$\nabla = B(r - p) \int_1^{1+rt/1+pt} \frac{e^{-ax}}{(r - px)^2} dx \quad (21)$$

Change variables again, letting

$$y = -\frac{a}{p}(r - px),$$

then

$$y^2 = \frac{a^2}{p^2}(r - px)^2, \quad x = \frac{y}{a} + \frac{r}{p}$$

and

$$dx = \frac{dy}{a}$$

Substitute the new variable y for x, then

$$\nabla = \frac{B(r - p)ae^{-ar/p}}{p^2} \int_{a-ar/p}^{a(1+rt)/(1+pt)-ar/p} \frac{e^{-y}}{y^2} dy \quad (22)$$

Let $m = \dfrac{ar}{p}$ and recall that $p = K + r$. Equation 22 can then be written

$$\nabla = -\frac{BaKe^{-m}}{p^2} \int_{a-m}^\infty \frac{e^{-y}}{y^2} dy$$

$$+ \frac{BaKe^{-m}}{p^2} \int_{a(1+rt)/(1+pt)-m}^\infty \frac{e^{-y}}{y^2} dy \quad (23)$$

The integrals in equation 23 are second-order exponential integrals, which can be rewritten

$$\nabla = \frac{BaKe^{-m}}{p^2}$$
$$\cdot \left[E_2\left\{a\left(\frac{1 + rt}{1 + pt}\right) - m\right\} - E_2(a - m) \right] \quad (24)$$

where $E_2(z)$ is as defined in equation 13.

When the individual characteristics for the various sterilization cases are substituted for the general parameters a, b, K, and r into equations 12, 18, and 24, these equations become the particular design equations for the respective cases. These equations are summarized in table 2. The parametric expressions were listed in table 1.

EVALUATION OF EXPONENTIAL INTEGRAL FUNCTIONS

The first and second order exponential integral functions, appearing in equations 12, 18, and 24 and defined in equations 13 and 19, have been numerically evaluated and graphically represented for positive arguments as large as 33 and negative arguments as large as -21 in reactor handbooks edited by Rockwell (1956) and Hogerton and Grass (1955). They appear tabulated in table 3 for positive arguments up to 15. Above this argument, the following approximation yields a satisfactory value:

$$E_n(z) = \frac{e^{-z}}{z^n} \quad (25)$$

Thus, for large arguments as are common in many sterilization operations, the integral can be evaluated by simple computation. Usually, too, where the func-

TABLE 2
Design equations for various temperature-time profiles

Type of Profile (Sterilizer)	Design Equation
Linear (figure 1b; 2a, b)	$\nabla = \dfrac{Ba}{K}\left[E_2\left(\dfrac{a}{1 + Kt}\right) - E_2(a) \right]$
Exponential (figure 1c, d; 2c, d)	$\nabla = \dfrac{B}{K}\left[E_1\left(\dfrac{a}{1 + b}\right) - E_1\left(\dfrac{a}{1 + be^{-Kt}}\right) \right] - \dfrac{Be^{-a}}{K}\left[E_1\left(\dfrac{a}{1 + b} - a\right) - E_1\left(\dfrac{a}{1 + be^{-Kt}} - a\right) \right]$
Hyperbolic (figure 1a)	$\nabla = \dfrac{BaKe^{-m}}{p^2}\left[E_2\left\{a\left(\dfrac{1 + rt}{1 + pt}\right) - m\right\} - E_2(a - m) \right]$

tions appear in a design equation as a difference, one is sufficiently larger than the other so that the smaller value can be neglected. A rule of thumb to follow here is that, whenever one argument is larger than another by six or more, its functional value will be small enough to neglect.

APPLICATION OF DESIGN EQUATIONS

To illustrate the facility of the equations developed, the following example of a batch sterilization is cited.

TABLE 3
*Exponential integral functions**

z	$E_1(z)$	$E_2(z)$	z	$E_1(z)$	$E_2(z)$
0.1	1.823	7.23	7.2	9.22×10^{-5}	1.150
0.2	1.223	2.87	7.4	7.36	8.96×10^{-6}
0.3	9.06×10^{-1}	1.563	7.6	5.89	6.99
0.4	7.02	9.74×10^{-1}	7.8	4.71	5.46
0.5	5.60	6.52	8.0	3.77	4.27
0.6	4.54	4.61	8.2	3.02	3.34
0.7	3.74	3.35	8.4	2.42	2.62
0.8	3.11	2.51	8.6	1.936	2.05
0.9	2.60	1.916	8.8	1.552	1.610
1.0	2.19	1.485	9.0	1.245	1.265
1.2	1.584	9.26×10^{-2}	9.2	9.99×10^{-6}	9.95×10^{-7}
1.4	1.162	5.99	9.4	8.02	7.82
1.6	8.63×10^{-2}	3.99	9.6	6.44	6.16
1.8	6.47	2.71	9.8	5.17	4.85
2.0	4.89	1.877	10.0	4.16	3.83
2.2	3.72	1.317	10.2	3.34	3.02
2.4	2.84	9.36×10^{-3}	10.4	2.69	2.39
2.6	2.19	6.72	10.6	2.16	1.887
2.8	1.686	4.86	10.8	1.740	1.492
3.0	1.305	3.55	11.0	1.400	1.180
3.2	1.013	2.60	11.2	1.127	9.35×10^{-8}
3.4	7.89×10^{-3}	1.925	11.4	9.08×10^{-7}	7.40
3.6	6.16	1.430	11.6	7.31	5.87
3.8	4.82	1.067	11.8	5.89	4.65
4.0	3.78	8.00×10^{-4}	12.0	4.75	3.69
4.2	2.97	6.02	12.2	3.83	2.93
4.4	2.34	4.54	12.4	3.08	1.848
4.6	1.841	3.44	12.6	2.49	1.469
4.8	1.453	2.62	12.8	2.01	1.165
5.0	1.148	1.993	13.0	1.622	1.025
5.2	9.09×10^{-4}	1.523	13.2	1.309	9.30×10^{-9}
5.4	7.20	1.166	13.4	1.057	7.40
5.6	5.71	8.95×10^{-5}	13.5	8.53×10^{-8}	5.89
5.8	4.53	6.88	13.6	6.89	4.69
6.0	3.60	5.30	14.0	5.57	3.74
6.2	2.86	4.10	14.2	4.50	2.98
6.4	2.28	3.17	14.4	3.63	2.38
6.6	1.816	2.45	14.6	2.94	1.896
6.8	1.448	1.903	14.8	2.37	1.513
7.0	1.155	1.479	15.0	1.919	1.207

* Adapted from Hogerton and Grass (1955).

The use of the equations in designing sterilization conditions in continuous sterilizers has also been described by the authors (1958).

Example. Large industrial fermentors containing raw medium often are sterilized by passing steam through the air sparger until a desired sterilization temperature is reached. The medium and fermentor are maintained at this temperature for a prescribed amount of time, and then cooled by passing water through coils located within the fermentor. Obviously, some contribution to the sterilization occurs during both heating and cooling the fermentor. The problem then is one of assessing the sterilization accomplished during these portions of the sterilizing cycle and calculating the necessary holding time at so-called sterilization temperature required to complete the desired sterilization. Consider this problem where the following conditions prevail.

1. The fermentor contains 12,000 gallons of a raw medium which in periodic laboratory checks has consistently shown a bacterial count in the neighborhood of 20×10^9 cells per gal. It is desired to reduce this population to such an extent that the chance for a contaminant surviving the sterilization is 1 in 1000.

2. During heating, 50 psig saturated steam is passed into the fermentor at a rate of 200 lb per min, until the temperature reaches 250 F, the desired sterilization temperature. The medium is initially at 130 F. The enthalpy of the steam relative to 130 F water is 1091 BTU per lb.

3. In cooling the fermentor, 4000 lb per min of 50 F water passes through coils until a process temperature of 85 F is reached. The coils have a heat transfer area of 400 ft² and for this operation the average over-all heat transfer coefficient for cooling is 120 BTU per $hr \times ft^2 \times °F$.

4. The most heat resistant bacterial spores in the medium are characterized by an Arrhenius coefficient of $1 \times 10^{36.2}$ sec^{-1} and an activation energy of 67,700 cal per gmol for thermal destruction.

The design criterion can be calculated from condition 1 using equation 2.

$$\nabla = \ln \frac{N_0}{N}$$

$$= \ln \frac{(2 \times 10^{10} \text{ cells/gal})(1.2 \times 10^4 \text{ gal})}{(1 \times 10^{-3} \text{ cell})} = 40.0$$

The constant addition of steam to the medium results in a hyperbolic temperature-time heating profile. Using the equation for this case listed in table 2, it is easily shown that 62 min heating time are required to heat the medium from 130 to 250 F. The sterilization design equation for a hyperbolic profile, listed in table 3, is shown below.

$$\nabla = \frac{BaKe^{-m}}{p^2}\left[E_2\left\{a\left(\frac{1+rt}{1+pt}\right)-m\right\} - E_2(a-m)\right]$$

The factors in this equation that are known include:

$$B = 1 \times 10^{36.2} \text{ sec}^{-1} = 6 \times 10^{37.2} \text{ min}^{-1}$$

$$t = 62 \text{ min}$$

The remaining factors are calculated as follows:

i. $a = \dfrac{\mu}{R\mathfrak{T}}$

$\mu = 67,700$ cal/gmol

$R = 1.10$ cal/gmol $\times$ °R

$\mathfrak{T} = T_0 = 130$ F $= 590$°R

$$a = \frac{(67,700 \text{ cal/gmol})}{(1.10 \text{ cal/gmol} \times \text{°R})(590\text{°R})} = 104.3$$

ii. $K = \dfrac{hs}{McT_0}$

$h = 1091$ BTU/lb

$s = 200$ lb/min

$M = (12,000 \text{ gal})(8.34 \text{ lb/gal}) = 100,000$ lb

$c = 1$ BTU/lb $\times$ °R

$$K = \frac{(1091 \text{ BTU/lb})(200 \text{ lb/min})}{(100,000 \text{ lb})(1 \text{ BTU/lb} \times \text{°R})(590\text{°R})}$$
$$= 3.70 \times 10^{-3} \text{ min}^{-1}$$

iii. $m = \dfrac{ar}{p}$

$r = \dfrac{s}{M} = \dfrac{(200 \text{ lb/min})}{(100,000 \text{ lb})} = 2.0 \times 10^{-3} \text{ min}^{-1}$

$p = K + r = 3.70 \times 10^{-3} \text{ min}^{-1} + 2.0 \times 10^{-3} \text{ min}^{-1}$
$= 5.70 \times 10^{-3} \text{ min}^{-1}$

$m = \dfrac{(104.3)(2.0 \times 10^{-3} \text{ min}^{-1})}{(5.70 \times 10^{-3} \text{ min}^{-1})} = 36.6$

$e^{-m} = e^{-3.66} = 1.26 \times 10^{-16}$

Then,

$$\frac{BaKe^{-m}}{p^2} = \frac{(6 \times 10^{37.2} \text{ min}^{-1})(104.3)}{(5.70 \times 10^{-3} \text{ min}^{-1})^2}$$
$$(3.70 \times 10^{-3} \text{ min}^{-1})(1.26 \times 10^{-16})$$
$$= 8.72 \times 10^{-25.2}$$

$a\left(\dfrac{1 + rt}{1 + pt}\right)$

$= 104.3 \left[\dfrac{1 + (2.0 \times 10^{-3} \text{ min}^{-1})(62 \text{ min})}{1 + (5.7 \times 10^{-3} \text{ min}^{-1})(62 \text{ min})}\right] = 86.6$

The exponential integral functions are, therefore,

$E_2\left\{a\left(\dfrac{1 + rt}{1 + pt}\right) - m\right\} = E_2(86.6 - 36.6) = E_2(50.0)$

$E_2(a - m) = E_2(104.3 - 36.6) \; E_2(67.7)$

The lowest of these is evaluated using the approximation of equation 25. Since $67.7 > 50.0 + 6.0$, E_2 (67.7) is not significant and need not be evaluated.

$$E_2(50.0) = 7.10 \times 10^{-25}$$

The extent of sterilization during heating is

$$\nabla = (8.72 \times 10^{25.2})(7.10 \times 10^{-25}) = 9.8$$

or almost 25 per cent of the desired sterilization.

Cooling of the medium by cold water in coils (batch cooling using a continuous nonisothermal heat sink) leads to an exponential cooling curve. Under the conditions described, it will take 4 hr to cool the fermentor from 250 to 85 F. The design equation for the exponential cooling curve, listed in table 3, is shown below.

$$\nabla = \frac{B}{K}\left[E_1\left(\frac{a}{1+b}\right) - E_1\left(\frac{a}{1 + be^{-Kt}}\right)\right]$$
$$- \frac{Be^{-a}}{K}\left[E_1\left(\frac{a}{1+b} - a\right) - E_1\left(\frac{a}{1 + be^{-Kt}} - a\right)\right]$$

The various factors in this equation that are known are:

$$B = 6 \times 10^{37.2} \text{ min}^{-1}$$

$$t = 240 \text{ min}$$

The remaining factors are calculated as follows:

i. $K = \dfrac{wc}{Mc}\left(1 - e^{-\frac{UA}{Wc}}\right)$

$w = 4000$ lb/min $= 24 \times 10^4$ lb/hr

$U = 120$ BTU/hr $\times$ ft$^2 \times$ °F

$A = 400$ ft^2

$$K = \frac{(4000 \text{ lb/min})(1 \text{ BTU/lb} \times \text{°F})}{(100,000 \text{ lb})(1 \text{ BTU/lb} \times \text{°F})}$$
$$\left(1 - e^{-\frac{(120 \text{ BTU/hr} \times \text{ft}^2 \times \text{°F})(400 \text{ ft}^2)}{(24 \times 10^4 \text{ lb/hr})(1 \text{ BTU/lb} \times \text{°F})}}\right)$$

$K = 7.24 \times 10^{-3}$ min^{-1}

ii. $a = \dfrac{\mu}{R\mathfrak{T}}$

$\mathfrak{T} = T_{c_0} = 50$ F $= 510$ °R

$$a = \frac{(67,700 \text{ cal/gmol})}{(1.10 \text{ cal/gmol} \times \text{°R})(510\text{°R})} = 120.8$$

$e^{-a} = e^{-120.8} = 3.16 \times 10^{-53}$

iii. $b = \dfrac{T_0 - T_{c_0}}{T_{c_0}}$

$T_0 = 250$ F $= 710$ °R

$b = \dfrac{710 \text{ °R} - 510 \text{ °R}}{510 \text{ °R}} = 0.392$

Then,

$$\frac{B}{K} = \frac{6 \times 10^{37.2} \text{ min}^{-1}}{7.25 \times 10^{-3} \text{ min}^{-1}} = 8.28 \times 10^{39.2}$$

$$\frac{Be^{-a}}{K} = (8.28 \times 10^{39.2})(3.16 \times 10^{-53}) = 2.62 \times 10^{-12.8}$$

$$be^{-Kt} = 0.392 \, e^{-(7.25 \times 10^{-3} \text{min}^{-1})(240 \text{min})} = 0.069$$

The exponential integral functions are, therefore:

$$E_1\left(\frac{a}{1+b}\right) = E_1\left(\frac{120.8}{1+0.392}\right) = E_1(86.8)$$

$$E_1\left(\frac{a}{1+be^{-Kt}}\right) = E_1\left(\frac{120.8}{1+0.069}\right) = E_1(113.0)$$

$$E_1\left(\frac{a}{1+b} - a\right) = E_1(86.8 - 120.8) = E_1(-34.0)$$

$$E_1\left(\frac{a}{1+be^{-Kt}} - a\right) = E_1(113.0 - 120.8) = E_1(-7.8)$$

Since $113.0 > 86.8 + 6.0$ and $-7.8 > -34.0 + 6.0$, the function of the larger argument of each of these pairs will not be significant in the sterilization calculation. The other functions are evaluated using equation 25.

$$E_1(86.8) = 2.01 \times 10^{-40}$$

$$E_1(-34.0) = -1.82 \times 10^{13}$$

The extent of sterilization during cooling is

$$\nabla = (8.28 \times 10^{39.2})(2.01 \times 10^{-40})$$
$$- (2.62 \times 10^{-12.8})(-1.82 \times 10^{13})$$
$$= 2.6 + 7.5 = 10.1$$

or slightly more than 25 per cent of the sterilization. Now since

$$\nabla_{\text{Total}} = \nabla_{\text{heating}} + \nabla_{\text{holding}} + \nabla_{\text{cooling}}$$

$$\nabla_{\text{Holding}} = 40.0 - 9.8 - 10.1 = 20.1$$

At 250 F, the velocity constant for thermal death is 1.83 min^{-1}. The required time at so-called sterilization temperature is therefore

$$t = \frac{\nabla}{k} = \frac{20.1}{1.83 \text{ min}^{-1}} = 11.0 \text{ min}$$

Thus, the sterilization time needed is shortened considerably (in this example almost 50 per cent) by long heating and cooling periods in the sterilization cycle.

Summary

The equations developed in this paper make it analytically possible to calculate the degree of sterilization accomplished during portions of sterilization cycles where temperature varies. Their incorporation in sterilization design procedures permits a simplified and rational approach to calculating the degree of sterilization in the over-all process. One should be aware that the basic assumption in their development is that spore destruction rates can be correlated by an Arrhenius-type relationship over the temperature range of the sterilization. This assumption is believed valid on the basis of the success of similar though approximate design procedures employed for heat sterilization calculations in the food industry. A suggested safety factor is introduced when the entire contaminant population is assumed to consist of the most heat resistant spore species.

NOMENCLATURE

a = parameter in design equations, equal to $\mu/R\Im$, dimensionless
A = surface area across which heat transfer occurs during sterilization, ft^2
b = temperature ratio in design equations, defined in table 1, dimensionless
B = constant in Arrhenius equation, sec^{-1}
c = specific heat of medium, sources and sinks, BTU/lb $\times$ °F
g = constant in absolute rate theory equation, sec^{-1} $\times$ °Rankine^{-1}
h = enthalpy of steam relative to raw medium temperature, BTU/lb
k = specific reaction rate for thermal spore destruction, sec^{-1}
K = time parameter in design parameter, defined in table 1, sec^{-1}
m = parameter in design equation, equal to ar/p, dimensionless
M = initial mass of medium in batch sterilizer, lb
N, N_0 = number of viable spores, number of viable spores initially present
p = time parameter in design equation, equal to $K + r$, sec^{-1}
q = rate of heat transfer, BTU/sec
r = time parameter in design equation, equal to s/M, sec^{-1}
R = universal gas constant, cal/gmol $\times$ °R
s = steam mass flow rate, lb/sec
t = time of heat exposure, sec
$T, T_c, T_{c_0}, T_H, T_{H_0}, T_0$ = absolute temperature of medium, heat sink, heat sink (initial), heat source, heat source (initial) and medium (initial), respectively, °Rankine
U = over-all heat transfer coefficient, BTU/sec $\times$ ft^2 $\times$ °F
w = coolant mass flow rate, lb/sec
W = mass of flowing medium in contact with surface area A in sterilizer, lb
ΔH^* = standard activation energy for thermal spore destruction, cal/gmol
ΔS^* = entropy of activation for thermal spore destruction, cal/gmol $\times$ °R
$\Im$ = temperature parameter in design equations, defined in table 1, °R
μ = activation energy for thermal spore destruction in Arrhenius equation, cal/gmol
∇ = design criterion for sterilization, equal to $\ln N_0/N$, dimensionless
$E_1(z), E_2(z), E_n(z)$ = exponential integrals of argument z, of first, second and the n-th order, respectively.

The use of sec as the unit time measure is arbitrary. Min or hr can also be used. Care should be exercised, however, to maintain consistency when employing units of time measure.

REFERENCES

BALL, C. O. 1923 Thermal process time for canned food. Bull. Natl. Research Council (U. S.), **7**, 1–76.

CHARM, S. E. 1958 The kinetics of bacterial inactivation by heat. Food Technol., **12**, 4–8.

DEINDOERFER, F. H. 1957 Calculation of heat sterilization times for fermentation media. Appl. Microbiol., **5**, 221–228.

DEINDOERFER, F. H. AND HUMPHREY, A. E. 1958 Principles in the design of continuous sterilizers. Fermentation Subdivision, American Chemical Society, 134th Meeting, Chicago, Illinois, September 10, 1958.

HOGERTON, J. F. AND GRASS, R. C. (Editors) 1955 *Reactor handbook: physics*, pp. 686–692. McGraw-Hill Book Co., New York, New York.

HUMPHREY, A. E. 1957 Dynamics of sterilization, Ch. E. Seminar, University of Pennsylvania, Philadelphia, Pennsylvania, October 7, 1957.

JOHNSON, F. H., EYRING, H., AND POLLISAR, M. J. 1954 *The kinetic basis of molecular biology*, pp. 187–285. John Wiley & Sons, New York, New York.

KERN, D. Q. 1950 *Process heat transfer*, pp. 626–635. McGraw-Hill Book Co., New York, New York.

LEVINE, S. 1956 Determination of the thermal death rate of bacteria. Food Research, **21**, 295–301.

POLLARD, E. C. 1953 *The physics of viruses*, pp. 103–109. Academic Press, Inc., New York, New York.

ROCKWELL, T. (Editor) 1956 *Reactor shielding design manual*, pp. 372–384. U. S. Atomic Energy Commission, Washington, D. C.

ERRATA

Page 261, column 2, line 46 should read: ". . . the equation for this case listed in table 1, . . ." and line 49 should read: ". . . listed in table 2, . . ."

Page 262, column 2, line 13 should read: ". . . listed in table 2, . . ."

Air Sterilization by Fibrous Media

ARTHUR E. HUMPHREY[1] AND ELMER L. GADEN, JR.

Department of Chemical Engineering, Columbia University, New York, 27, N. Y.

STERILE or highly purified air is now required in large volumes for a variety of operations in the chemical process industries. While individual volumetric demands were relatively limited, air sterilization was a fairly simple matter with a variety of techniques applicable to the task. The rapid growth, in the last decade, of commercial submerged fermentation processes involving intensive aeration and pure culture operation has, however, greatly amplified the problem and many solutions have been proposed.

Steam jets, ultraviolet radiation, incinerators, insulated compressors, and electrostatic precipitators have all been used to destroy or remove the microorganisms in air. None of these has, however, been fully satisfactory for large scale applications like fermentation. In such cases filtration of air through deep beds of fibrous or granular materials has been simpler and economically preferable. The most common of these materials are glass fibers, activated carbon, stainless steel wool, and cotton.

Air filters must satisfy two at least partially incompatible standards; high efficiency for removal of organisms and low pressure drop. The strictness of the sterility standard is determined by the application involved. For the aeration of fermentation broths and aseptic canning, sterility requirements are extremely high; whereas, for packaging areas and sterile rooms they are considerably lower. In the strictest bacteriological sense, the term "sterile" refers to air absolutely free from viable microorganisms. In practice, however, absolute sterility can rarely be guaranteed and instead a standard of practical or commercial sterility, involving some statistical probability, is applied. For example, in discussing the sterilization of fermentation media, Finn (10) noted that the examination of small samples from a large tank, even if the available sterility tests are perfect (and they are certainly not), can only positively establish that the degree of contamination is below a certain level. Similarly an air filter can generally be designed only so that the probability of penetration by a single contaminant (or some minimal number of them) during the operating period of the filter is so small that for all practical purposes the air stream may be considered sterile.

As a matter of fact it is impossible for any air filter in which the interstices of the bed are larger than the least characteristic dimension of organisms being collected to be 100% efficient; only a statistical prediction of its collection can be made. So-called absolute filters, like the millipore assay filters, are formed of a membrane or matrix with openings smaller than the particles to be removed. Consequently, theirs is a sieving rather than an impacting action.

Filter design and the choice of filtering media so far seem to have been made with little reference to filtration fundamentals. Rather, they have been based on performance tests and accumulated experience, with design procedures amounting essentially to the extrapolation of existing units whose performance had been found satisfactory. In particular, the effect of air stream velocity on filtering efficiency has been neglected. The purpose of this investigation was to supplement practical knowledge with a fundamental understanding of the mechanics of fibrous filters removing microorganisms from air streams. The results may shed some useful light on the behavior of deep bed filters in general.

Studies are noted in which microorganisms are the aerosol particles

While a great deal of attention has been given to aerosol filtration in general, only a few studies have been reported in which microorganisms were used as the specific aerosol particles. Terjesen and Cherry (28) were among the early experimenters testing fibrous media for removing microorganisms from air streams. They reported that 3 inches of slag wool were 99.9998+% efficient, as determined by the Bourdillon slit sampler (2), in removing *Bacillus subtilis* spores from air streams at low velocities.

Cherry, McCann, and Parker (6) examined the applicability of slag and glass wool filters to industrial penicillin fermentations of 7 to 10 days duration. At superficial air velocities of 0.2 to 0.5 foot per second, the filters they tested performed satisfactorily for industrial uses and gave efficiencies of 99.89 to 100%. They concluded that filters compounded of 3 inches of slag wool with a packing density of 25 pounds per cubic foot gave the best performance characteristics.

Decker and coworkers (9) investigated the removal of bacteria and bacteriophage particles from air using electrostatic precipitators and spun glass filter pads. As high as 99 to 100% removal of *Serratia indica* and *E. coli bacteriophage* could be obtained with Corning glass wool filters number 25-FG, 50-FG, and 100-FG by combining each filter with an electrostatic precipitator. Two glass wool filters in series would remove 99% of the organisms from inlet air streams.

Kluyver and Visser (15) have reported on a means for testing such filter material as cotton wool, stilbite, and carbon for removing *Bacillus cereus* spores from streams of artifically contaminated air. More recently McDaniel and Long (20) have described laboratory equipment used by a commercial antibiotic producer for testing filter materials that are to be used in sterilizing air streams.

Commercial tests have been reported on the effectiveness of the Cambridge absolute filter (3) in removing cells of *Serratia marcescens* from air streams. This filter was reported to have an average leakage of 0.0031% when filtering air containing 23,000 to 1,730,000 organisms per cubic foot and operating at a resistance of 0.8 to 1.0 inch of water. The assays were obtained with an electrostatic air sampler.

Membrane filters, first developed in Germany during World War II and later perfected by Goetz (11), are reported to retain virtually 100% of all bacteria cells on their surfaces when filtering air. These filters, sometimes referred to as millipore filters, are prepared from stabilized, dry cellulose-ester gel membranes, 120 to 160 microns thick. They have a uniform porosity of approximately 85%. The pores are less than 0.5 micron in diameter and number 10^6 to 10^8 per square centimeter. The delicate mechanical structure of the membranes limits their use as industrial air filters; nevertheless, they have wide application as an assay tool for hydrosols and aerosols. In the experimental work reported in this paper the membrane filters were used for the assays, because they gave better quantitative results than the impinger type described by Rosebury (26) and used by the Chemical Corps.

---

[1] Present address, School of Chemical Engineering, University of Pennsylvania, Philadelphia, Pa.

Some years ago it was discovered that fibrous filters impregnated with certain resinous materials were, because of their ability to develop an electrostatic surface charge, more efficient in collecting aerosol particles (*21*). This observation started many investigators testing the increased adhesiveness and collecting power of fibrous filters impregnated with various agents. One of these investigators, Hopper (*12*), reported that fibrous filters coated with triethylene glycol had increased bacterial collection action.

Much of the early theory of aerosol filtration and behavior—particularly that for the deposition of aerosols by inertial forces—has come from the work of two German investigators, Albrecht (*1*) and Sell (*27*). Using a mathematical model for potential flow of a fluid around an object, they have described the impaction of a particle under the influence of inertial forces only. Langmuir and Blodgett (*19*) further examined this potential flow model and, with the aid of a differential analyzer, they obtained a complete characterization of inertial impaction of spherical particles collected on cylindrical, spherical, and ribbon objects. This inertial action has been described for viscous flow by Landahl and Herrman (*17*) and Davies (*8*) at Reynolds number flows of 10 and 0.2, respectively. All these investigators reported that the inertial action could be partially characterized by a unit Ranz and Wong (*24*) called the inertial parameter—the ratio of the distance a particle will travel in still air, given the initial velocity of the air stream, to a characteristic dimension of the collector.

Of course forces other than inertial ones can affect particle motion and impaction. A discussion of the various mechanisms operating in a fibrous filter to cause aerosol collection has been presented in the Handbook on Aerosols (*25*). Inertial, electrostatic, gravitational, diffusional, and thermal forces are all considered capable of contributing to impaction. Also direct interception has been proposed as an important means of collection. This latter action arises from the fact that an aerosol particle possesses a finite size. Because of this finite size there exists a certain chance of the particle brushing against a collector and being collected.

All these mechanisms of impaction are currently being exhaustively investigated by many research workers. Of these, Johnstone and Roberts (*14*) were among the earlier investigators to examine the collection arising from Brownian motion. They characterized this action, often referred to as the diffusional effect, by analogy to heat transfer and concluded that the diffusional mechanism is not important in the collection of particles larger than 0.5 micron in diameter.

Ranz and Wong (*24*) have examined, both theoretically and experimentally, the action of aerosol particles in air streams and have outlined the important mechanisms that contribute to aerosol impaction by cylindrical and spherical collectors. According to their interpretations, four primary mechanisms contribute to impaction in ordinary fibrous filters. They are inertial forces, interception, settling, and electrostatic attraction. Ranz and Wong used dimensionless parameters—each essentially a comparison of the force tending to cause impaction with the fluid resistance force opposing particle motion—to characterize the inertial, settling, and electrostatic attraction mechanisms. Table I lists these parameters as they defined them. A definition of symbols may be found at the end of the paper.

At first, direct interception was believed to be solely a function of the particle and collector size ratios and completely characterized as shown in Table I. However, Ranz (*23*) and Davies (*8*) have shown that interception is also a function of the Reynolds number and that its effects decrease with a decrease in velocity. Finn (*10*) has suggested that interception is the important mechanism operating to collect bacteria in industrial air filters where the air streams are moving at reasonable velocities. However, the work of Davies (*7*) has shown that this probably is not true for cases where the Reynolds number is less than 0.1 and

Table I. Parameters of Collection

Mechanism	Parameter
Inertial	$\dfrac{C'\rho_p V_0 D_p^2}{18\mu D_c}$
Settling	$\dfrac{C'\rho_p g D_p^2}{18\mu V_0}$
Interception	D_p/D_c
Electrostatic attraction by charge differences	$\dfrac{C' q_p q_{ac}}{3\pi\mu\epsilon_0 D_p V_0}$
Electrostatic attraction by induction	$\dfrac{2C'(\epsilon_p - \epsilon_a) q_{ac}^2 D_p^2}{18\mu\epsilon_0 D_c V_0}$

the fiber diameter is several times larger than that of the aerosol particle.

In all the theory of aerosol collection it has been assumed sufficient for a particle only to brush the surface of the collector to impact. Whether this fact is true or not would be difficult to prove particularly since so little is known about the forces operating to bring the particle in contact with the collector. Terjesen and Cherry (*28*) have shown that once a particle adheres to a collector it remains attached and changes in air stream velocity do not seem to dislodge it. One could ask, however, how hard a particle must impact on a collector or how close a contact is necessary between the particle and collector for impaction to occur.

Simplified solutions relating the various impaction mechanisms to the efficiency of a single collector have been presented in a number of papers (*1, 7, 13, 17, 19, 24, 27, 30*). Incorporation of this single fiber efficiency in an operating efficiency for a fibrous filter bed is infinitely complex. Not only may several mechanisms of collection operate simultaneously, but the distribution of the fibers in the bed is difficult to describe mathematically for they occur randomly rather than uniformly. Portions of the filter exist where, due to this random distribution, the fibers are sufficiently close to interfere with each other's action. As such they may disturb the flow field that surrounds them and, in certain cases, they may be sufficiently close to give rise to a sieving rather than an impacting action.

A few papers, particularly those of Langmuir (*18*), Wong and Johnstone (*30*), and Davies (*7*), have attempted to incorporate single fiber efficiency in over-all operating efficiency for a fibrous filter. These investigators have considered the combined effects of several mechanisms including those of inertial, interception, diffusional, and settling action. Also Ramskill and Anderson (*22*) have presented a qualitative picture for the simultaneous action of the inertial, interception, and diffusional mechanisms in a fibrous filter.

One expression for the over-all efficiency of a filter bed can be obtained by assuming that the fibers are uniformly distributed and that each differential thickness of filter removes the same average fraction of aerosol particles. Such a relation will have the form

$$\ln \frac{C_{\text{out}}}{C_{\text{in}}} = -4\eta(\rho_b X/\rho_c \pi D_c) \qquad (1)$$

In this equation C_{in} is the concentration per unit volume of particles in the incoming air stream, C_{out} is the concentration in the air stream at X depth in the bed, ρ_b is the bulk density of the bed, ρ_c is the density of the fiber medium, and η is the single fiber efficiency obtained from some relationship that characterizes the collecting action. Equation 1 is often referred to as the log penetration law. In reality it is an expression for the probability of collection.

Davies (*7*) has reported that the fraction of aerosol particles penetrating a filter may be given by

$$\ln \frac{C_{\text{out}}}{C_{\text{in}}} = -4\eta\alpha X/\pi D_c (1 - \alpha) \qquad (2)$$

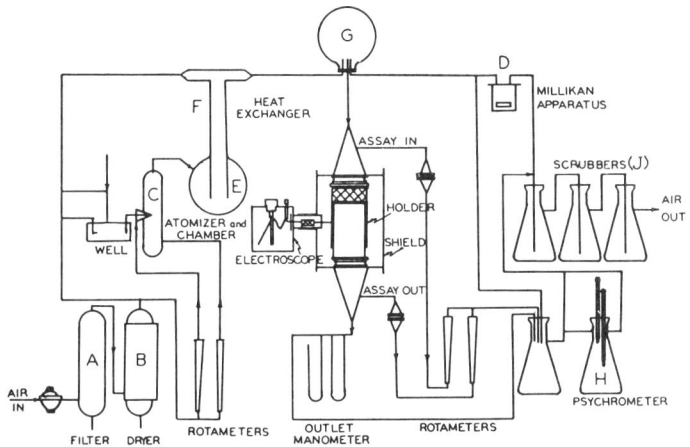

Figure 1. Flow diagram

where α is equivalent to the ratio of ρ_b/ρ_c. For small values of α, Equations 1 and 2 are the same.

Since the completion of this work, unpublished experimental studies at another institution have shown that aerosol penetration of fibrous media can be represented as

$$\ln \frac{C_{\text{out}}}{C_{\text{in}}} = -\frac{4}{\pi} \eta k' \left(\frac{1-p}{p}\right) \frac{D_c}{(D_c)_S{}^2} X \quad (3)$$

In Equation 3 p is the porosity of the fibrous filter, D_c is the average fiber diameter, $(D_c)_S$ is the surface average fiber diameter, and k' is a constant related to the distribution of fibers in a given filter bed.

For a more extensive review of aerosol filtration, the reader is referred to the papers of Wong and Johnstone (30), Thomas (29). or Chen (5).

Aerosol preparation, apparatus, assaying, and test filters are described

The experimental work was designed to determine
1. Whether the log penetration law could be applied to glass wool filter beds with random fiber distribution
2. The dominant mechanisms operating to collect bacterial aerosols in a fibrous medium similar to that employed in commercial air sterilization operations. To achieve these objectives it was decided to use test filters constructed from thin mats of glass wool. In the early phases of the work $^3/_{16}$-inch lengths of Pyrex wool fibers were slurried in a solution containing a carboxymethylcellulose thickener and settled on a 200-mesh screen to form the differential filter thicknesses. It was extremely difficult to fabricate reproducible mats in this way; therefore, an industrially prepared glass fiber mat was employed instead.

Owens-Corning Industrial Mat, a glass fiber material 0.02 inch thick and impregnated with a furfural resin binder, was the most satisfactory mat commercially available. By carefully selecting individual thicknesses from a roll of the mat, sufficiently uniform sections were obtained so that reproducible filters could be prepared.

Figure 1 shows a flow sheet of the experimental equipment. Basically, it includes a means for dispersing bacteria in the air, a holder for the test filter, and equipment for assaying the bacterial concentrations in the air streams. Filter efficiency was obtained from air stream assays while penetration, a measure of bacterial collection with depth, was determined by enumerating the number of organisms collected on individual filter layers. These data, along with air stream velocity and pressure drop across the filter provided the essential experimental results.

Aerosol Preparation. The bacterial aerosol was generated by spraying a buffered suspension of *B. subtilis* spores in an air stream. These spores were obtained by growing *B. subtilis* in trypticase soy broth (BBL) containing 5 micrograms of manganese per milliliter as $MnCl_2$ (4). When complete sporulation had occurred, the organisms were resuspended in a phosphate buffer solution (pH 7.0). An atomizer (C, Figure 1) taken from a Beckman flame spectrophotometer was used for generating the aerosol. This atomizer was similar to those employed in the work at Camp Detrick (26). By directing the spray against the atomization chamber wall, the quality of the bacterial aerosol could be controlled. It depended both on the velocity at which the spray impinged against the chamber walls and the velocity of air flowing past the spray. When properly adjusted, the larger aerosol particles, containing clumps of bacteria and water, were removed by impingement and only single spores and small water particles were carried away by the air stream.

To check homogeneity and particle size, the settling action of several samples was observed in a modified Millikan oil drop apparatus (D). Table II gives the results of 65 observations on various aerosol particles produced by the spray. The nature and extent of surface change was also determined on 34 of these particles.

Table II. Aerosol Particle Size and Charge

	Number of Observations	Particle Radius, Micron
Average particle radius	65	0.575
Average deviation		0.082
Maximum positive deviation		0.295
Maximum negative deviation		0.175
		Charge, Electron Units
Particle surface charges measured	34	
Negative charge	25	1 to 60
Positive charge	5	5 to 14
No charge	4	

Microscopic examination of the spore particles revealed them to be elipsoidal in shape, 1.0 by 1.2 microns in size; the results shown in Table II, therefore, indicate that the aerosol was composed of monodispersed spores. Not all the spores exhibited a negative charge.

Test Apparatus. Air for aerosol generation was obtained from a constant pressure (15 pounds per square inch gage) source. It was filtered through a 12-inch bed of glass wool and dried in a silica gel column to an average relative humidity of 15% at a temperature of 80° F. Dry air was essential for reproducible filtration results. Air having a high humidity or containing water droplets invariably gave erratic results; apparently wet air decreases the particle's ability to adhere to filter fibers. To allow for evaporation of water droplets formed during aerosol generation and for drying of the spore particles, the aerosol was passed through the settling chamber (E) and through a heat exchanger (F). From the heat exchanger the aerosol passed to a distribution chamber (C) from which portions of the flow were directed to various sections of the test apparatus.

Two telescoping lengths of brass tubing machined to a concentric fit formed the filter holder. Test filters were held in place by two brass rings clamped between the ends of the concentric tubes. By varying the hole size of the rings, it was possible to vary the air velocity through the filter while keeping the volume of air tested constant during the experimental runs. Conical adapter sections preceded and followed the filter holder section. For purposes of observing any electrostatic charges that might accumulate on the filter and its holder they were isolated electrically and connected to a gold leaf electroscope. No particular effect was noted.

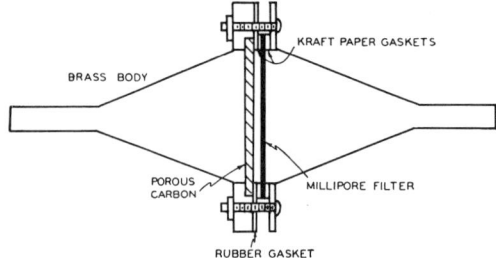

Figure 2. Millipore filter holder

Humidity measurements were obtained with a dew point meter and later with a psychrometer (H). Pressure drop across the test filter was measured with an inclined iso-octane manometer which could be read to 0.01 inch of water. During test runs exit air streams were bubbled through phenol solution (J) before being released to the atmosphere. At the conclusion of each run the atomizer was disassembled and cleaned, and the silica gel bed was regenerated. Before each run the entire system, including the new test filter, was sterilized by purging with ethylene oxide vapors for 3 hours.

Assay Methods. Millipore filters, marketed by the Lovell Chemical Co. of Cambridge, Mass., were used to assay bacterial concentrations in the air streams. Holders (Figure 2) for these filters consisted of two brass funnels bolted together through their lips. Each holder had a $1/4$-inch porous carbon disk backing to prevent rupture of the weak filters. Air seal on the holders was maintained by brown kraft paper gasketing. This material was used because electrostatic charges readily built up on the filters. It caused them to adhere to almost everything else, making their aseptic removal almost impossible.

Bacterial concentrations in the air streams were determined by passing a sample, sufficiently large for accurate measurement, through the Millipore assay filter. After sampling was completed each Millipore filter was removed from its holder with sterile forceps and placed in 100 ml. of sterile, distilled water. The filter was then disintegrated by beating for two minutes in a Waring Blendor. This treatment accomplished the separation of collected spores from filter material, and they could then be counted by the usual microbiological plating techniques. After a test run, the filter bed itself was similarly handled, each differential thickness being removed aseptically and disintegrated in the blendor. By counting the recovered organisms from each layer, penetration data was obtained.

The spore suspensions resulting from the Blendor treatment were diluted to the proper concentration to give an estimated bacterial count of 30 to 300 per ml. and plated out on nutrient agar. Since *B. subtilis* is a very motile organism, it was necessary to use four layers of agar in the plate; the suspension was included in the second layer from the bottom. This procedure minimized the surface growth of the spores and made counting easier and more accurate.

Special Test Filters. To determine the effect, if any, of fiber surface characteristics on the collection of aerosols, filter pads were compounded from similarly sized Pyrex wool, cotton, nylon, and Dacron fibers, 3 mm. in length. These were prepared by settling the fibers from a stabilized slurry to a 200-mesh screen. Characteristics of these filter pads are presented in Table III. Nylon and Dacron fibers gave quite uniform mats. The cotton mats were very dense, yet uniform, while Pyrex wool mats were generally nonuniform. The apparent reason for this nonuniformity was the fact that air bubbles were readily entrapped among the fibers leaving little voids after drying the mats.

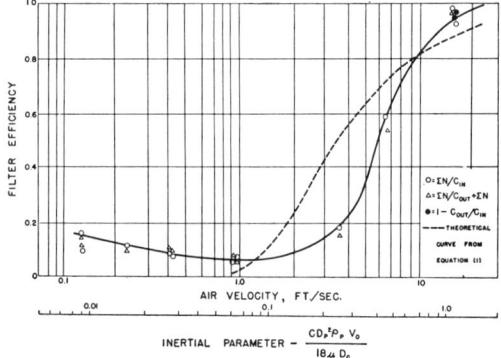

Figure 3. Velocity-efficiency relationships of IMF filters

Filter thickness = 0.12 inch

Velocity-efficiency relationships are established

The effect of air stream velocity on the filtration efficiency of filters compounded from six thicknesses (0.12 inch) of the IMF mats is shown in Figure 3. Note that the efficiency of the filters,

Table III. Physical Data of Test Filter Pads

Code Names	Material	Source	Binder	Pad Weight, Gram	Fiber Density, Grams/Cc.	Av. Fiber Diameter of 50 Samples, Microns	Variation in Fiber Diameter Microns
IMP	Industrial Mat, glass fibers	Owens-Corning	Polyethylene	0.180	2.54	13.8	±4.0
IMS	Industrial Mat, glass fibers	Owens-Corning	Starch	0.180	2.54	13.8	±4.0
IMF	Industrial Mat, glass fibers	Owens-Corning	Furfural	0.271	2.54	16.0	±3.5
C	Cotton	Johnson & Johnson	None	0.556	1.50	13.9	±5.5
GW	Pyrex wool	Owens-Corning	None	0.272	2.54	7.5	±0.7
N	Nylon	Du Pont	None	0.300	1.15	11.6	±1.3
D	Dacron	Du Pont	None	0.300	1.24	11.2	±1.7

determined from air stream and test filter assays, can be expressed by any one of the forms

Efficiency = $\Sigma_N/C_{in} = \Sigma_N/(C_{out} + \Sigma_N) = 1 - (C_{out}/C_{in})$ (4)

Σ_N is the total number of spores collected on all thicknesses of the test filter, C_{in} is the number in the entering air stream, and C_{out} is the number in the exit air stream, all based on the total volume of air filtered. For purposes of illustrating the inertial action, the efficiencies are also plotted against the inertial parameter and compared to the theoretical curve which would be predicted if only inertial impaction were involved. This theoretical curve is obtained from the single fiber data of Langmuir and Blodgett (19), for the case where potential flow is assumed around the fiber. When these single fiber efficiencies are incorporated in Equation 1, the over-all filter efficiency may be computed. In using Equation 1 the assumption of uniform fiber distribution was made; of course this is not quite true. Furthermore, the observed filter density of 2.29 pounds per cubic foot was assumed to be the actual density of the bed.

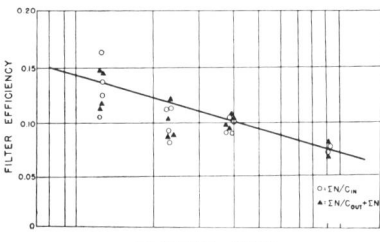

Figure 4. Efficiency data for low velocity region

Figure 3 indicates that there are two velocity regions which exhibit different effects; a high velocity region, where efficiency increases with velocity, and a low velocity region where efficiency decreases with velocity. Comparison to the theoretical curve leaves little doubt but that this high velocity effect results from the dominance of the inertial mechanism. Its effects disappear close to an inertial parameter value of $1/16$, that predicted by Langmuir and Blodgett. The low velocity action could be due to a settling, diffusional, or electrostatic action. It could not be due to interception, because the collection efficiency due to this mechanism increases with velocity. Furthermore, interception contributes little to the collection even in the high velocity region. Here the Reynolds numbers for the flow around the filter fibers range from 0.1 to 1.0. Also in these tests the ratio of the particle to collector was less than 0.07. For such conditions the Davies theory would predict almost no interception action.

Low Velocity Region. Details of the data taken in the low velocity region are shown in Figure 4. From this curve it is evident that there truly exists an inverse velocity effect. Although there is considerable scattering of the data, the upward trend is quite definite. This action is due to settling, diffusion,

or electrostatic attraction. The diffusional mechanism can be eliminated immediately. Observations of the spore particle in the Millikan chamber have shown that the lateral motion as the *B. subtilis* particle falls is considerably less than the settling motion, because it falls in nearly a straight line. Therefore, this low velocity action results either from gravitational or electrostatic forces or a combination of both.

Industrial Mat furfural filter is most practical of different filter mediums compared

As a possible means for evaluating electrostatic surface effects, the collection efficiencies of several different fibrous media were investigated. These results are reported in Figure 5 for 0.045 inch of the various Industrial Mat filters and in Table IV for the filters compounded by the investigators. Results for the Industrial Mat furfural filter are also reported in Table IV for comparison purposes.

The primary difference among the Industrial Mat filters reported in Figure 5 is the bonding agent. The starch bonded (IMS) filter gave lower efficiencies in the low velocity region than either the furfural (IMF) or polyethylene (IMP) bonded filters. The difference found among all filters is within the experimental error. Although one would like to draw the conclusion that this small effect may be an electrostatic action, it is impossible to justify such a conclusion with limited data. However, the data of Figure 5 because of their similarity certainly justify the use of the inertial parameter for characterizing filter efficiencies at high velocities.

The fibers of the filters whose tests are reported in Table IV are of the same approximate size (see also Table III). As a consequence any radical differences in filter operation at low velocities, where electrostatic action is most effective, could be attributed to differences in surface charges of the fibers. In order to compare these filters, it was necessary to compound them so that each had the same total volume of fibrous material. This is equivalent to the same number of fibers—i.e., collectors—in each pad if all the fibers have the same length.

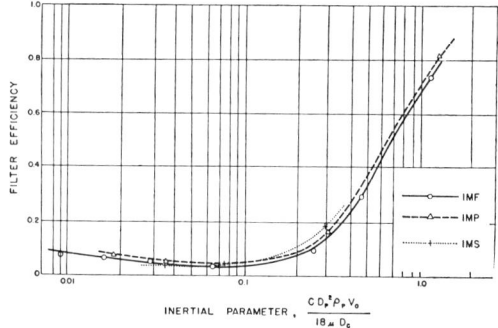

Figure 5. Comparison of various glass mat filters

Filter thickness = 0.045 inch

Since none of the filters were similar in density or porosity, it was decided to correlate the efficiencies with pressure drop across the filter. Actually the efficiency per unit of pressure drop is a measure of utility or practicality of a given filter. It is also proportional to the filter thickness. All filters reported in Table IV were compounded from six differential pads of each material and tested at a superficial air velocity of 0.421 foot per second.

From these data the utility is apparently about the same for all filters except the one compounded from furfural pads. The Industrial Mat furfural filter is evidently more practical in its

Table IV. Comparison of Efficiency for Different Filter Mediums

Material	IMF	C	GW	D	N
Efficiency of 6-pad filter, $1 - (C_{out}/C_{in})$	0.104	0.594	0.274	0.333	0.375
Pressure drop across filter; inch of water	0.05	0.75	0.25	0.283	0.361
Pressure drop across a sufficient amount of filter to remove 90% of spores, inches of water	1.05	1.92	1.79	1.62	1.75
Utility, efficiency/pressure drop	0.96	0.52	0.56	0.62	0.57

Air Sterilization by Fibrous Media

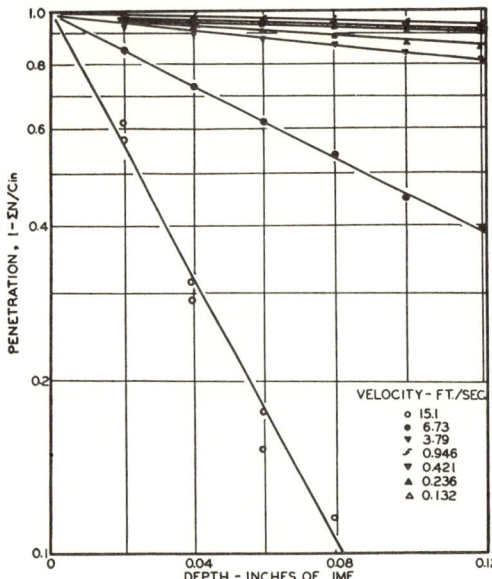

Figure 6. Penetration of spores in IMF filters

Table V. Individual Pad Efficiencies

	Spore Fraction Retained from Air Entering	
No. Pads/Filter	First pad	Last pad
6	0.376	0.403
12	0.441	0.384
25	0.448	0.610

thicker filters (to 25-pad thickness or 0.5 inch) were used to accentuate the penetration variation.

In these thicker filters a greater number of organisms were retained in the end sections than predicted by the log penetration law. Careful checking of the curves show that the amount retained was less than $1/_{100}$th of the total number and could be attributed to the collection of organisms from the air leaking around the filter edges. This edge or radial leakage could not be avoided in the experimental filter design because of the problem of removing and analyzing the differential filter thicknesses. To support the view that leakage around the edges caused these erratic values, it can be shown that essentially the same fraction of aerosol particles are retained from the air streams passing through the first and the last differential thicknesses where radial and edge leakage would not be encountered. Table V gives a breakdown of individual efficiencies for the first and last pads (0.02-inch thicknesses) of each test filter.

In all cases the fraction of particles retained from the passing air stream by each pad is nearly the same. The high value on the last pad of the 25-pad filter was due to difficulties in enumeration; the low concentration of spores occurring on this pad approached the limit of detection by plate count. As a matter of interest,

operation even though it is less efficient per volume of fiber than the other filters. However, the Industrial Mat furfural filter differed from the others in several important features. It had a resinous binder, its fibers were more uniformly distributed; they were fixed in place; and the filter was considerably less dense.

The only generalization obtainable from the results seems to be a relation of utility to bulk density of the filters. C and GW filters, the denser ones, had the lowest utility while IMF and D filters, the less dense, had the highest utility. Also, the fibers of D and N filters and particularly those of the IMF filter were more uniformly distributed. The more uniform the distribution of fibers, the more effective, on a pressure drop basis, are individual fibers in their collection action. Unfortunately any electrostatic effects seem to be masked by these density and uniformity factors. Some of the utility and hence collection action of the IMF filters can probably be attributed to its resinous binder.

Minimal Filter Efficiency. Because of the two inverse velocity effects, there existed a velocity for each of the Industrial Mat filters at which the penetration of *B. subtilis* spores was a maximum and the efficiency was a minimum. This type of action has been reported previously by other investigators including La Mer (*16*). La Mer reported that a maximum penetration of the Chemical Corps No. 5 (CC-5) filter paper by dioctyl phthalate aerosol particles smaller than 0.66 micron in diameter occurred at superficial air velocities of 13 cm. per second (0.043 foot per second).

This minimal filter efficiency of approximately 0.06 is less than that of 0.33 predicted for direct interception only from a potential flow model and greater than that of 0.016 predicted from the Ranz (*23*) viscous flow model.

Penetration. Figure 6 shows penetration of spores to different depths of the test IMF filter. The parameter is the superficial air velocity (feet per second) through the filter. As anticipated the penetration is logarithmic. Figure 6 is only a different presentation of the data given earlier in Figure 3, all obtained for shallow filter beds with fairly high penetrations. It seemed desirable to check the penetration of deeper filters with considerably greater efficiencies. Figure 7 presents the results of runs where the highest air velocity (15.1 feet per second) and

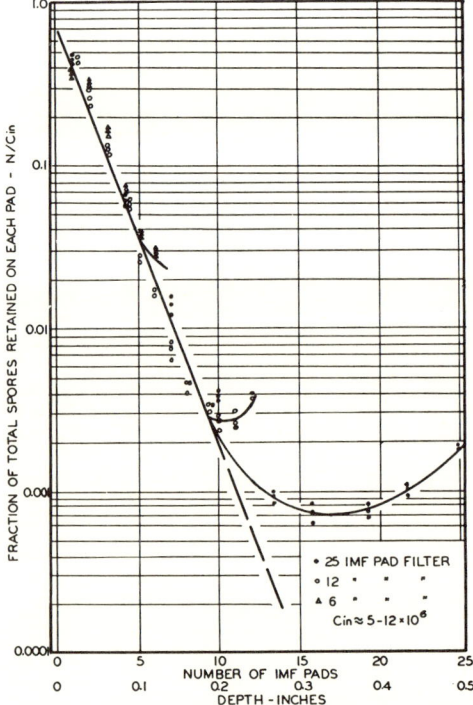

Figure 7. Retention of spores on each differential filter pad

logarithmic penetration patterns were found in the filter tests reported in Figure 5 and Table IV.

Conclusions

These experiments for filter design are applicable only to homogeneous bacterial aerosols with particles similar in nature and size to *B. subtilis* spores. The majority of microorganisms in air do not exist as single cells but rather as clumps of cells adhering to other organic matter. It is conceivable that the collection of such a heterogeneous aerosol might have different implications than considered here. *B. subtilis* spores do, however, represent about the smallest airborne contaminants detrimental to commercial fermentation and food packaging operations, with the exception of phages and viruses. Consequently, the results can be used to determine the minimum bacterial collection of a filter and so have much wider application than the limitations just cited might suggest.

With these provisos in mind it can be concluded that

1. Penetration of single bacterial cells through a fibrous filter is logarithmic; that a logarithmic expression can be used for filter design. It has already been noted that any edge leakage will destroy this penetration pattern.
2. Inertial mechanism operates in the manner predicted by Langmuir and Blodgett with their potential flow model.
3. Since inertial parameter does depict the collection action at high velocities, filter efficiency will increase as fiber diameter decreases in this region.
4. Interception effect is not so great as predicted by the potential flow model, because at low velocities potential flow is a poor assumption; rather the viscous flow model should be used.
5. In low velocity regions, where inertial impaction is negligible, some inverse velocity effect is important to the collection action. It is probably an electrostatic or a settling action.
6. Two opposing velocity effects observed give rise to a minimal filter efficiency at some intermediate velocity. For fibrous mediums having fibers 16 microns in diameter, the superficial velocity that results in the minimum collection efficiency for *B. subtilis* spores is in the range 1 to 2 feet per second.
7. Bulk density and fiber distribution in filter beds are more important to the collection action than is the kind of fibrous mediums used.

In general more work must be done before the full nature of impaction is understood. The question of which mechanism is most important at low velocities has not been answered fully. Furthermore, the observations that wet air streams give inconsistent results certainly points to a relationship between humidity and some important filtration factor not considered in this experimental work.

From personal observations it would seem that a great deal more consideration should be given to the mechanical problems of compounding uniform filters which avoid the difficulties of edge leakage and which are capable of withstanding repeated steam sterilization. These factors are certainly as important as the physical problem of collection. The complexities of air filtration are many. Some day the various mechanisms of impaction will be more fully understood so that filters can be designed from basic relationships rather than from extensive and costly test programs.

Nomenclature

C' = empirical correction factor for fluid resistance to movement of small particles same order of magnitude as mean path of fluid molecules
C_{in} = concentration of bacteria in air stream entering filter
C_{out} = concentration of bacteria in air stream leaving filter
D_c = average diameter of filter fibers
$(D_c)_s$ = surface average diameter of filter fibers
D_p = diameter of bacterial aerosol particles
g = local acceleration of gravity
k' = constant pertaining to a particular filter medium
N = number of bacteria retained by each thickness or layer of filter

ΣN = sum of bacterial concentrations retained by thicknesses of filter under consideration
p = porosity of filter
q_{ac} = electrical charge/unit area of fiber surface
q_p = electrical charge on bacterial particle
V_0 = superficial air stream velocity through filter
X = filter thickness
η = single fiber efficiency
μ = viscosity of air
α = ρ_b/ρ_c
ρ_p = density of bacterial particle
ρ_b = bulk density of filter bed
ρ_c = density of filter fiber material
ϵ_p = dielectric constant of bacterial particle
ϵ_a = dielectric constant of air
ϵ_0 = permittivity of free space

Literature cited

(1) Albrecht, F., *Physik. Z.*, **32**, 48 (1931).
(2) Bourdillon, A. G., "Studies in Air Hygiene," Medical Research Council, Special Rept. **262**, Her Majesty's Stationery Offices, London (1951).
(3) Bristol Laboratories, Inc., Syracuse, N. Y., "Summary of Report Showing Efficiency of Cambridge Absolute Filter in Filtering Bacteria from Air," 1951.
(4) Charney, J., Fisher, W. P., and Hegarty, C. P., *J. Bacteriol.*, **62**, 145 (1951).
(5) Chen, C. Y., "Filtration of Aerosols by Fibrous Media" (to be published).
(6) Cherry, G. B., McCann, E. P., and Parker, A., *J. Appl. Chem.*, **1**, Suppl. 2, S103 (1951).
(7) Davies, C. N., *Proc. Inst. Mechn. Engrs.*, London, **B1**, 185 (1952).
(8) Davies, C. N., *Proc. Phys. Soc.*, London, **B63**, 268 (1950).
(9) Decker, H. M., Geile, F. A., Moorman, H. E., and Glick, C. A., *Heating, Piping, Air Conditioning*, **23**, 125 (1951).
(10) Finn, R. K., Bio-Engineering Symposium, Rose Polytechnic Institute, Terre Haute, Ind., May 23, 1953.
(11) Goetz, A., *Am. J. Public Health*, **43**, 150 (1953).
(12) Hopper, S. H., *J. Am. Pharm. Assoc.*, **39**, 291 (1950).
(13) Humphrey, A. E., Doctoral Dissertation, Columbia University, 1953.
(14) Johnstone, H. F., and Roberts, M. H., IND. ENG. CHEM., **41**, 2417 (1949).
(15) Kluyver, A. J., and Visser, J., *Antonie van Leeuwenhoek J. Microbiol. Serol.*, **16**, 311 (1950).
(16) La Mer, V. K., "Studies on Filtration of Monodispersed Aerosols," Columbia University, Final Report NYO-512, AEC Contract AT(30-1)-651 (1951).
(17) Landhal, H. D., and Herrman, R. G., *J. Colloid Sci.*, **4**, 103 (1949).
(18) Langmuir, I., "Filtration of Aerosols and the Development of Filter Materials," OSRD 865 (1942).
(19) Langmuir, I., and Blodgett, K., "Smoke and Filters," OSRD 3460 (1944).
(20) McDaniel, L. E., and Long, R. A., *Applied Microbiol.*, **2**, 240 (1954).
(21) Murphy, H. C., *Heating, Piping, Air Conditioning*, **25**, 108 (1953).
(22) Ramskill, E. A., and Anderson, W. L., *J. Colloid Sci.*, **6**, 416 (1951).
(23) Ranz, W. W., "Role of Particle Diffusion and Interception in Aerosol Filtration," University of Illinois, Engineering Experiment Station, Technical Report 8, AEC Contract No. AT(30-3)-28, Serial No. SO-1009 (1953).
(24) Ranz, W. E., and Wong, J. B., IND. ENG. CHEM., **44**, 1371 (1952).
(25) Rodebush, W. E, Handbook on Aerosols, Atomic Energy Commission, Washington 25, D. C., 1950.
(26) Rosebury, T., "Experimental Airborne Infection," The Williams & Wilkins Co., Baltimore, 1947.
(27) Sell, W., *Forsch. Gebiete Ingenieurw.*, **2**, *Forschungsheft*, 347 (1931).
(28) Terjesen, S. G., and Cherry, G. B., *Trans. Inst. Chem. Engrs.*, (London), **25**, 89 (1947).
(29) Thomas, D. J., *J. Inst. Heating Ventilating Engrs.* (London), **20**, 35 (1952).
(30) Wong, J. B., and Johnstone, H. F., "Collection of Aerosols by Fiber Mats," University of Illinois, Engineering Experiment Station, Technical Report 11, AEC Contract AT(30-3)-28 (1953) (to be published).

RECEIVED for review October 2, 1954. ACCEPTED January 28, 1955. Contribution No. 42, Chemical Engineering Laboratories, Engineering Center, Columbia University.

Part VII

OXYGEN TRANSFER: AERATION AND AGITATION

Editor's Comments
on Papers 16 and 17

16 BARTHOLOMEW et al.
Oxygen Transfer and Agitation in Submerged Fermentations

17 MAXON
Aeration-Agitation Studies on the Novobiocin Fermentation

Treating aeration and agitation together and offering only two papers in this group should not be interpreted as a failure to recognize the importance of these aspects of fermentation conduction. On the contrary, the importance of proper measurement and control of oxygen mass transfer, mixing, and their interactions in characterizing fermentation processes, translating findings from one scale to another, and improving fermentor productivity as well as reducing raw material consumption cannot be overemphasized.

Paper 16 and a companion paper by Bartholomew et al. (1950) were among the first extensive quantitative aeration-agitation studies with biological systems reported. Appearing simultaneously was a paper that had considerable theoretical depth but discussed a less complex system (yeast) (Hixson and Gaden 1950). Prior to the studies by Bartholomew et al. (1950), Cooper, Fernstrom, and Miller (1944) had described the sulfite oxidation method for studying air oxidation, a method employed by Bartholomew and many others who followed. Another nonbiological study of lasting impact was that reported by Rushton, Costich, and Everett (1950). A large number of studies on aeration and agitation appeared in the literature in the 1950s. Paper 17 was selected because it was outstanding in that decade and in the 1960s.

Several papers on mixing that measured only physical forces and effects but contributed to advancing the design of fermentors and fermentation processes were written by Rushton and Oldshue (1959), Calderbank and Moo-Young (1959), Metzner and Taylor (1960), and Deindoerfer and Humphrey (1961).

The fundamental problem of mass transfer of oxygen within a mycelial pellet was first attacked seriously by Yano (1961) and most recently by Bhavaraju and Blanch (1975).

Two excellent reviews of aeration-agitation theory and design of systems are those by Gaden (1968) and Wang and Humphrey (1968).

REFERENCES

Bartholomew, W. H., E. O. Karow, M. R. Sfat, and R. H. Wilhelm. 1950. Effect of air flow and agitation rates upon fermentation of *Penicillium chyrsogenum* and *Streptomyces griseus*. *Ind. Eng. Chem.*, **42**:1810–1816.

Bhavaraju, S. M., and H. W. Blanch. 1975. Mass transfer in mycelial pellets. *J. Ferment. Technol.*, **53**:413–415.

Calderbank, P. H., and M. B. Moo-Young. 1959. The prediction of power consumption in the agitation of non-Newtonian fluids. *Trans. Inst. Chem. Engrs.* (London), **37**:26–33.

Cooper, C. M., G. A. Fernstrom, and S. A. Miller. 1944. Performance of agitated gas-liquid contactors. *Ind. Eng. Chem.*, **36**:504–509.

Deindoerfer, F. H., and A. E. Humphrey. 1961. Mass transfer from individual gas bubbles. *Ind. Eng. Chem.*, **53**:755–759.

Gaden, E. L., Jr. 1968. Aeration and agitation in fermentation. *Sci. Repts. Inst. Super. Sanita,* **1**:161–176.

Hixson, A. W., and E. L. Gaden, Jr. 1950. Oxygen transfer in submerged fermentation. *Ind. Eng. Chem.*, **42**:1792–1800.

Metzner, A. B., and J. S. Taylor. 1960. Flow patterns in agitated vessels. *A I CH E Journal,* 109–114.

Rushton, J. H., and J. Y. Oldshue. 1959. Mixing of liquids. *Chem. Eng. Prog.*, Sym. Ser., **55**:181–198.

———, E. W. Costich, and H. J. Everett. 1950. Power characteristics of mixing impellers. Parts I and II. *Chem. Eng. Progr.,* **46**:395–404 and 467–476.

Wang, D. I. C., and A. E. Humphrey. 1968. Developments in agitation and aeration of fermentation systems. *Progr. Ind. Microbiol.,* **8**:1–34.

Yano, T., T. Kadana, and K. Yamada. 1961. Fundamental studies on the aerobic fermentation. Part VIII. Oxygen transfer within a mold pellet. *Agr. Biol. Chem.*, **25**:580–584.

Oxygen Transfer and Agitation in Submerged Fermentations

MASS TRANSFER OF OXYGEN IN SUBMERGED FERMENTATION OF *Streptomyces griseus*

W. H. BARTHOLOMEW, E. O. KAROW, AND M. R. SFAT
Merck & Co., Inc., Rahway, N. J.,
R. H. WILHELM, *Princeton University, Princeton, N. J.*

A theory of oxygen absorption by suspended mycelia in aerated nutrient broth is proposed. Diffusion mechanism steps include an oxygen transfer resistance at bubbles and other air-liquid interfaces, a resistance through cell clumps and liquid films around individual cells, and a mechanism which involves direct contact between cells and air bubbles. Direct and indirect experimental evidence is presented in support of the mechanisms. Air supply and agitation rate affect oxygen transfer resistances in various ways and determine whether the local oxygen concentrations at the cells lead to a state of oxygen saturation or deficiency.

OXYGEN transfer and agitation are important in maintaining a desirable environment for mycelial growth and antibiotic synthesis in submerged cultures. The design of deep tank fermentors and the establishment of optimum fermentation conditions depend upon proper choice and control of these variables. The problem of supplying adequate oxygen to fermentors arises because of the limited solubility of this element in water. A program was started several years ago in this laboratory to explore, measure, and analyze fermentation variables such as aeration and agitation from a biological-engineering point of view. The work was undertaken with the object of acquiring increased understanding of mechanisms by attempting to distinguish between and study separately the physical and biological rate variables involved in oxygen transfer. A background was thereby to be provided for logical procedures of translation from one scale of fermentation to another. Strains of *Penicillium chrysogenum* and *Streptomyces griseus* were the organisms used, but the techniques that were developed and conclusions that have been reached may also be of interest in other fermentations.

An initial step in the program was to develop a versatile, laboratory scale fermentor, which in multiple units would provide a rapid and economic means of investigating fermentation variables. Such a bench scale type of fermentor was constructed with a capacity of 5 liters (*2*). Fermentations in the unit were found to be consistent with pilot plant and factory fermentations. The concepts and techniques set forth in this and the succeeding paper (*3*) have been used with success in translation of results among the three scales of operation. The laboratory fermentor was found to reduce the extent of research and development effort necessary on the pilot plant scale.

The present paper is concerned with the transfer of oxygen from air to the submerged mycelium. Component rate steps and their interrelations were studied in a quantitative manner. Experiments were limited to studies with *Streptomyces griseus* and with uninoculated mediums. A second paper (*3*) deals with the effect of aeration and agitation upon mycelial growth, sugar utilization, and biosynthetic formation of penicillin and streptomycin.

The literature (*1, 7, 11, 14, 16, 17, 23*) dealing with oxygen transfer and agitation in various fermentation systems is extensive, but the number of quantitative studies in the field is limited. The aeration of water alone by means of spargers in agitated vessels has been studied by Cooper, Fernstrom, and Miller (*10*). Becze and Liebmann (*8*) reviewed the literature on aeration in compressed yeast manufacture. Studies in aeration and agitation in penicillin fermentation have recently been reported by Brown and Peterson (*9*), and in yeast fermentation by Olson and Johnson (*25*).

THEORY OF OXYGEN ABSORPTION BY SUSPENDED MYCELIA IN AERATED NUTRIENT BROTH

Oxygen for mycelial growth and biosynthesis in industrial fermentations frequently is supplied by diffusion from air bubbles suspended and rising through the broth. The air usually is introduced through spargers near the bottom of the fermentor. Mechanical agitation is supplied by rotating propellers, turbine impellers, and similar devices. There is present a three-phase system of liquid, gas, and suspended solids.

As a suggested theory, oxygen transfer from gas to organism may be divided into a sequence of primary steps, which are illustrated in a schematic manner in Figure 1. Within a bubble or at any other air-broth interface there exists a gas film through which oxygen must diffuse. On passing through the interface, a liquid film resistance is encountered at the surface. A liquid phase resistance between the gas-liquid interface and the vicinity of the cells is postulated to follow. This liquid phase diffusion resistance becomes particularly important when cells are entangled or clumped to form structures or temporary groupings. Diffusion must take place to the centers of such aggregates. Finally, a liquid film diffusion resistance around each cell must be traversed before the cell is entered. The ultimate rate of oxygen passage into the cell is a characteristic of the cell in its immediate environment. The preceding diffusion steps are assumed to be in series in the sense that oxygen is not utilized in parallel processes such as direct oxidation of broth constituents.

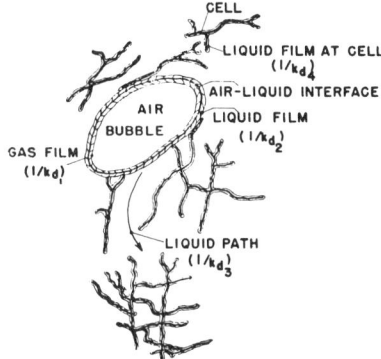

Figure 1. Schematic Representation of Possible Diffusional Resistance External to Cells

The appropriately numbered diffusion resistances in Figure 1 are indicated as reciprocals of k's, the values of the different step conductances. An over-all resistance, $1/k_{d_0}$, between gas bubble and cell wall may be written as the sum of the individual resistances as follows:

$$1/k_{d_0} = 1/k_{d_1} + 1/k_{d_2} + 1/k_{d_3} + 1/k_{d_4} \quad (1)$$

where
k_{d_1} = gas film conductance at air-liquid interface
k_{d_2} = liquid film conductance at air-liquid interface
k_{d_3} = liquid path conductance
k_{d_4} = liquid film conductance adjacent to cell surface

It is a characteristic of steps in series that if any conductance is markedly larger than other conductances, the mechanism step may be neglected in comparison with the others. For example, it will be shown experimentally that the gas film resistance, $1/k_{d_1}$, is unimportant compared to the liquid film resistance. The succeeding discussion will therefore not be burdened by carrying along $1/k_{d_1}$.

The transfer of oxygen from air to cell is complex in detail, with series and parallel paths simultaneously operative. It is convenient to distinguish three combinations of the individual mechanism steps as components of the total transfer process. To varying degrees, the properties of these rate steps are experimentally separable.

The resistances set forth in Equation 1 are divided into two groups, characterized by the conductance symbols, k_d and k'_d. The corresponding resistances are defined as follows:

$$1/k_d = 1/k_{d_2} \quad (2)$$

$$1/k'_d = 1/k_{d_3} + 1/k_{d_4} \quad (3)$$

The diffusion resistance associated with gas bubbles or other interfaces is represented as $1/k_d$. The diffusion resistance constant for oxygen transfer through cell aggregates and through cell wall liquid films is $1/k'_d$.

The rate equation for diffusion in which these constants appear is as follows:

$$r_d = k_d(p - p^*) = k'_d(p^* - p_c) \quad (4)$$

where
r_d = differential rate of oxygen absorption, gram moles of O_2/(ml. of soln.)(hour)
k_d = diffusion rate coefficient over partial pressure interval, $(p - p^*)$, gram moles of O_2/(ml. of soln.)(hour)(atm.)
k'_d = diffusion rate coefficient over partial pressure interval, $(p^* - p_c)$
p = partial pressure of oxygen in gas stream, atm.
p^* = partial pressure of oxygen in equilibrium with oxygen in main body of solution (C^*) at any moment, atm.
p_c = partial pressure of oxygen in equilibrium with oxygen in solution at mycelial cell surface, atm.
C^* = average oxygen concentration in main body of broth, gram moles of O_2 per ml. of solution

In addition to the oxygen diffusion path between gas and cell wall, represented by Equation 4, there is evidence of a shorter parallel path between the same concentration limits by direct contact between the cells and the gas-liquid interface. The conductance constant for this arrangement is designated as k''_d. Direct contact eliminates a path through cell clumps and because some cells are adsorbed at the interface, a single liquid surface film is assumed to be formed at the interface and k_{d_2}, k_{d_4} become merged. Thus k''_d is defined as

$$1/k''_d \cong 1/Ak_{d_2}$$

where A = fraction area of contact between cells and interface
The diffusion rate equation for direct contact is:

$$r_d = k''_d (p - p_c) \quad (5)$$

Turning from diffusion processes, a second type of rate process of importance concerns the oxygen absorption rate of the organism. Except at the lowest partial pressures of oxygen, this rate ordinarily is constant. Thus,

$$r_r = k_r C_m \quad (6)$$

where
r_r = differential rate of oxygen absorption, gram moles of O_2/(ml. of soln.)(hour)
k_r = specific uptake rate of oxygen by the organism, gram moles of O_2/(grams of mycelium)(hour)
C_m = concentration of mycelium at any time, grams of mycelium per ml. of solution

The value of k_r is constant for short time intervals in the total time of fermentation, but it changes as age of cells and conditions in the growth cycle vary. Normal biological processes take place when p_c, the cell surface oxygen partial pressure, is in the range in which the oxygen uptake rate is independent of oxygen concentration. When p_c falls to low values, the uptake rate becomes concentration-dependent, and growth and biosynthesis processes are unfavorably affected.

A final equation expresses the fact that in the normal, slow mycelial growth cycle the system, in any short time interval, may be considered in steady-state condition or dynamic balance. It is:

$$r_r = r_d \quad (7)$$

During fermentative growth, diffusion and cell absorption rates for oxygen are related. For deep tank fermentors, the interrelation of variables may be shown by combining Equations 4, 6, and 7 and integrating. The resulting solution is as follows:

$$k_r = \frac{l2GH(C_i - C^*)}{PC_m V\left(\dfrac{2G}{Vk_dP} + 1\right)} \quad (8)$$

where
- H = Henry's law constant, (atm.) (ml. of soln.)/gram moles of O_2
- C_i = saturation concentration of oxygen in the liquid, gram moles of O_2/ml. of solution ($p_i = HC_i$)
- G = total gram moles of gas entering fermentor per hour (assumed constant)
- V = volume of fermentor, ml.
- P = total gas pressure, average, atm.

Rewriting Equation 4 as the log mean expression between p_i and p_0,

$$r_d = \frac{k_d(p_i - p_0)}{\ln\frac{(p_i - p^*)}{(p_0 - p^*)}} \tag{A}$$

p_i = partial pressure of oxygen in gas to fermentor, atm. If air is used, $p_i = 0.21$
p_0 = partial pressure of oxygen in gas leaving fermentor, atm.

From a material balance around the fermentor,

$$r_d = \frac{G(p_i - p_0)}{PV} \tag{B}$$

Combining Equations A, 6, and 7

$$k_r C_m = \frac{k_d(p_i - p_0)}{\ln\frac{(p_i - p^*)}{(p_0 - p^*)}} \tag{C}$$

Because it is not convenient to measure the exit oxygen concentration, p_0, this variable may be eliminated mathematically.

Combine Equations B, 6, and 7 and solve for p_0,

$$p_0 = p_i - \frac{V k_r C_m P}{G} \tag{D}$$

Eliminate p_0 between Equations C and D

$$\frac{PVk_d}{G} = \ln \frac{(p_i - p^*)}{\left(p_i - p^* - \frac{PVk_rC_m}{G}\right)} \tag{E}$$

Through Henry's law, $p = HC$, convert Equation E from partial pressures to oxygen concentrations

$$\frac{PVk_d}{G} = \ln \frac{(C_i - C^*)}{\left(C_i - C^* - \frac{PVk_rC_m}{GH}\right)} \tag{F}$$

It is convenient under certain conditions to write Equation F in terms of arithmetical rather than logarithmic mean concentration differences. This may be done by writing Equation A in terms of arithmetical differences:

$$r_d = k_d \frac{(p_i - p^*) + (p_0 - p^*)}{2} = \frac{k_d}{2}(p_i + p_0 - 2p^*) \tag{G}$$

Combining Equations G, 6, and 7

$$k_r C_m = \frac{k_d}{2}(p_i + p_0 - 2p^*) \tag{H}$$

Eliminating p_0 between Equations D and H, the alternate to Equation F is obtained:

$$k_r = \frac{2GH(C_i - C^*)}{PC_m V\left(\frac{2G}{Vk_dP} + 1\right)} \tag{8}$$

The above equation takes into account variation of oxygen concentration in air as it passes through a fermentor. In shake flask operation it may be assumed that liquid is exposed to air having a uniform oxygen partial pressure at all points in the interface. For this case, the working equation is:

$$k_r = \frac{k_d H(C_i - C^*)}{C_m} \tag{9}$$

When the group $(2G/Vk_dP)$ is large compared to unity, which frequently is the case in tank fermentors, Equation 8 becomes identical with Equation 9.

It is of interest to consider certain aspects of the oxygen absorption theory outlined above. The diffusion rate constant, k_d, for the gas-liquid interface is capable of direct experimental determination through measurement of the absorption rate, r_d, and the oxygen partial pressures in gas and liquid phases, p and p^*, respectively. The measurement may be independently performed in a fermentation medium in the absence of cells. The rate constant is proportional to the interfacial area per unit volume of broth and so will depend upon size and number of bubbles per volume of liquid. The constant is a convenient measure of the capacity of fermentation equipment for absorbing oxygen.

The diffusion rate constant for oxygen transport through cell clumps and films, k_d', is not directly determinable because p_e at cell walls is not ordinarily measured. The constant probably represents a diffusion barrier not only for oxygen, but also nutrient components which enter cells and for products which leave them, particularly those present in small concentrations. Fully populated broths have been found to be non-Newtonian fluids with plastic yield values indicative of physical interlacing and structure formation by the cells. Agitation to break down cell clumps is presumed to increase the value of k_d' favorably. The constant is important because it, together with k_d, determines whether the oxygen concentration at cell walls, p_e, lies above or below an approximate value that distinguishes between independence and dependence in the oxygen uptake rate of the cells. The mechanism step, represented by the constant, serves as an indirect means of explaining the effects of agitation on mycelial growth and biosynthesis (3).

Although the cell oxygen concentration, p_e, may under certain conditions be at a low and limiting value, one may anticipate that the normally measured oxygen concentration in solution, p^* (or C^*), on the other side of the diffusion barrier ($1/k_d'$) may well be at a substantially high value. Thus, if one bases conclusions upon concentration of oxygen in solution, the possibility exists of a misinterpretation as to the condition of oxygen saturation or deficiency in the system.

The two processes discussed above form one type of series diffusion path. Direct interface-mycelial contact with conductance, k_d'', is an alternate path which depends upon the probabilities of contact. Analogous to molecular or turbulent motion, the number of contacts will depend upon the mean length of path of bubbles in a fermentor and mean relative velocity between bubbles and mycelia. This conductance path is favored by a large population of small bubbles having large interfacial area, coupled with thorough mixing to give relative motion between bubbles and cells. In shake flask operation, it is likely that aeration is predominantly through this mechanism at the continuously reformed interface.

Equations 8 and 9 show the interrelation of rate variables over short time intervals of the growth cycle. If p_e at the cell wall is such that k_r is independent of oxygen concentration, then variation in other variables in the equations leads primarily to changes in the concentration of dissolved oxygen, C^*. Strictly speaking, C^* should be defined as the oxygen concentration in liquid exterior to clumps and adjacent to bubble interfaces. Practically, it is taken as the mixed average oxygen concentration in a fermentation sample in which the cells are destroyed rapidly. The equations serve as a convenient means of checking the validity of assumptions in the oxygen absorption theory. Used in a reverse manner, they give one means of determining the oxygen absorption constant, k_d, in large fermentors.

EXPERIMENTAL

For purposes of checking Equations 8 and 9, several well aerated and agitated fermentation experiments were performed in 5-liter fermentors and shake flasks. The 5-liter fermentors were fully baffled, equipped with impeller agitators, and provided with spargers. The shake flasks were 250-ml. Erlenmeyer flasks on rotary shakers turning at 220 r.p.m. on a 1.5-inch (3.75-cm.) diameter throw.

Experiments were performed with two strains of *Streptomyces griseus*, Waksman No. 1 and a mutant strain developed in these laboratories. Mediums used were similar to the soybean medium described by Dulaney *et al.* (*12*) and the meat extract medium of Waksman (*31*). The temperature of fermentation was 27° C. and the total pressure was constant at 1 atmosphere.

Mycelial weight, dissolved oxygen and oxygen saturation concentrations, and air flow rate were measured at time intervals. The specific oxygen uptake constant, k_r, of the organism and the interface diffusion coefficient, k_d, of the medium were also determined.

The cell dry weights were determined gravimetrically. Cell weights were obtained by a differential centrifugation technique when other insoluble solids were present. Details of the measurements for dissolved oxygen, cell oxygen uptake rate, and oxygen absorption follow.

Determination of Dissolved Oxygen Concentration, C^*. The oxygen concentration in solution, C^*, was determined amperometrically by measuring the diffusion current attributable to oxygen at a constant applied potential. Vitek (*30*) first employed the polarographic technique in the determination of dissolved oxygen, and Baumberger and Müller (*6*) were the first to apply the technique to biological systems. Others have successfully used the dropping mercury electrode to determine dissolved oxygen in biological systems (*19, 20, 22*).

A procedure was developed for rapid and accurate determination of dissolved oxygen in a fermenting broth.

The bulk of the cells were removed by filtration and pressing through cheesecloth until the filtrate completely filled a 10-ml. volumetric flask previously blown with nitrogen and containing 1 ml. of deaerated phenol. This process required approximately 5 seconds from the time the sample was withdrawn from the fermentor. The flask was stoppered to eliminate any air bubbles and shaken to mix the phenol with the filtrate to ensure quick killing of any remaining organisms. The dissolved oxygen was determined, correcting for the dilution by phenol. Phenol does not interfere with the dropping mercury electrode measurement of oxygen. Next the saturation concentration of the broth, C_i, was determined after saturation of the sample with air. Finally the sample was always deaerated with nitrogen to determine the zero concentration reading. The oxygen concentration has been shown to be a linear function of the galvanometer deflection. The electrode constant was checked frequently against a standard.

Determination of Specific Oxygen Uptake Rate for Organism, k_r. This section describes the independent determination of the cell uptake rate constant, k_r, as defined by Equation 6. The method used was essentially that of Baumberger and Müller (*6*), using the dropping mercury electrode. Preliminary investigations showed that the viability of the cells was not affected by the mercury over much longer periods of time than that used for the determination. Because it was desired to determine the actual oxygen uptake rate of the organism in the fermentor at a specific time, it was necessary to utilize a portion of the fermenting broth for the determinations.

A sample of fermenting medium was removed from the fermentor, and a portion was filtered, placed in the electrode cell, and saturated with oxygen. An appropriate amount of the remaining sample containing the organisms was added to the electrode cell in such concentration that the organism utilized the dissolved oxygen over a period of 5 to 15 minutes, during which time any effect due to settling of the cell suspension was negligible. The current as measured by the maximum galvanometer swing at a constant applied potential of -0.4 volt across the dropping mercury electrode was read. Readings were taken at half-minute intervals.

Because the variation of galvanometer deflection with time is linear, the slope of the line was determined and the specific oxygen uptake rate, k_r, was then calculated by means of the equation

$$k_r = \frac{(b)(I.C._{(t)})(t)}{(d)(C_m)(f)} \quad (10)$$

where k_r = specific oxygen-uptake rate, gram moles of O_2/ (grams of mycelium)(hour)
b = slope, scale divisions per minute
$I.C._{(t)}$ = instrument constant, γ of O_2/(ml.)(scale division) at $t°$ C.
d = dilution ratio of whole broth to total sample placed in cell
C_m = cell dry weight, grams of mycelia per ml. of solution
t = 60 minutes per hour
f = $32 \times 10^6 \gamma$ of O_2 per gram mole

Determination of Oxygen Absorption Rate Coefficient, k_d. The oxygen absorption rate coefficient, k_d, is defined through Equation 4. The coefficient was determined independently of the fermentations by two procedures. The first is a modification of the sulfite method of Cooper, Fernstrom, and Miller (*10*) and the second was developed in this laboratory, employing an amperometric technique with nonsteady-state operation. Each of these methods is described.

MODIFIED SULFITE METHOD. Johnstone (*21*) observed that the oxidation of sulfite to sulfate when catalyzed by iron or manganese was a measure of the rate of absorption of oxygen. Sherwood (*29*) used Johnstone's data to calculate rate absorption coefficients for oxygen. This method was extensively used with success by Cooper, Fernstrom, and Miller for a water system; however, the basic assumptions of a catalyzed reaction and a zero oxygen back-pressure do not hold for a system containing organic anticatalysts (*24*) which are found in biological mediums. In the present work it was therefore necessary to correct for the back-pressure of oxygen in the liquid. This was accomplished by determining the actual oxygen concentration in the liquid.

The medium, made up as for a fermentation, was adjusted to approximately 0.5 N sodium sulfite and 0.0001 M sodium cobaltoquinolinate. The cobaltoquinolinate acts as a catalyst and does not interfere with subsequent polarographic procedures.

After the conditions of air flow, agitation, and temperature of the experiment had been established, 5 ml. of solution were withdrawn in a pipet previously flushed with nitrogen in order to determine sulfite oxidation. This sample was transferred into 50 ml. of 0.1 N iodine reagent and back-titrated with standard 0.1 N sodium thiosulfate. This procedure was repeated at several intervals.

Also, during the run, 15-ml. samples for dissolved oxygen determinations were withdrawn, quickly transferred into standard 15-ml. graduated centrifuge tubes containing 5 ml. of saturated lead acetate solution which had been deoxygenated with nitrogen, and stoppered with a cork. After replacement of the cork, making certain that no gas remained in the tube, the tube was thoroughly shaken to ensure immediate precipitation of the sulfite and the precipitate was centrifuged. The supernatant liquid was analyzed amperometrically for dissolved oxygen and corrected for dilution of the sample by the lead acetate solution. At the applied potential of operation, -0.4 volt, the lead wave does not interfere with the oxygen wave.

The saturation dissolved oxygen of the test medium was determined by adjusting a portion of medium, which was not previously treated with sulfite, to 0.5 N sodium sulfate. After saturating with air, the oxygen content was determined amperometrically.

The over-all specific oxygen absorption rate coefficient, k_d, was calculated by means of the equation

$$k_d = \frac{n(R_2 - R_1)}{mvp_i\left(\dfrac{C_i - C^*}{C_i}\right)(\theta_2 - \theta_1)} \quad (11)$$

where n = normality of thiosulfate solution
R_1, R_2 = titers of thiosulfate at times θ_1 and θ_2, ml.
v = volume of sample, ml.
C_i = saturation oxygen concentration of sample, gram moles of O_2 per ml.
C^* = corrected oxygen concentration of sample, gram moles of O_2 per ml.
p_i = partial pressure of oxygen in gas stream, atm.
m = 4×10^3, milliequivalents per gram mole of O_2
θ = time, hours

NONSTEADY-STATE AMPEROMETRIC METHOD. The method consists of continuously measuring the dissolved oxygen of the medium during nonsteady-state absorption by means of an amperometric technique. Two advantages of the method are the elimination of effects due to the presence of sulfite and increased speed of manipulation. A working equation, based on Equation 4, is

$$k_d = \frac{2.303\, C_i[\log(R_i - R_1) - \log(R_i - R_2)]}{p_i f(\theta_2 - \theta_1)} \quad (12)$$

where C_i = saturation oxygen concentration at p_i, micrograms of O_2 per ml.
p_i = partial pressure of oxygen in gas phase, atm.
R_i = galvanometer deflection when liquid is saturated with oxygen, scale divisions
R_1, R_2 = galvanometer deflection at times θ_1 and θ_2, scale divisions
f = 32×10^6, micrograms O_2 per gram mole
θ = time, hours

The apparatus, illustrated in Figure 2, consists of a 5-liter fermentor fitted with a side-entering calomel half-cell and a top-entering dropping mercury electrode shielded by a small diameter cylindrical stainless steel screen. A Fisher Elecdropode was connected across the electrodes.

The design and installation of the dropping mercury electrode are critical features. It was found that 80-mesh stainless steel screening allowed sufficiently rapid flow of liquid through the screen to eliminate screen resistance as a controlling factor, but not rapid enough to cause unduly erratic drop formation. It was also found desirable to keep the screen diameter to a minimum and to provide an air relief

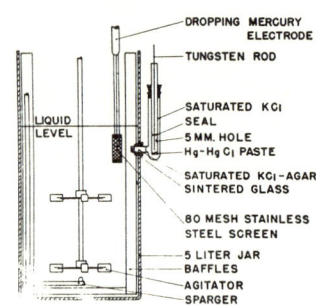

Figure 2. Apparatus for Direct Amperometric Determination of k_d

passage from the inside of the shield. A small stopper in the bottom of the screen served to collect the mercury and eliminate vertical liquid flow effects on drop formation. To eliminate the effects of vibration on drop formation, the electrode assembly was mounted independent of the fermentor and cushioned on rubber supports. Although agitation of the liquid increases diffusion around the mercury drop, a series of calibration runs showed that the variation of diffusion current with dissolved oxygen concentration is linear.

After the fermentor had been filled with the desired volume of medium, maintained at the proper temperature, the medium was deoxygenated by blowing with nitrogen. The electrode shield was cleaned and the dropping mercury electrode submerged into the liquid near the calomel half-cell. Mercury dropping time was adjusted to a range of 2 to 4 seconds. According to English (13), this range gives an approximately constant diffusion current for a given applied potential. A constant potential of -0.4 volt was applied across the cell. The fermentor agitator and air flow were started and galvanometer readings were taken every 5 seconds until the readings reached a constant maximum value equivalent to an oxygen-saturated system. The absolute saturation value of dissolved oxygen of the medium, C_i, was determined amperometrically.

Typical data for the logarithm of $(R_i - R)$ versus time at different agitator speeds and constant air flow are plotted in Figure 3. The theoretically linear plot is obtained.

This method has several limitations which must be kept in mind. If the medium contains a substance whose potential of polarization is the same as that for oxygen, the method is not directly applicable. Furthermore, solids in the liquid may clog the electrode shield, thereby introducing a diffusion resistance which may be the controlling rate mechanism. At extremely rapid rates of oxygen absorption, the inertia of the galvanometer

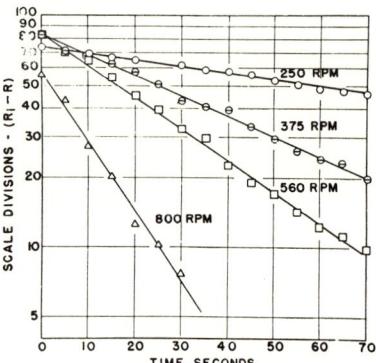

Figure 3. Amperometric Determination of k_d for Corn Steep Medium

Log $(R_i - R)$ vs. time
Conditions. 24° C.; 3200 ml.; constricted pipe sparger; 2 turboimpellers; 6 liters per minute air flow

may give readings that lag behind actual values. When this situation is met, it is possible to determine the desired result by reducing the partial pressure of oxygen in the gas stream, thereby slowing down the observed absorption process to a rate which will lie within the range of the available instruments. Care was taken in the present work to avoid the above possible difficulties.

The two methods for determining k_d agree within 0 to 30% at the various conditions imposed upon the mediums employed in this work.

RESULTS AND DISCUSSION

The results are presented under the subject headings of interfacial oxygen transfer coefficient, k_d; adsorptive oxygen transfer coefficient, k_d''; specific organism oxygen uptake rate, k_r; oxygen diffusion through cell aggregates, k_d'; and experimental verification of over-all rate equation.

Interfacial Oxygen Transfer Coefficient, k_d. Oxygen absorption experiments were performed with uninoculated mediums. Because it was desired to distinguish between gas phase and liquid phase diffusion at the gas-liquid interface, absorption experiments were performed at different temperatures to permit establishment of the rate determining step. The gas film resistance at a bubble interface was assumed unimportant compared to the liquid film at that site. The effect of temperature upon absorption rate permits selection between these mechanisms. In addition, evidence for the interface as the prime site of the diffusion step is given through air holdup experiments and through the effects of introducing surface-active agents into an air-water system.

TEMPERATURE. The amperometric method of measuring oxygen absorption was used in determining the effect of temperature in uninoculated medium, aerated and agitated in a fermentor. Triplicate experiments were performed at 27° and 32° C. The data are presented in Table I. An energy of activation, E, of 4200 calories per gram mole is obtained from the measurements applied to the Arrhenius equation for an activated process:

$$k_d = C e^{-E/RT} \quad (13)$$

The absolute value of the energy of activation lies well within the characteristic range of from 3000 to 5000 calories per gram mole for diffusional processes in liquid water. This range has been established through theoretical considerations (18) and measurements with other processes involving liquid phase dif-

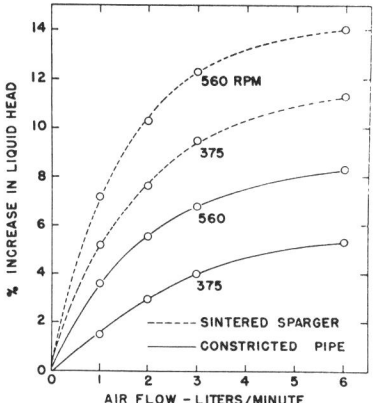

Figure 4. Air Holdup in Laboratory Fermentors vs. Air Flow Rate

R.p.m. as parameter. Streptomycin soybean medium

fusion. In contrast, it is well known (28) that the temperature dependency in gas phase diffusion is very small. If a bubble in the present case is nonagitated internally with respect to gas movement, the gas film coefficient, k_d, may be expected to vary as $\sqrt{T}$, T being the absolute temperature. This temperature variation of k_d is of the order of one tenth that for a liquid phase diffusion process when expressed in terms of E in Equation 13. It is apparent from the temperature measurements of oxygen absorption in this cell-free system that the rate-controlling step is diffusion through water rather than air. This is consistent with past experience for absorption of oxygen in water under other than immersed bubble conditions.

BUBBLE SIZE AND HOLDUP. By definition, the oxygen diffusion coefficient, k_d, is directly proportional to the area through which the diffusion takes place, or

$$k_d = ak \quad (14)$$

where a is the effective transfer area per unit volume of broth, and k is the absolute diffusion coefficient which is independent of area. This last coefficient would be difficult to measure in the system under consideration. However, exploratory measurements were made to determine whether k_d for oxygen diffusion might be affected by an obvious area, that of the bubble cloud in the fermentor. Total bubble area was changed experimentally in two ways. The size of individual bubbles was changed markedly by the use of two different spargers, constricted pipe and sintered metallic disk. In addition, the holdup population of bubbles was varied by means of changing the rate of air flow through the fermentor and the degree of agitation.

In general, as bubbles are caused to stream from a sparger the size of the individual bubbles varies with gas flow rate. Their size is affected by the type of sparger and by the intensity of agitation in the system. After bubbles have left the sparger their rate of travel upward depends upon their size and upon the density and viscosity of the liquid. The total instantaneous bubble population or holdup depends upon a balance between the rate at which bubbles are formed and the rate at which they can escape upward. Mechanical agitation can affect holdup because bubbles are swept downward after they have risen part way, thus affecting residence time.

Total bubble holdup may be measured through changes in liquid height as aeration rate, type of sparger, or degree of mechanical agitation is changed. Figure 4 presents typical air holdup curves as functions of air rate and agitation speed. Sintered and constricted pipe spargers are represented. It is first desirable to fix attention upon curve shape as percentage increase in head is related to linear air flow rate at constant revolutions per minute. The head rises rapidly at first with increasing air rate. In this region of behavior the bubbles are more or less uniform in size and of a size characteristic of the sparger—that is to say, very small for a sintered sparger and larger for a constricted pipe sparger. With increasing air flow rate the curve tends to become flat, and little added holdup results from further increase in air rate. This state of affairs results from the fact that very large bubbles eventually are observed to form in addition to a complement of small ones. The large ones rise very rapidly through the broth and do not contribute as strongly to holdup and surface area as the small ones. Operation under these conditions is one of "flooding" of the apparatus. A further examination of Figure 4 shows a strong differentiation in holdup characteristics between fine and larger bubbles as normally produced by sintered and open pipe spargers. The effect of increasing rate of agitation to increase holdup may be noted.

The effective time of contact (15) may be more appropriately represented by (volume holdup)/(volume air/minute) rather than by (volume of fermentor charge)/(volume air/minute) as it has on occasion been reported.

For comparison with the above holdup results, data for k_d at different air rates, agitator speeds, and sparger types are presented in Figure 5. The similarity in curve shapes between holdup and over-all k_d and similarity in the relationships between sparger and r.p.m. variables are apparent. A direct quantitative relation between k_d and holdup has not been presented because additional data are required before quantitative conclusions of a general nature are justified.

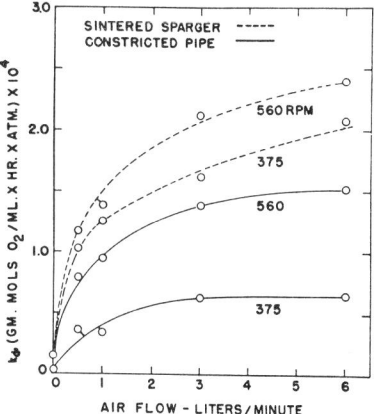

Figure 5. Oxygen Diffusion Rate Coefficient vs. Air Flow Rate

k_d determined in laboratory fermentor using amperometric method. R.p.m. as parameter. Streptomycin soybean medium

Table I. Effect of Temperature of Gas Diffusion in a Laboratory Fermentor

Medium, meat-peptone
Volume, 3200 ml.
Air flow, 6 liters per minute
Agitator speed, 375 r.p.m.

Test	Temperature, °C.	k_d, Gram Moles of O_2/Ml./Hour/Atm. $\times 10^4$	
1	27	0.722	
2	27	0.732	0.73
3	27	0.730	
4	32	0.820	
5	32	0.804	0.82
6	32	0.829	

SURFACE-ACTIVE AGENTS. Experiments concerning the effect of surface-active components serve to trace diffusion activity to bubble interfaces at which they congregate, and the experiments serve also to indicate differences in oxygen absorption between an ideal system, water, and a fermentation broth in which numerous agents of this type are present.

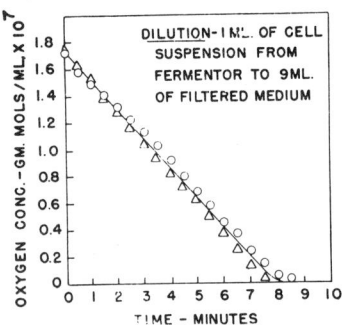

Figure 6. Oxygen Removal from Solution by *S. griseus* Cells vs. Time
Typical determination

The effect of a surface-active agent on the oxygen diffusion rate constant, k_d, is shown in Table II. The addition of the surface-active agent lowers the value of k_d from that in water, in spite of the fact that the total area is extended through the effect of these agents upon bubble size. The congregation of the agent in the surface to form a high diffusion resistance is more important in its effect on k_d than is the increase in total area upon this constant. These results give added evidence for the association of k_d with the interface.

Adsorptive Oxygen Transfer Coefficient, k_d''. The oxygen transfer by direct contact between organism and air interface is considered on qualitative rather than quantitative bases.

It is known (27) that microorganisms can form an intimate adsorptive contact with air bubbles much in the manner of mineral froth flotation. Indeed, microorganisms can be separated through froth flotation. For example, in this laboratory, it was found that a characteristic concentration isotherm between main liquid and froth concentrations for yeast cells may be obtained. As observed under the microscope, yeast cells are seen to adhere and slide about at a bubble surface. *S. griseus* and *P. chrysogenum* in small clusters are also observed to adhere and have mobility in the surface. Larger clusters of these organisms also adhere, but tend to break away because of the shear applied by the cluster mass. These microscopic observations were in dilute suspensions. In normal, heavy non-Newtonian suspensions of the organisms it is likely that much direct contact with the bubble takes place with the aid of mechanical pressure from the mycelial mass.

Organism Oxygen Uptake Rate Coefficient, k_r. Figure 6 is a typical curve showing the disappearance of oxygen with time in the polarograph. The curve is similar to those reported previously for yeast cells and other microorganisms by Baumberger (4), Petering and Daniels (26), and Winzler (32). It can be seen that over a large portion of the curve the rate of oxygen utilization by the organism at any time is constant and is independent of oxygen concentration. However, at very low values of oxygen concentration, the linearity ceases, the absorption rate becomes oxygen-dependent, and growth rates as well as antibiotic production rates are adversely affected.

Oxygen Diffusion through Cell Aggregates, k_d'. Evidence has been gathered regarding the presence of liquid diffusion processes through the following oxygen absorption experiments performed in the absence of air bubbles.

Specific oxygen absorption rates for *S. griseus*, expressed by k_r, were measured in a laboratory fermentor rather than in the polarograph as above. The bulk of the medium was saturated with air and the supply was turned off. A suspension of cells in the remaining medium was then added to obtain the desired concentration. No bubbles were provided and the fermentor was filled completely to eliminate a surface source of oxygen. The rate of oxygen depletion was measured by an appropriately installed polarograph cell located directly in the fermentor. The specific oxygen uptake rate, k_r, was computed from the depletion rate of the dissolved oxygen.

The depletion curves were similar to those given in Figure 6. Figure 7 presents the value of k_r as a function of mycelium concentration at zero agitation and at 375 r.p.m. The marked effect of cell concentration on k_r in the unagitated vessel is noted. Baumberger's (5) data for yeast cells show a similar effect (Table III). The explanation is offered that the adverse diffusion gradients caused by resistance, $1/k_d'$, are dissipated through agitation or dilution.

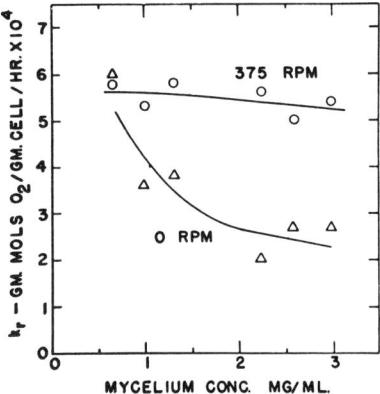

Figure 7. Specific Oxygen Uptake Rate vs. Mycelium Concentration
Medium, meat. Organism, *S. griseus*. Temperature, 24° C. Amperometric technique in 5-liter fermentor

Table II. Effect of Medium Constituents on Specific Oxygen Diffusion Rate, k_d

Equipment, 5-liter fermentor
Technique, amperometric
Sparger, constricted pipe
Air flow, 6 liters per minute
Agitator speed, 560 r.p.m.

Medium	$k_d \times 10^4$	Observed Bubble Size
Water	2.4	Coarse
Corn steep	0.97	Fine
Soybean	1.53	Fine
0.5% peptone	0.88	Fine

Table III. Specific Oxygen Uptake Rate vs. Yeast Cell Concentration

Yeast Concentration, %	Slope of O_2 Uptake Curve	Specific O_2 Uptake Rate
(1)	(2)	(2)/(1)
0.033	0.096	2.9
0.16	0.49	3.1
0.5	1.19	2.4
1.0	2.14	2.1

Calculated from data of Baumberger (5).

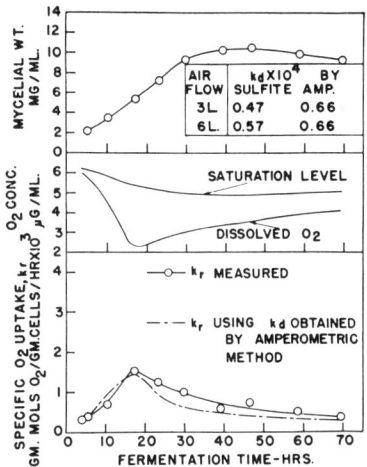

Figure 8. Oxygen Uptake Rates during Fermentation in 5-Liter Fermentor

Medium, soybean. Air flow, 3 liters per minute to 8 hours, 6 liters per minute to end. Inoculum mutant strain of *S. griseus*. Volume. Initial 3900 ml., final 3300 ml. Agitator, 375 r.p.m.

Experimental Verification of Over-all Rate Equation. Equations 8 and 9 were developed on a mechanism hypothesis that oxygen diffusion and organism rate processes are in series and that each of these processes may be independently characterized. The assumption also was made that fermentations are so slow that, at any given instant, steady-state conditions obtain.

These equations for tank and shake flask fermentations may be checked experimentally by alternate procedures. During the course of a research fermentation it is customary to measure C_m, the mycelium concentration, C^*, the instantaneous oxygen concentration in the medium, and H, Henry's law constant, as these depend upon time. In addition, G, the air flow rate, V, the batch volume, and the inlet oxygen partial pressure are established. The oxygen diffusion and organism oxygen uptake rate coefficients, k_d and k_r, are determined experimentally independent of each other. A check of the oxygen rate equations may therefore be accomplished by inserting either of these independently determined rate constants into the appropriate equation, 8 or 9, and computing the other through the equation, all other variables having been identified through fermentation measurements. The rate constant so computed then is compared with its independent counterpart.

Such comparisons are made for shake flasks and laboratory fermentors in Figures 8 through 11. Experiments using *S. griseus* with two mediums and two cultures in laboratory fermentors and one medium and culture in shake flasks are represented. Agitation was sufficiently great in all experiments to associate oxygen transfer primarily with k_d at the air-liquid interface. In these figures it was elected to compare "experimental" values of k_r as determined independently in the polarograph with "computed" values of k_r obtained through the equations. The k_d as measured by the modified sulfite method and by the direct amperometric method is represented in the laboratory fermentor results; only the former could be used with shake flasks. Although the dissolved oxygen concentration and mycelial weights varied widely and in a complex manner, "computed" and "experimental" values of k_r are in satisfactory agreement with respect to absolute value and particularly with respect to curve shape, thereby lending support for the equations. Although k_r is constant over a short time interval, the cell activity varies considerably during the course of a growth cycle.

Auxiliary measurements also were performed with *S. cerivisiae* fermenting under nitrogen in stoppered shake flasks. One series of flasks was quiescent; the other was shaken. A 40% increase in cell yield was obtained in the shaken flasks. The greater yield with agitation may be attributed to an increased diffusion of nutrients through a favorable increase in k_d'.

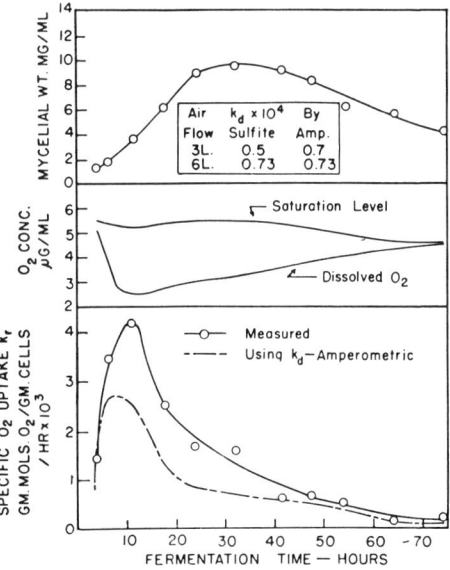

Figure 9. Oxygen Uptake Rates during Fermentation in 5-Liter Fermentor

Medium, meat. Air flow, 3 liters per minute to 8 hours, 6 liters per minute to end. Inoculum, Waksman No. 1 strain of *S. griseus*. Volume initial 3900 ml., final 3000 ml. Agitator, 375 r.p.m.

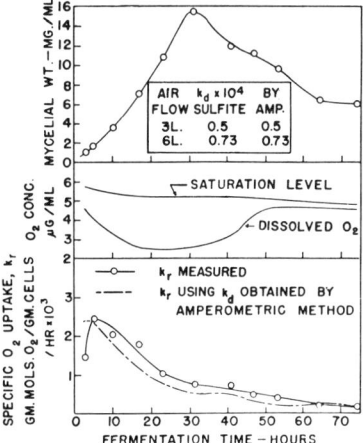

Figure 10. Oxygen Uptake Rates during Fermentation in 5-Liter Fermentor

Medium, meat. Air flow, 3 liters per minute to 8 hours, 6 liters per minute to end. Inoculum, mutant strain of *S. griseus*. Volume. Initial 3900 ml., final 3300 ml. Agitator, 375 r.p.m.

213

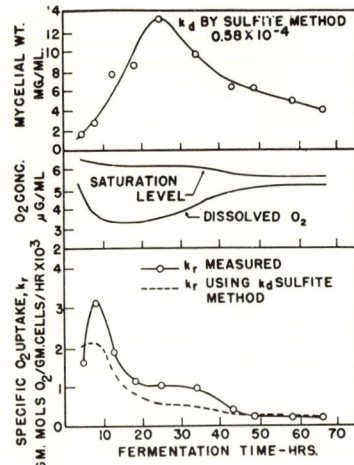

Figure 11. Oxygen Uptake Rates during Fermentation in Shake Flasks

Medium, meat. Agitator shaker 1.5-inch throw, 220 r.p.m. Inoculum, Waksman No. 1 strain of *S. griseus*

through their life cycle and fermentation conditions changed with time. The condition of limiting oxygen level at the cell wall under which the absorption rate becomes dependent upon oxygen concentration is of particular interest because cell growth and antibiotic formation are impaired. Whether or not cells, on the average, are in a condition of oxygen sufficiency or dearth depends upon the interplay of the diffusion rate processes, which in turn are dependent upon variables—agitation rate, air flow rate, interfacial area, and oxygen partial pressure in the gas phase.

Fermentation systems may, over short time intervals, be considered to be in dynamic equilibrium, and diffusion and cell oxygen absorption rate equations may be combined for conditions of deep tank and shake flask fermentations. The elements of the oxygen absorption theory were verified through an appropriate experimental check of such equations. An analysis of the rate steps indicates that the average measured oxygen concentrations in the liquors may vary widely as experimental conditions vary, but that an adequate average concentration does not necessarily exist at the all-important site of the cell interface.

ACKNOWLEDGMENT

The authors express their appreciation to P. K. Frolich, who initiated the formation of the biological-engineering group, and to L. E. McDaniel, H. E. Silcox, and J. C. Woodruff for encouragement and advice during this study. Acknowledgment is due J. B. Conn for advice on polarographic techniques, and the technicians and operators who helped experiments and analyses.

CONCLUSIONS

A theory for oxygen absorption by suspended mycelia in aerated nutrient broths involves a complex pattern of diffusional rate steps between air and organisms coupled with the rate of oxygen uptake by the cells.

One diffusion step, characterized by a rate constant k_d, is associated with the transfer of oxygen through air bubbles or other air-liquid interfaces. The resistance has been found to be located in the liquid phase. Its magnitude depends upon the extent of interfacial area in a unit volume of liquid and upon presence of adsorbed surface-active material in the interface. The rate constant, which is determined independent of the presence of cells in a broth, serves as a valuable index of the oxygen transfer performance of fermentors.

A second diffusion step, in series with the above and having a rate constant k_d', is concerned with the diffusion or eddy transfer of oxygen through clumps or groupings of cells and through film diffusion barriers around individual cells. The presence of cell aggregates is indicated by the non-Newtonian character of fermentation broths. Although the constant was not evaluated because oxygen concentration at the cell wall was not readily measurable, indirect experimental evidence was obtained for the presence of this diffusion resistance. It was found that the resistance could be dissipated to a large extent by diluting the broth or by agitating it strongly.

Evidence of a third mode of oxygen transfer, concurrent with the above, was obtained by frothing experiments and by microscopic observations. The rate step, designated by k_d'', occurs as cells form a momentary adsorptive contact at the gas-liquid interface. The number of such contacts presumably depends upon statistical considerations and varies with bubble size and population density as well as relative motion between bubbles and mycelia.

The oxygen uptake rate for a strain of *S. griseus* has been found to be constant and independent of oxygen concentration in the broth except at the lowest concentrations. The rate also is directly proportional to cell concentration under such conditions that diffusion gradients through cell clusters are not limiting the supply of oxygen. The specific rate constant was found not to vary over brief time intervals in which measurements were made, but to change during the course of fermentation as cells went

LITERATURE CITED

(1) Achorn, G. B., Jr., and Schwab, J. L., *Science*, **107**, 377 (1948).
(2) Bartholomew, W. H., Karow, E. O., and Sfat, M. R., IND. ENG. CHEM., **42**, 1827 (1950).
(3) Bartholomew, W. H., Karow, E. O., Sfat, M. R., and Wilhelm, R. H., *Ibid.*, **42**, 1810 (1950).
(4) Baumberger, J. P., *Am. J. Phys.*, **123**, 10 (1938).
(5) Baumberger, J. P., "Symposia on Quantitative Biology," Vol. VII, p. 200, New Bedford, Mass., Darwin Press (1939).
(6) Baumberger, J. P., and Müller, O. H., presented at Winter Meeting, Western Society of Naturalists, Stanford Univ. (1935).
(7) Becze, G. de, Hung. Patent 110,202 (1934).
(8) Becze, G. de, and Liebmann, A. J., IND. ENG CHEM., **36**, 882 (1944).
(9) Brown, W. E., and Peterson, W. H., *Soc. Am. Bact., Proc. Meetings*, **1**, 46 (1948).
(10) Cooper, C. M., Fernstrom, G A., and Miller, S. A., IND. ENG. CHEM., **36**, 504 (1944).
(11) Daniel, H. S., and Stahley, G. L., *J. Bact.*, **52**, 351 (1946).
(12) Dulaney, E. L., *et al.*, *Mycologia*, **41**, 388 (1949).
(13) English, F. L., *Anal. Chem.*, **20**, 889 (1948).
(14) Feustel, I. C., and Humfeld, H., *J. Bact.*, **52**, 229 (1946).
(15) Foust, H. C., Mack, D. E., and Rushton, J. H., IND. ENG. CHEM., **36**, 517 (1944).
(16) Gee, L. L., and Gerhardt, P., *J. Bact.*, **52**, 271 (1946).
(17) Glassman, H. N., and Elberg, S., *Ibid.*, **52**, 423 (1946).
(18) Glasstone, S., Laidler, K. J., and Eyring, H., "Theory of Rate Processes", New York, McGraw-Hill Book Co., 1941.
(19) Ingols, R. S., IND. ENG. CHEM., **14**, 256 (1942).
(20) Ingols, R. S., *Sewage Works J.*, **13**, 1097 (1941).
(21) Johnstone, H. F., *Combustion*, **5**, No. 2, 19 (1933).
(22) Lewis, V. M., and McKenzie, H. A., ANAL. CHEM., **19**, 643 (1947).
(23) Locke, F. G., *et al.*, U. S. Dept. Commerce, Washington, D. C., *FIAT Final Rept.* 499, 17 (1945).
(24) Mellor, J. W., "Inorganic and Theoretical Chemistry," pp. 10, 263, New York, Longmans, Green & Co., 1930.
(25) Olson, B. H., and Johnson, M. J., *J. Bact.*, **57**, 235 (1949).
(26) Petering, H. G., and Daniels, F., *J. Am. Chem. Soc.*, **60**, 2796 (1938).
(27) Shedlovsky, L., *Ann. N. Y. Acad. Sci.*, **49**, 279 (1948).
(28) Sherwood, T. K., "Absorption and Extraction," New York, McGraw-Hill Book Co., 1937.
(29) Sherwood, T. K., "Chemical Engineers' Handbook," 2nd ed., p. 1187, New York, McGraw-Hill Book Co., (1941).
(30) Vitek, V., *Collection Czechoslov. Chem. Communs.*, **7**, 537 (1935).
(31) Waksman, S. A., Schatz, A., and Reilly, H. C., *J. Bact.*, **51**, 753 (1946).
(32) Winzler, R. J., *J. Cellular Comp. Physiol.*, **17**, 263 (1941).

RECEIVED November 14, 1949.

17

Copyright © 1959 by John Wiley and Sons, Inc.
Reprinted from *J. Biochem. Microbiol. Tech. Eng.*, **1**(3), 311–324 (1959)

Aeration-agitation Studies on the Novobiocin Fermentation

W. D. MAXON

The Upjohn Company, Kalamazoo, Michigan

Summary. Novobiocin fermentations in 20 l. baffled fermentors were studied as a function of the size and speed of the dual four-bladed flat-blade turbines used. Power input and sulphite oxidation rate measurements were made. The courses of pH, sugar utilization, mycelial dry weight, carbon dioxide evolution and antibiotic titer were determined. Optimum antibiotic yields were achieved at a power input of 0·5 h.p./100 gal, equivalent to a sulphite oxidation rate of 110 mmoles O_2/l.h, when the impeller diameters were 29 per cent or 39 per cent of the tank diameter. A power input of 0·75 h.p./100 gal, equivalent to a sulphite oxidation rate of 160 mmoles O_2/l.h, was required for equivalent results with the impeller diameter 49 per cent of the tank diameter. Some explanations of the lack of equivalent results with the large impeller are discussed.

Introduction

Of those methods used for the estimation of oxygen transfer characteristics of fermentation vessels, the most easily applied are the sulphite oxidation rate and the power input per unit volume. The latter measurement is extremely indirect and depends to a considerable degree on maintaining geometric similitude and constant superficial air velocity between the scales of operation being compared. Its successful application in the penicillin fermentation has been reported by Wegrich and Shurter.[1]

The sulphite oxidation rate, first used by Cooper, Fernstrom and Miller[2] for study of oxygen transfer in agitated vessels, has been investigated extensively. Many workers have determined the effect of aeration-agitation in fermentors upon it[3,4] and many have shown that it gives a good prediction of oxygen transfer rate in actual fermentations. The correlation is best with the non-viscous yeast and bacterial systems,[5–8, 34] and in some of these instances a close agreement between sulphite oxidation rate and oxygen uptake in aeration-limited fermentations was actually observed.[9-11] Adequate scale-up of viscous, mycelial fermentations has also been achieved.[12, 13] Some of the limitations of the method have been demonstrated in publications by Wise,[14] Phillips and Johnson,[15] Pirt *et al.*[16] and Steel and Brierley.[17] The difficulty is that the chemical reaction under certain circumstances causes deviation from the diffusionally determined oxygen transfer rate. The degree of this deviation varies with agitation and may be quite minimal under normal conditions for antibiotic fermentations.[18]

Another limitation of the sulphite method is that it is measured in dilute aqueous solution, while fermentations are often highly viscous and non-Newtonian. The marked influence of suspended materials on oxygen transfer has been shown by Deindoerfer and Gaden,[19] Chain *et al.*[18] and Brierley and Steel.[20] The success of the method must be due to a maintained ratio of sulphite oxidation rate to oxygen transfer rate actually available in the fermentation under the aeration-agitation conditions being studied.

The interference of the chemical reaction may be avoided by measuring oxygen transfer rates polarographically. This method is not considered in the present report but it has been successfully applied by many workers.[18-24]

An excellent review of the subject of fermentation aeration-agitation has been published by Finn.[25]

The present paper deals with the effect of aeration-agitation in pilot equipment on the novobiocin fermentation.* This is an aerobic fermentation using a mycelial organism. The beers are pronouncedly non-Newtonian. The parameters used to measure the aeration-agitation are power input and sulphite oxidation rate. The variations in tank geometry, agitation rate and air flow are well within the normal range employed for antibiotic fermentations, yet it has been possible to demonstrate clearly some limitations of these parameters in predicting fermentation behaviour.

Equipment and Procedures

Fermentor

The fermentor, seed vessel, and auxiliary equipment used in these experiments have been described previously.[26] The geo-

---

* Smith[33] has recently completed some shaken flask studies on aeration effects in the novobiocin fermentation. He studied the interaction of mycelium concentration and aeration with respect to its influence on novobiocin titers.

metry of the fermentor agitation system was considerably modified, however. Fig. 1 illustrates the significant dimensions: two four-bladed flat-blade turbine impellers and four symmetrically spaced baffles were employed. Three sizes of impeller were available, their over-all diameters being 29 per cent, 39 per cent and 49

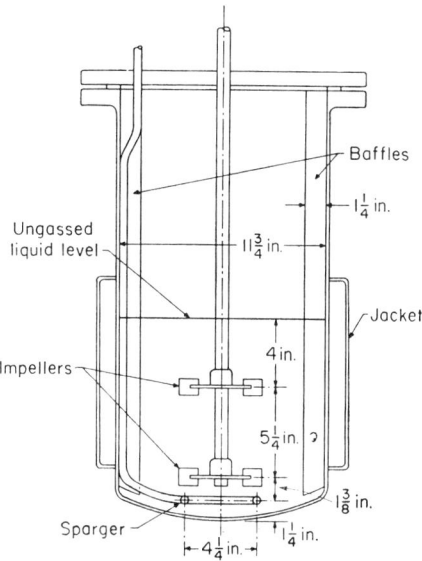

Fig. 1. Fermentor interior

Small impeller: over-all diameter 3·44 in.; blades 0·72 × 0·86 in.
Standard impeller: over-all diameter 4·63 in.; blades 0·94 × 1·19 in.
Large impeller: over-all diameter 5·81 in.; blades 1·19 × 1·44 in.

per cent of the inside tank diameter. The blade dimensions were proportional to the diameter in each case.

All runs were made with a liquid volume of 20 l. in the fermentor after inoculation. The air-flow rate was held constant at 2·2 SCFM (standard ft³/min humidified air measured at 14·7 lb/in² and 70°F) and the pressure in the tank maintained at 5 lb/in² gauge for all runs.

Fermentation

Strain 3R of the novobiocin-producing culture *Streptomyces niveus* was used in a medium containing (after sterilization and inoculation) 40 g/l. of Brown–Forman distillers' solubles and 40 g/l. of Cerelose (Corn Products Refining Co., glucose monohydrate). Sufficient sodium hydroxide solution was added to raise the pH before sterilization to 8·0. Vegetative inoculum, 5 per cent by volume, grown for 3 days on the same medium (but with the pH *after* sterilization readjusted to 8·0) was used. The fermentations were continued for 115 h. Lard oil containing 1 per cent octadecanol was added automatically as a defoaming agent; about 50 ml was the average consumption.

Analyses

The antibiotic titer was determined by the ultraviolet method described elsewhere.[27] Results are expressed as μg/g of whole beer (weighed sample). All components of the novobiocin complex are included. The microbiological assay procedure[27] was used as a check and ordinarily gave slightly lower, more variable, results.

The sugar determination, a modified Somogyi method,[28] was run on unhydrolyzed filtered beer against a glucose standard.

For mycelial dry weight measurement, an aliquot of filtered beer was centrifuged under reproducible conditions in a tared, calibrated tube. The percentage sediment was measured and the solids washed two or three times and dried *in vacuo* at 60°C to constant weight. Since, during the fermentation, mycelium and medium solids both contribute significantly to the total dry weight, an equation involving both percentage sediment and dry weight was used to calculate the approximate *mycelial* dry weight:

$$\text{mycelial dry wt.} = \frac{S - (S/D)_0 D}{(S/D)_{\max} - (S/D)_0}$$

where S = sediment, ml/l., D = total dry weight, g/l. The subscript '0' refers to the initial pre-inoculation conditions and 'max' refers to the maximum achieved during the fermentation.

The method's validity is based on two assumptions:

(1) Dry weight and sediment due to insoluble material from the medium are negligible late in the fermentation. Visual and microscopic observation indicated that this was true.

(2) The ratio of dry weight to percentage sediment remains constant for the mycelium and the medium solids independently through the run. This assumption could not easily be verified.

This method is of dubious accuracy but no more satisfactory procedure for the determination of mycelial weight in the presence of medium solids is available.

The method for sulphite oxidation rate determination was modified somewhat, for the sake of convenience, from that used by Cooper, Fernstrom and Miller.[2] In our procedure the fermentor was filled to the 20 l. level with a solution of sodium sulphite of proper concentration (1·0 N was usual), to which was then added a concentrated solution of cupric sulphate to a level of 0·005 M. After air and agitation were started, a sample was withdrawn into a flask. A measured aliquot of this sample (usually 5 ml) was pipetted into a test tube containing a small piece of dry ice, which blanketed the sample with carbon dioxide to terminate further oxidation. The sulphite solution was then titrated to the starch endpoint with an iodine solution of appropriate normality. The sampling was repeated at timed intervals. A plot of titration against time was prepared, and the slope of a straight line determined by four or five points (expressed as mmoles O_2/l.h) was taken as the sulphite oxidation rate.

Instrumentation

The torsion dynamometer used for power measurements and the recording infrared analyzer used for determination of carbon dioxide were described in an earlier publication.[26] A modified Stormer viscometer was used for rheological measurements. This measured the rate of turn of a two-bladed fork-shaped paddle in the fluid in question under the application of different amounts of torque.

Results

Physical Measurements

The effect of agitator speed and size on the power input to the water in the fermentor is shown in Fig. 2. The slope of the lines on this log-log plot was drawn equal to 3·0 ($P = KN^3$). This relationship is expected from the work of previous investigators.[29]

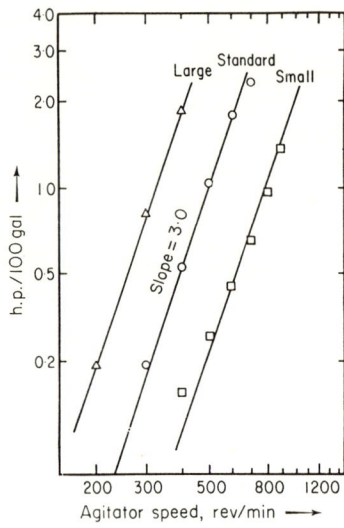

Fig. 2. Power input to water in the 20 l. fermentor for three impeller sizes

Experimental points fall, within error, on the lines. A cross plot shows that power input varies, as expected, with agitator diameter to the 5th power ($P = K'D^5$).

Power inputs to actual fermentation beers were also measured in several instances. In spite of the much higher viscosities of these beers the results were equivalent to those with water within the accuracy of the measurement (± 10 per cent). Furthermore, no significant variation in power during the course of a fermentation was noted. Published power correlations predict no effect of viscosity under conditions of high turbulence.[29] The rheological behaviour of beer samples taken from a typical fermentation is shown in Fig. 3: a shear diagram obtained from measurements in

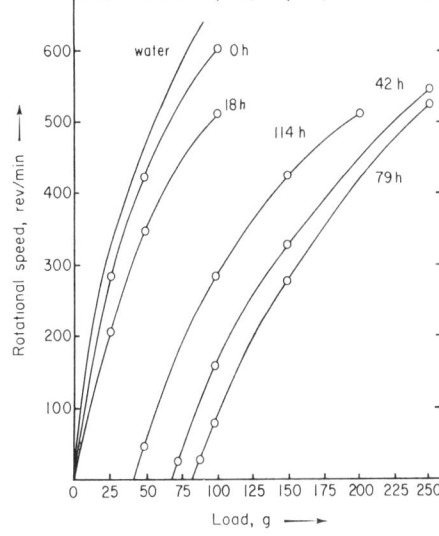

Fig. 3. Shear diagram of typical novobiocin beers

Age of fermentation at time of sampling is given in hours.

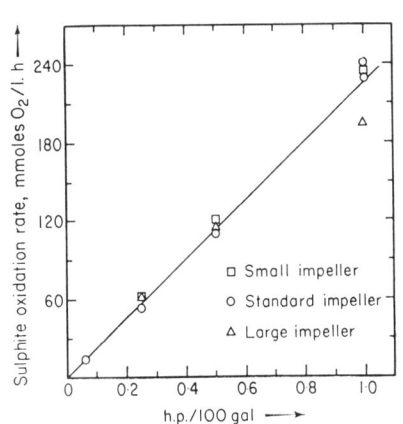

Fig. 4. Effect of power input and impeller size on sulphite oxidation rate

the Stormer viscometer. The differences in apparent viscosity from one time to another during the fermentation can be seen to be due more to the changes in yield point than to the changes in the slope of the lines. The curvature of the lines is real, but exaggerated by frictional resistances in the viscometer, as is evidenced by the curvature of the line for water.

Fig. 4 shows the sulphite oxidation rates obtained as a function of the h.p. input using the three agitator sizes. Except for

11*

one point with the large impellers at 1·0 h.p./100 gal, the rates are directly proportional to power input regardless of agitator size. Thus, sulphite number and power input are equivalent parameters for correlation of fermentation behaviour within the range where antibiotic yield effects occur. The one point of deviation falls outside this range.

Fermentation Behaviour

A series of novobiocin fermentations was run over a range of power inputs from 0·1–1·0 h.p./100 gal for each of the three agitator sizes. Measurements were made of pH, sugar, mycelial

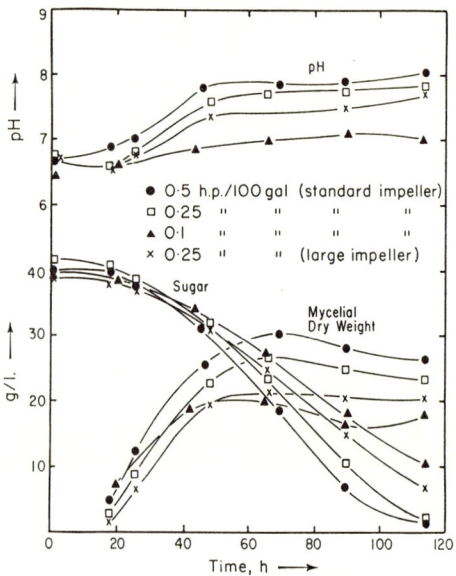

Fig. 5. Changes in pH, mycelial dry weight, and sugar during the fermentation, as a function of agitation

dry weight, novobiocin titer and carbon dioxide evolution rate. The results of typical runs are plotted in Figs. 5 and 6, giving concentration or rate against fermentation time. A key index, the novobiocin titer at 115 h, is plotted against the agitation conditions, power input and impeller size, in Fig. 7.

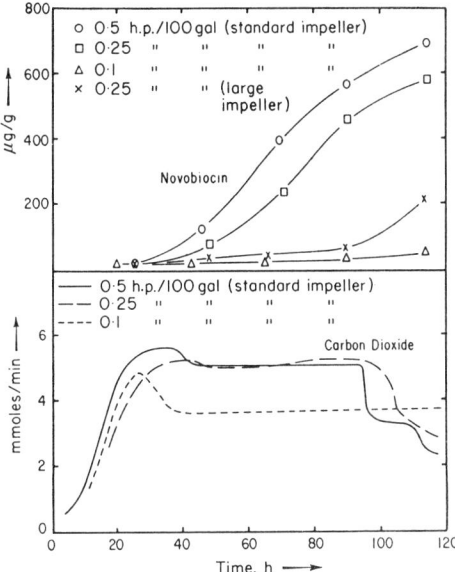

Fig. 6. Changes in novobiocin titer and carbon dioxide evolution rate during the fermentation, as a function of agitation

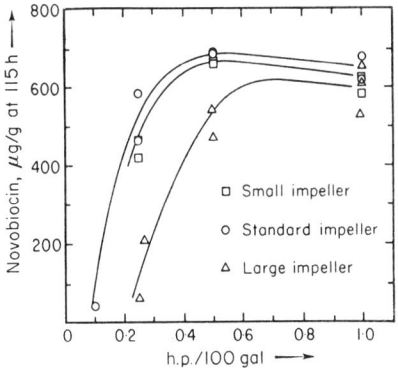

Fig. 7. Effect of power input and impeller size on harvest yield of the novobiocin fermentation

For the small and standard impellers the effect of power input (or of sulphite oxidation rate) on novobiocin titer is similar. Antibiotic yield increases from near zero at 0·1 h.p./100 gal to an optimum at 0·5 h.p./100 gal. Examination of Figs. 5 and 6 shows that antibiotic production is the most sensitive to aeration-agitation of those factors measured. Limiting conditions cause not so much a change in rate, as a delay in the beginning of production. This results in a striking effect on the 115 h yield level. Less pronounced are the effects on the other factors; insufficient agitation-aeration is manifested by a lower pH level during the antibiotic production phase, a less rapid utilization of sugar, and a smaller amount of mycelial dry weight. Effects on carbon dioxide evolution rate are even less pronounced. No reduction is noted at 0·25 h.p./100 gal even though the antibiotic yield is about 70 per cent of optimum. At 0·1 h.p./100 gal carbon dioxide evolution rate is reduced only by about 30 per cent, but the yield is near zero. Even this effect can be noted only in the period after antibiotic production begins. In the most active period of growth carbon dioxide evolution reaches the same level regardless of aeration-agitation. The maintenance of carbon dioxide evolution may be due to the presence of anaerobic dissimilation pathways. One might speculate that the shift from aerobic to anaerobic metabolism has a pronounced effect on the elaboration of the rate-limiting enzyme for novobiocin synthesis.

While 0·5 h.p./100 gal is sufficient power for optimum yield with the small and standard impellers, the yield at this power level is only 75 per cent of optimum when the large impellers are used. This difference is statistically significant at the 5 per cent level. The reduced yield is even more noticeable at 0·25 h.p./100 gal. Interpolation on the curves in Fig. 7 indicate that 0·75 h.p./100 gal is required for optimum yield with the large impellers. Fermentations at 1·0 h.p./100 gal show that when this power level is reached the antibiotic yields are approximately equivalent for all three impeller sizes.

While the evidence is only presumptive, it would appear that the reason for reduced yield with the large impellers at less than 0·75 h.p./100 gal is inadequate aeration. The curves in Figs. 5 and 6 indicate the same effects as occur with the smaller impellers under insufficient aeration-agitation, i.e. a lower pH course, less mycelial dry weight, reduced rate of sugar consumption and a delay in time of antibiotic production rather than a decrease in rate. If this assumption is correct, it must be concluded that neither power input per unit volume nor sulphite oxidation rate, the two factors being directly interconvertible, according to Fig. 4, in the region below 0·5 h.p./100 gal, is a satisfactory index of aeration-agitation with mycelial fermentations under all conditions of agitator geometry.

There is an indication, though not statistically significant at the 5 per cent level, that yields of novobiocin are reduced under conditions of excessive agitation. Dion et al.[30] have shown such a phenomenon to occur in the penicillin fermentation. They attribute the effect to mechanical damage to the mycelium, but observe similar damage at lower agitation levels when air is replaced by oxygen. Excessive agitation results in lower yields in a kojic acid fermentation according to Camposano et al.,[31] and Steel[32] has shown reduced yields with high agitation in an actinomycete fermentation under conditions of constant oxygen supply.

Discussion

There may be several reasons for the observed lack of correlation between sulphite oxidation rate or power input on the one hand and antibiotic yield on the other. Basically, most of these resolve to the simple proposition that these parameters are not always proportional to the oxygen transfer rate actually available in a non-Newtonian fermentation beer. Just where the deviation occurs is open to speculation. Some of the alternatives are:

(1) Sulphite oxidation rate is not a true, or even proportional, measure of the maximum oxygen transfer rate, even in a dilute aqueous system as measured, for example, by polarographic procedures. Wise,[21] and more recently Phillips and Johnson,[15] have shown how the chemical reaction may cause higher oxygen transfer than occurs by purely diffusional means. The degree of this discrepancy is dependent on the agitation characteristics and may not be appreciable under the conditions employed in most antibiotic fermentations. It is, however, possible that the large impeller gives a higher ratio of sulphite-measured oxygen transfer to purely diffusional oxygen transfer than the two smaller sizes.

(2) Oxygen transfer rate in a dilute aqueous system is not a true, or even proportional, measure of the oxygen transfer rate in a viscous

non-Newtonian fermentation beer. Several workers have shown how added mycelium or other fibrous materials reduces polarographically measured oxygen transfer. For this to explain the present observations, however, requires that the effect of mycelium is *greater* proportionally with the large impeller than with the two smaller impellers. Such might well be the case since the flow component of the power consumption (proportional to ND^3) is greater and the turbulence component (proportional to N^2D^2) is less with the large impeller. This reduced turbulence may have a more pronounced effect on oxygen transfer in the viscous than the non-viscous system.

(3) Non-homogeneity of oxygen supply may exist in the fermentor and its degree may be influenced by the impeller size. It is likely that the larger impeller with its higher flow component would cause a more uniform distribution of oxygen supply throughout the vessel. With the smaller impellers regions of local high turbulence and oxygen transfer could be visualized even though the over-all transfer would remain unchanged. This situation might cause the observed effect on antibiotic yield.

Which, if any, of these three alternatives is operating in the present case is not demonstrated. Perhaps it is a combination of them all. Effort to elucidate the phenomenon is continuing in these laboratories. The studies reported here serve to illustrate that, while sulphite oxidation rate and power input are useful parameters for estimating oxygen transfer rate in fermentations, they are applicable only within certain limits. This work partially indicates these limits.

References

[1] Wegrich, O. G. and Shurter, R. A., Jr. Development of a Typical Aerobic Fermentation. *Industr. Engng. Chem.*, 45 (1953), 1153–1160

[2] Cooper, C. M., Fernstrom, G. A. and Miller, S. A. Performance of Agitated Gas-Liquid Contactors. *Industr. Engng. Chem.*, 36 (1944), 504–509

[3] Chain, E. B., Paladino, S., Callow, D. S., Ugolini, D. and Van der Sluis, J. Studies on Aeration. *Bull. World Hlth Org.*, 6 (1952), 73–97.

[4] Elsworth, R., Williams, V. and Harris-Smith, R. A Systematic Assessment of Dissolved Oxygen Supply in a 20 l. Culture Vessel. *J. appl. Chem.*, 7 (1957), 261–268

[5] Elsworth, R., Williams, V. and Harris-Smith, R. The Effect of Oxygen Supply on the Rate of Growth of *Aerobacter cloacae*. *J. appl. Chem.*, 7 (1957), 269–274

[6] Olson, B. H. and Johnson, M. J. Factors Producing High Yeast Yields in Synthetic Media. *J. Bact.*, 57 (1949), 235–246

[7] Benedict, R. G., Koepsell, H. J., Tsuchiya, H. M., Sharpe, E. S., Corman, J., Kemp, C. E., Meyers, G. B. and Jackson, R. W. Studies on the Aerobic Propagation of *Serratia marcescems*. *Appl. Microbiol.*, 5 (1957), 308–313

[8] Lumb, M., Mercer, C. K. and Wilken, G. D. The use of *Aerobacter aerogenes* in Scale-up. Abst. VIIth Intern. Congr. Microbiol., p. 412, Stockholm (1958)

[9] Maxon, W. D. and Johnson, M. J. Aeration Studies on Propagation of Baker's Yeast. *Industr. Engng. Chem.*, 45 (1953), 2554–2560

[10] Pirt, S. J. The Oxygen Requirement of Growing Cultures of an Aerobacter Species Determined by Means of the Continuous Culture Technique. *J. gen. Microbiol.*, 16 (1957), 59–75

[11] Owen, S. P. and Johnson, M. J. Continuous Shake-Flask Propagator for Yeast and Bacteria. *Ag. and Food Chem.*, 3 (1955), 606–608

[12] Karow, E. O., Bartholomew, W. H. and Sfat, M. R. Oxygen Transfer and Agitation in Submerged Fermentations. *Ag. and Food Chem.*, 1 (1953), 302–306

[13] Roxburgh, J. M., Spencer, J. F. T. and Sallans, H. R. Factors Affecting the Production of Ustilagic Acid by *Ustilago zeae*. *Ag. and Food Chem.*, 2 (1954), 1121

[14] Wise, W. S. The Aeration of Culture Media. A Comparison of the Sulphite and Polarographic Methods. *J. Soc. chem. Ind., Lond.*, 61 (1950), 540–541

[15] Phillips, D. H. and Johnson, M. J. Oxygen Transfer in Agitated Vessels. *Industr. Engng. Chem.*, 51 (1959) 83–88

[16] Pirt, S. J., Callow, D. S. and Gillett, W. A. Oxygen Absorption Rates in Sodium Sulphite Solutions. Comparison of Cu^{2+} and Co^{2+} as Catalysts. *Chem. and Ind.*, 730–731 (June 8, 1957)

[17] Steel, R. and Brierley, M. R. Agitation-Aeration in Submerged Fermentation—I. A Comparative Study of the Sulfite and Polarographic Methods for Measuring Oxygen Solution Rates in a Fermentor. *Appl. Microbiol.*, 7 (1959), 51–56

[18] Chain, E. B. and Gualandi, G. Aeration Studies. *Rendiconti Istituto Superiore di Sanita*, 17 (1954), 5–60

[19] Deindoerfer, F. H. and Gaden, E. L., Jr. Effects of Liquid Physical Properties on Oxygen Transfer in Penicillin Fermentation. *Appl. Microbiol.*, 3 (1955), 253–257

[20] Brierley, M. R. and Steel, R. Agitation-Aeration in Submerged Fermentation—II. Effect of Solid Disperse Phase on Oxygen Absorption in a Fermentor. *Appl. Microbiol.*, 7 (1959), 57–61

[21] Wise, W. S. The Measurement of the Aeration of Culture Media. *J. gen. Microbiol.*, 5 (1951), 167–177

22 Bartholomew, W. H., Karow, E. O., Sfat, M. R. and Wilhelm, R. H. Oxygen Transfer and Agitation in Submerged Fermentations. *Industr. Engng. Chem.*, 42 (1950), 1801–1815

23 Hixon, A. W. and Gaden, E. L., Jr. Oxygen Transfer in Submerged Fermentation. *Industr. Engng. Chem.*, 42 (1950), 1792–1801

24 Strohm, J., Dale, R. F. and Peppler, H. J. Polarographic Measurement of Dissolved Oxygen in Yeast Fermentations. Abst. 134th Meeting Amer. chem. Soc., Chicago, Illinois (Sept. 1958)

25 Finn, R. K. Agitation-Aeration in the Laboratory and in Industry. *Bact. Rev.*, 18 (1954), 254–274

26 Nelson, H. A., Maxon, W. D. and Elferdink, T. H. Equipment for Detailed Fermentation Studies. *Industr. Engng. Chem.*, 48 (1956) 2183–2189

27 Smith, R. M., Perry, J. J., Prescott, G. C., Johnson, J. L. and Ford, J. H. Assay Methods for Novobiocin. Antibiotics Annual 1957–1958 Medical Encyclopedia, Inc., New York

28 Somogyi, M. A New Reagent for the Determination of Sugars. *J. biol. Chem.*, 160 (1945), 61–68

29 Rushton, J. H., Costich, E. W. and Everett, H. J. Power Characteristics of Mixing Impellers. *Chem. Engng. Progr.*, 46 (1950), 395–404, 467–476

30 Dion, W. M., Carilli, A., Sermonti, G. and Chain, E. B. The Effect of Mechanical Agitation on the Morphology of 'Penicillium Chrysogenum' Thom in Stirred Fermentors. *Rendiconti Istituto Superiore di Sanita*, 17 (1954), 185–205

31 Camposano, A., Chain, E. B. and Gualandi, G. The Effect of Mechanical Agitation on Morphology of *Aspergillus flavus* (Link) on the Production of Kojic Acid by this Organism in Submerged Culture. VIIth Intern. Congr. Microbiol., Stockholm (Aug. 1958)

32 Steel, R. Some Effects of Agitation on Antibiotic Production in Submerged Fermentation. International Botanical Congress, Montreal (Aug. 1959)

33 Smith, C. G. Effect of Aeration on the Novobiocin Fermentation. To be published

34 Smith, C. G. and Johnson, M. J. Aeration Requirements for the Growth of Aerobic Microorganisms. *J. Bact.*, 68 (1954), 346–350

Part VIII

METABOLIC REGULATION

Editor's Comments
on Paper 18

18 DEMAIN
Cellular and Environmental Factors Affecting the Synthesis and Excretion of Metabolites

    The newer knowledge of coordination of microbial metabolism has only begun to be applied to problems of synthesis of primary and secondary metabolites of microorganisms that are useful to human beings. That is not to say that the phsyiological regulatory mechanisms of industrial microorganisms have not been altered by selection and adaptation to conditions imposed by humans; instead, it is simply an expression of confidence that fermentation technology is reaching a new level of sophistication as our understanding of regulation and coordination of metabolism improves.

    In Paper 18, Demain discusses factors affecting synthesis and excretion of microbial metabolites from the point of view of one who has dedicated himself to integrating theoretical advances in molecular biology, genetics, and biochemistry into a unified theory of the physiology of fermentation processes. (See also Demain 1971 and 1974.)

REFERENCES

Demain, A. L. 1971. Overproduction of microbial metabolites and enzymes due to alteration of regulation. *Adv. Biochem. Eng.*, **1**:113–142.

———. 1974. How do antibiotic-producing organisms avoid suicide? *Ann. N. Y. Acad. Sci.*, **235**:601–612.

18

Copyright © 1972 by Arnold L. Demain

Reprinted from *J. Appl. Chem. Biotechnol.*, 22, 345–362 (1972)

Cellular and Environmental Factors Affecting the Synthesis and Excretion of Metabolites[a]

A. L. Demain

Department of Nutrition and Food Science, Massachusetts Institute of Technology, Cambridge, Mass. 02139, U.S.A.

1. Introduction

Pirt[1] concluded his address at the 19th Symposium of the Society of General Microbiology this way:

> "The wild-type of organism possesses a host of characters which probably are essential to the organism's survival in the natural habitat but are unnecessary in pure culture where the most favorable conditions for growth can be artificially contrived. The aim of microbial culture for efficient product formation should be to ensure that the autosynthetic processes of the organism are limited to reproduction of only that system (the 'minimal system') which performs the required function . . . Such induced degeneration of the organism is the antithesis of the normal aim of the biologist, which is concerned with preserving as many as possible of the organism's characters."

A growing microbial cell breaks down high molecular weight carbon and energy sources, brings the smaller derivatives into the cell, degrades them to smaller molecules, converts these to amino acids, nucleotides, vitamins, carbohydrates and fatty acids, and finally builds these basic materials into proteins, coenzymes, nucleic acids, mucopeptides, polysaccharides and lipids. Hundreds of enzymes must be made and must act in an integrated manner to avoid total chaos. Thus, regulatory mechanisms have evolved that enable a species to efficiently compete with other forms of life and survive in nature. These control mechanisms allow only necessary enzymes to be made, in the correct amounts; they then control the action of these enzymes. Thus, the ideal cell does not overproduce metabolites no matter what its environment. Some of the important control mechanisms are listed in Table 1, along with pertinent references.

Let us now shift our attention from the point of view of the micro-organism to that of the fermentation microbiologist or bioengineer. He is not interested in efficient organisms; on the contrary, he desires a wasteful strain which will overproduce and excrete a particular compound which can then be isolated. Usually, he first empirically

---

[a] Publication No. 1749 from the Department of Nutrition and Food Science, Massachusetts Institute of Technology, Cambridge, Mass. 02139, U.S.A.

screens organisms from culture collections and from nature for their ability to overproduce the desired product. Without necessarily realising it, he is usually searching for the organism with the weakest regulatory mechanisms. Once the desired strain is found, he begins a development programme to improve yields by modification of culture conditions and by mutation. These two facets of process development go hand

TABLE 1. Important regulatory mechanisms in micro-organisms.

Mechanism	References
A. Substrate induction	Clarke and Lilly[2]
B. Feedback regulation	
1. Repression	Clarke and Lilly[2]
2. Inhibition	Stadtman[3]
3. Modifications used in branched pathways	
a. Isoenzyme feedback regulation	Stadtman[4]
b. Concerted (multivalent) feedback regulation	Datta[5]
c. Cumulative feedback regulation	Stadtman[3]
d. Compensatory feedback regulation	Datta[5]
e. Sequential feedback regulation	Datta[5]
C. Catabolite regulation	
2. Repression	Paigen and Williams[6]
2. Inhibition	Paigen and Williams[6]
D. Cross-pathway regulation	Rebello and Jensen[7]
E. Energy-charge regulation	Atkinson[8]
F. Amino acid regulation of RNA synthesis	Edlin and Broda[9]

in hand, since each superior mutant responds to a specific environment for optimum product formation. The microbiologist and bioengineer are actually modifying the regulatory controls remaining in the original culture so that its "inefficiency" can be further increased. Until very recently, these manipulations were done in the total absence of any understanding of the basic factors involved. Due to the increase in our knowledge of microbial biochemistry and genetics, we now have some idea of these factors and, indeed, new fermentation processes for primary products, such as amino acids and purine nucleotides, are being developed on a more rational basis. On the other hand, development of fermentations for secondary metabolites (those which have no general function in life processes) still relies mainly on the empirical approach because of our ignorance of the pathways and the regulatory circuits involved in secondary metabolism. This situation may be remedied in the near future.

2. Subnormal regulation in fermentation organisms

Only within the last few years has evidence started to accumulate to support the thesis that the organisms selected by our screening procedures are, in fact, subnormally regulated. For example, *Corynebacterium glutamicum* (previously called *Micrococcus glutamicus*), a species used for the commercial product of L-lysine, possesses an initial enzyme of the lysine branch (dihydropicolinate synthetase) which is resistant to feedback inhibition by lysine.[10] The superior ability of *Bacillus subtilis* to produce hypoxanthine and inosine probably results from the poor ability of the nucleotide derivative, IMP, to inhibit the first enzyme of purine biosynthesis (PRRP amidotransferase) in this organism.[11] In *Claviceps paspali*, the production of alkaloids is markedly stimulated by a precursor, tryptophan; in this organism the first enzyme of the tryptophan biosynthetic branch (anthranilate synthetase) is resistant to feedback inhibition by tryptophan.[12] Chloramphenicol production by *Streptomyces* sp. occurs by a shunt pathway from shikimic acid, a normal intermediate in aromatic amino acid biosynthesis. In the producing organism, the first enzyme of aromatic biosynthesis (DAHP synthetase) resists feedback inhibition by phenylalanine and tyrosine.[13] In fact, Jensen and Rebello[14] tested six different species of *Streptomyces* and found that the enzyme in all six was resistant to feedback inhibition by tyrosine, phenylalanine and tryptophan. It is tempting to believe that the ability of streptomycetes to produce antibiotics derived from the aromatic amino acid pathway (e.g. chloramphenicol, antimycin, indolmycin, telomycin, bottromycin, actinomycin) results from this lack of regulation of DAHP synthetase.

3. Other changes in fermentation micro-organisms

Some cultures are selected in screening programmes because they have poor permeability barriers or are deficient in certain enzymes. Both characters can be illustrated by the glutamic acid-producing bacteria, all of which are deficient in α-ketoglutarate dehydrogenase and have a nutritional requirement for biotin.[15] It is clear that an "efficient" micro-organism would possess this important TCA cycle enzyme and would encase itself in an effective cytoplasmic membrane to retain its important intermediary metabolites. The deficiency of the enzyme in *C. glutamicum*, however, blocks the TCA cycle at α-ketoglutarate and the flow of carbon is shunted to L-glutamic acid. The natural biotin requirement renders the organism biotin-deficient in a low biotin environment; because biotin is necessary in fatty acid synthesis, an altered plasma membrane is made which is incapable of retaining high concentrations of glutamate. Thus, the accumulated glutamate pours out of the cell. In contrast, when the permeability barrier is effective (as in the case of *C. glutamicum* grown in the presence of sufficient biotin), the glutamate stops its own synthesis when it reaches an internal level of 25 to 50 mg/g cells (dry weight).[16]

Even when grown in the presence of sufficient biotin, the cytoplasmic membrane of *C. glutamicum* and of other glutamate producers appears to differ from that of most micro-organisms.[17] Thus, purine nucleotides, which accumulate as a result of certain mutational blocks (see below), are excreted by glutamate producers, whereas in a species such as *B. subtilis*, the nucleotides must first be degraded to nucleosides and free bases before excretion takes place. This ability to excrete intact nucleotides is crucial to the direct fermentation of the flavour enhancers, IMP, GMP and XMP.

4. Feedback regulation in biosynthetic pathways

Once an organism is selected through screening, an improved culture environment (culture medium, oxygen transfer, temperature, pH, etc.) is devised. Occasionally potential precursors and probable intermediates are found to stimulate product formation and, if their cost is not too high, they are included in the medium. Often they prove useful not only for increasing product yields but for directing production towards one member of a family of related products. Such "directed biosynthesis" has been exploited in production of particular penicillins (precursors: phenylacetic acid, phenoxyacetic acid, etc.), actinomycins (valine, isoleucine, sarcosine, hydroxyproline, etc.), tyrocidines (phenylalanine, tryptophan, isoleucine), novobiocins (various derivatives of benzoic acid) and cobalamines (5,6-dimethylbenzimidazole).

After this investigational phase ends, the development process moves, predominantly, to alteration of the residual control mechanisms.

4.1. Primary metabolites

In biosynthetic pathways leading to primary metabolites, the main regulation is through feedback, involving *inhibition* of an early biosynthetic enzyme or *repression* of one or more of the biosynthetic enzymes by the final product or its derivative. In fact, most processes designed to produce primary metabolites work by decreasing

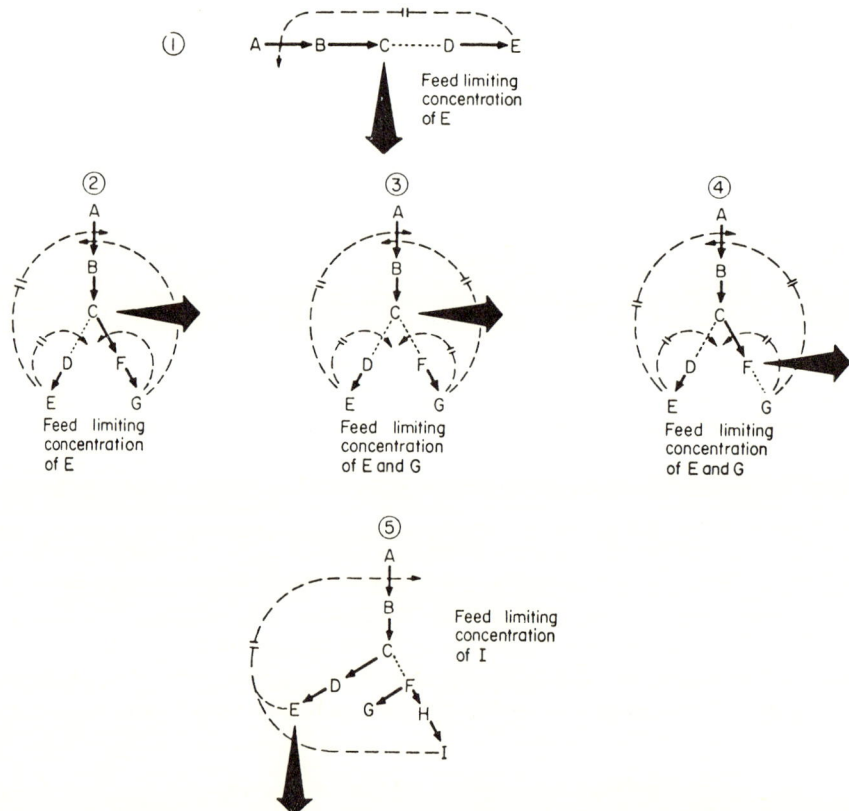

Figure 1. Overproduction of primary metabolites by decreasing the concentration of a repressing or inhibiting end product. ····, Site of auxotrophic mutation; ----, feedback regulation; →, overproduced product.

the intracellular concentration of such feedback inhibitors or repressors. Usually, one employs an auxotrophic mutant blocked in the further conversion of the desired product. Since such an organism requires the end product for growth, one can limit the intracellular concentration of this inhibitory or repressive end product by feeding growth-limiting levels of the end product to the culture; feedback regulation is then bypassed, and high levels of the desired intermediate accumulate. Various types of pathways can be handled by this technique (Figure 1 and Table 2). In the simplest case (1), an intermediate such as ornithine is produced in high yields when grown with limiting levels of arginine. In case 2, an intermediate of a branched pathway, such as IMP, is accumulated by an adenineless mutant. The double mutant depicted in case 3 produces a better IMP fermentation since feedback regulation by both end

TABLE 2. By-passing feedback regulation to produce primary metabolites.

Case	Type of pathway	Type of product	Product	Auxotrophic requirement	Reaction blocked (enzyme)
1	Simple	Intermediate	Ornithine	Arginine	Ornithine → citrulline (ornithine transcarbamylase)
2	Branched	Intermediate	IMP	Adenine	IMP → adenylosuccinate (adenylosuccinate synthetase)
3	Branched	Intermediate	IMP	Adenine + guanine (or adenine + xanthine)	IMP → adenylosuccinate (adenylosuccinate synthetase) IMP → XMP (IMP dehydrogenase)
4	Branched	Intermediate	XMP	Adenine + guanine	IMP → adenylosuccinate (adenylosuccinate synthetase) XMP → GMP (XMP aminase)
5	Branched	End product	Lysine	Homoserine (or threonine + methionine)	Aspartate semialdehyde → homoserine (homoserine dehydrogenase)

products (AMP, GMP) is by-passed. In case 4, a different type of double mutant is used to accumulate an intermediate such as XMP. Case 5 depicts the production of an end product by the same principle, of which the commercial lysine fermentation is a good example. The main obstacle to lysine (E) accumulation is feedback inhibition of the first enzyme (aspartokinase) by the concerted effect of lysine + threonine (E + F). By employing an auxotroph blocked in homoserine dehydrogenase, we make the intracellular level of threonine (F) dependent on the concentration of threonine in the medium. As long as the threonine concentration is low, concerted feedback inhibition cannot occur, and extremely high levels of lysine (over 40 g/l are excreted. In other branched pathways, valine can be overproduced by isoleucineless mutants, tyrosine by phenylalanine-requiring auxotrophs, phenylalanine by tyrosineless mutants and threonine by double mutants requiring lysine plus methionine. In every case, the required end product must be added in growth-limiting concentrations.

A second way to eliminate feedback regulation and accumulate primary metabolites is to alter the structure of the enzyme subject to inhibition or to modify the regulatory genes so that the system is no longer repressible. This modification is done by selecting mutants which resist the toxic effects of an analogue of the desired product. Many of the resistant mutants overproduce and excrete the natural end product. A compilation of the antimetabolites which have been successfully used to accumulate particular metabolites appears in Table 3. The best threonine fermentation developed to date (14 g/l) employs a single-step, α-amino-β-hydroxyvaleric acid-resistant mutant of *Brevibacterium flavum*.[18] Certain feedback-resistant overproducing mutants possess desensitised enzymes, while others have derepressed enzyme-forming systems. Strains in which both types of mutation are combined usually show synergistic excretion of the natural metabolite.

TABLE 3. Overproduction of primary metabolites by analogue-resistant mutants.

Accumulated product	Analogues for selection
Phenylalanine	*p*-Fluorophenylalanine; thienylalanine
Tyrosine	*p*-Fluorophenylalanine; D-tyrosine
Tryptophan	5-Methyltryptophan; 6-methyltryptophan
Histidine	2-Thiazolealanine; 1,2,4-triazole-3-alanine
Proline	3,4-Dehydroproline
Valine	α-Aminobutyric acid
Isoleucine	Valine
Leucine	Trifluoroleucine; 4-azaleucine
Threonine	α-Amino-β-hydroxyvaleric acid
Methionine	Ethionine; norleucine; α-methylmethionine; L-methionine-D,L-sulphoximine
Arginine	Canavanine
Adenine	2,6-Diaminopurine
Uracil	5-Fluorouracil
Hypoxanthine	5-Fluorouracil
Guanosine	8-Azaxanthine
Nicotinic acid	3-Acetylpyridine
Pyridoxine	Isoniazid
p-Aminobenzoic acid	Sulphonamide
Thiamine	Pyrithiamine

An alternate means of desensitising an enzyme to feedback inhibition is by removing it via one mutation and replacing it by a second ("reverting") mutation. Certain revertants apparently possess an active enzyme whose amino acid sequence is modified so that feedback inhibition is eliminated or is much less severe. Thus, removal and replacement of threonine deaminase leads to overproduction of isoleucine,[19] whereas the same procedure applied to IMP dehydrogenase results in GMP overproduction.[20]

4.2. Secondary metabolites

The major obstacle in applying the above clear-cut principles to the rational development of secondary metabolite fermentations is our ignorance; we know little of the pathway intermediates, less about the enzymes, and almost nothing about the feedback regulation of secondary metabolism. Only in the case of gramicidin S biosynthesis[21] are the enzymes and pathway known. In certain biosynthetic processes, mutant methodology has been and is being used to reveal the pathways. The most advanced system, in terms of established intermediates, is the tetracycline system of McCormick,[22] but its enzymes have not been studied. Other pathways, such as those for novobiocin,[23] polymyxin,[24] prodigiosin[25] and erythromycin[26] are being explored with non-producing mutants but are less clearly understood.

Although there is very little firm evidence, there are indications that secondary metabolites exert feedback regulation on their own formation. Chloramphenicol appears to limit its own synthesis.[27] 6-Methylsalicylic acid, a key intermediate in the formation of many phenolic compounds (including the antibiotic, patulin), inhibits its own production.[28] Selection of mutants resistant to high levels of streptomycin and ristomycin has resulted in superior producers of these antibiotics.[29,30] A chlortetracycline producer, mutated to non-productivity and then "reverted" to productivity, has become a superior producer.[31]

In secondary pathways where a primary metabolic end product is a precursor of a secondary metabolite, elimination of feedback regulation in the primary pathway appears to increase production of the secondary metabolite. For example, in a penicillin-producing mutant, the first enzyme of valine synthesis (acetohydroxy acid synthetase) is less sensitive to feedback inhibition by valine than the enzyme in its ancestral strain, *Penicillium chrysogenum* Wis. Q-176; furthermore, the enzyme content is doubled in the superior mutant.[32] In the pyrrolnitrin fermentation, optimum production requires addition of tryptophan as a precursor. Mutants selected for resistance to 5-fluorotryptophan and 6-fluorotryptophan appear to have a tryptophan biosynthetic path resistant to feedback regulation; the overproduced tryptophan is

TABLE 4. Branched pathways yielding primary and secondary metabolites.

Intermediate	Primary end products	Secondary end products
Shikimic acid	Tryptophan Phenylalanine Tyrosine *p*-Aminobenzoic acid	Chloramphenicol Pyocyanine
Malonyl-CoA	Fatty acids	Griseofulvin Tetracyclines Patulin Cycloheximide
Mevalonic acid	Sterols	Gibberellins Helvolic acid Fusidic acid β-Carotine Terpenes Ergot alkaloids
α-Aminoadipic acid	Lysine	Penicillins Cephalosporins
Acetolactate	Valine Leucine Pantothenic acid	Tetramethylpyrazine

incorporated into the antibiotic.[33] These superior mutants no longer require tryptophan supplementation. Another example involves the precursor role of methionine in chlortetracycline biosynthesis. *Streptomyces viridifaciens*, a producer of this antibiotic, was mutated to methionine auxotrophy and then reverted to prototrophy. Many of the revertants were superior chlortetracycline producers.[31] In contrast, four other auxotrophs which required nonprecursor amino acids failed to yield a high frequency of improved chlortetracycline producers upon reversion.

Many secondary biosynthetic pathways share intermediates with primary pathways; taken together they constitute branched sequences (Table 4). In such cases, feedback regulation of an early enzyme(s) by the primary end product would be expected to diminish production of the secondary metabolite. Thus, addition of lysine to *P. chrysogenum* reduces penicillin formation.[34,35] The site of action is unknown, but current studies in my laboratory indicate that we are dealing with feedback inhibition by lysine, not feedback repression.[36] Since branched pathways are so often involved in synthesis of secondary products it appears that auxotrophic mutation in the primary branch might shunt the flow of metabolites into the secondary branch. Indeed, an isoleucine-valine-leucine-pantothenic acid auxotroph of *C. glutamicum* excretes large amounts of tetramethylpyrazine into the medium.[37] Similarly, an aromatic amino acid auxotroph of *Neurospora crassa*, blocked between dehydroshikimic acid and shikimic acid, excretes protocatechuic acid, whereas the parent culture does not.[38] This compound, which might be considered a secondary metabolite, is formed by reduction of dehydroshikimate, a simple shunt reaction induced by blockage of the primary biosynthetic pathway.

TABLE 5. Fermentations sensitive to inorganic phosphate.

Streptomycin	Nystatin	Prodigiosin
Neomycin	Oleandomycin	Chlortetracycline
Viomycin	Amphotericin B	Monensin
Nebramycin		

Some fermentations must be conducted in the presence of a level of inorganic phosphate suboptimal for growth (Table 5). The diminution in product formation caused by normal phosphate concentrations in some of these fermentations might involve the well-known feedback repression and inhibition of phosphatase by inorganic phosphate. Since many biosynthetic intermediates of certain secondary pathways are phosphorylated, whereas the ultimate products are not, phosphatases must participate in biosynthesis. Streptomycin biosynthesis, markedly inhibited by phosphate, includes at least three phosphate-cleaving steps in formation of the streptidine moiety alone.[39,40] The penultimate compound in streptomycin production appears to be streptomycin phosphate (phosphate esterified to a hydroxyl group of the streptidine moiety).[41] Miller and Walker[42] recently reported that a partially purified preparation of "streptomycin phosphatase" was nearly 90% inhibited by 17 mM-inorganic phosphate. When fermentations were conducted in complex medium containing a 10 mM-phosphate supplement, streptomycin production decreased while streptomycin phosphate accumulated extracellularly. When phosphate was added after streptomycin began to form, production of streptomycin was stopped, and streptomycin phosphate accumulated.

5. Catabolite regulation in biosynthetic pathways

The inhibition, repression or inactivation of enzymes by catabolism of a rapidly used carbon source, usually glucose, does not influence fermentations yielding primary metabolites. There is little doubt, however, that some type of catabolite control affects secondary metabolism. After years of empirical development, most fermentations are now conducted with sources of carbon and energy other than glucose.[43] The unfavourable effect on penicillin production of rapid growth in glucose is well known, and glucose adversely affects even resting cells of *P. chrysogenum*. Lactose manufacturers benefited greatly from this effect until the discovery that continuous feeding of *low* glucose concentrations allowed maximal penicillin formation. Gibberellin fermentation also responds well to limited glucose. Glucose markedly decreases alkaloid production by *C. paspali*; here, polyols and organic acids are the preferred carbon and energy sources. Resting cells of *Streptomyces sioyaensis* are adversely affected in siomycin production by glucose. Glucose also supresses production of actinomycin, violacein, mitomycin, bacitracin, neomycin, coumermycin and enterotoxin B. The most basic discovery in this area is that production of phenoxazinone synthetase, an obligatory enzyme of actinomycin synthesis, is repressed by commercial galactose preparations,[44] the actual repression being exerted by the contaminating glucose.[45] Although production of streptomycins employs glucose as carbon and energy source, the sudden appearance of the mannosidase which catalyses the desirable conversion of mannosidostreptomycin to streptomycin very late in the fermentation is due to glucose exhaustion and release of catabolite repression.[46] Since bacterial sporulation appears to be controlled by catabolite repression,[47] it is a fairly safe bet that carbon sources such as glucose can repress formation of all polypeptide antibiotics produced in association with sporulation of bacilli.

In Gram-negative bacteria, catabolite repression is mediated by cyclic AMP, i.e. rapid catabolism of a carbon source drastically lowers intracellular levels of cyclic AMP, a compound specifically needed for transcription of DNA into messenger RNA.[48] The significance of cyclic AMP in other microbial groups is unknown, and it is important to determine whether cyclic AMP plays any role in the fermentative production of desirable metabolites. Much basic work on catabolite regulation must be done in the applied area if we are to utilize the recent findings of molecular biology. Experimentation with mutants resistant to catabolite repression as fermentation organisms would appear to be of primary importance.

6. Induction in biosynthetic pathways

In certain secondary metabolite fermentations, the response to stimulatory additives resembles the phenomenon of enzyme induction. One example involves tryptophan, a stimulatory precursor of ergoline alkaloids in *Claviceps*. Although the stimulatory action of tryptophan was once thought to result solely from its precursor activity, the following observations indicate that it also acts as an inducer of one or more idiophase (production phase) enzymes: (i) tryptophan analogues that are not incorporated into alkaloids nevertheless stimulate alkaloid production;[49] (ii) stimulation of alkaloid biosynthesis requires tryptophan addition during trophophase (growth phase), later addition has little or no effect;[50,51] (iii) the added tryptophan is removed from the medium during growth and reaches two- to threefold the normal intracellular concentration just prior to alkaloid production;[52] (iv) the molar yield of alkaloid is greater than the concentration of tryptophan added, but without exogenous tryptophan, production is negligible; (v) mycelium grown in a tryptophan medium and shifted to a fresh, tryptophan-free medium during trophophase produces little alkaloid. A similar shift during idiophase results in a high rate of alkaloid formation. Apparently, the added tryptophan accumulates in the mycelium, induces the enzyme(s) of alkaloid synthesis, and is used as a precursor; after depletion, endogenous synthesis supplies sufficient tryptophan.

A similar induction effect is seen in the methionine stimulation of cephalosporin C biosynthesis.[34] Methionine acts best when added during trophophase, its effect in defined medium can be duplicated by its analogue, norleucine, and methionine is completely removed from the medium before idiophase production ensues.

7. Energy charge regulation in biosynthetic pathways

Chlortetracycline formation is markedly reduced by inorganic phosphate; in batch fermentations, idiophase begins when phosphate in the medium is exhausted.[53] Since chlortetracycline biosynthesis involves no known phosphorylated intermediates, the detrimental effect of phosphate probably does not involve regulation of phosphatases. There is a distinct possibility that the mechanism of phosphate inhibition of chlortetracycline biosynthesis involves regulation of energy charge. This type of control, characterised by Atkinson,[8] involves activation and inhibition of enzymes of primary metabolism by the relative levels of ATP, ADP and AMP in the cell. Energy charge is quantitatively measured as

$$\frac{(ATP) + \frac{1}{2}(ADP)}{(ATP) + (ADP) + (AMP)}.$$

High energy charge inhibits some enzymes and activates others. A high phosphate concentration in the medium might increase ATP formation and lead to a high energy charge in the cell. If chlortetracycline biosynthesis were inhibited by high energy charge, the link between phosphate concentration and antibiotic synthesis would be established. Janglová, Suchý and Vanek[54] examined the ATP content of two strains of *Streptomyces aureofaciens*: a low producer (200 µg/ml) and a high producer (2000 µg/ml) of chlortetracycline. In both strains, ATP concentration rapidly increased during growth, then rapidly decreased and remained at a lower level for most of the fermentation cycle. The low producer was found to have two to four times as much ATP as the high producer throughout.

8. The trophophase–idiophase relationship

The most intriguing problem of secondary metabolism centres on the mechanism by which formation of secondary products is usually retarded until the trophophase approaches completion.[55] The presence of a distinct idiophase has been observed in a large number of fermentations, some of which are listed in Table 6. One might also add to this list bacterial spores and spore-specific products such as dipicolinic acid and the polypeptide antibiotics associated with sporulation of various bacilli.

Clearly, these metabolites fail to appear during growth because the enzymes responsible for their formation are repressed during trophophase. Among those key enzymes derepressed after growth are the amidinotransferase of streptomycin biosynthesis,[39] penicillin acyl transferase and the phenylacetate-activating enzyme of penicillin biosynthesis,[56,57] enzymes I and II of gramicidin S biosynthesis[58] and phenoxazinone synthetase of actinomycin biosynthesis.[59] In all cases examined, the derepression involves protein synthesis, not precursor activation, since inhibitors of protein synthesis block formation of the enzyme and subsequent antibiotic synthesis. Idiophase genes are clearly repressed during normal growth, but we have no idea what type of repression is involved. The following five possibilities merit serious consideration. At present, there is no reason to expect that only one of these will emerge as the "true" mechanism controlling all fermentations of secondary metabolites. More likely, any one biosynthetic process is influenced by several regulatory controls.

TABLE 6. Some metabolites produced during the idiophase.

Streptomycin	Gramicidin S	Edeine
Chlortetracycline	Bacitracin	Prodigiosin
Polymyxin	Gibberellic acid	Kojic acid
Circulin	Hadacidin	6-Methylsalicylic acid[a]
Erythromycin	Novobiocin	Diphtheria toxin
Tyrocidine	Actinomycin	Staphylococcal enterotoxin
Mitomycin	Aflatoxin	Botulinus toxin
Penicillin	Pyrocyanine	Tetanus toxin
Ergot alkaloids	Mycobacillin	Neomycin

[a]Also the phenolic derivatives produced from this compound, including the antibiotic, patulin.

(i) An inducer must accumulate after growth or be added exogenously for idiophase genes to be derepressed.

(ii) A primary end product exerts feedback repression on the secondary pathway. Exhaustion of this compound derepresses idiophase genes.

(iii) Growth on a readily utilisable carbon source represses idiophase genes by catabolite repression. Depletion of these catabolites derepresses idiophase genes.

(vi) The idiophase pathway is repressed by a high energy charge. Derepression occurs when ATP formation is diminished.

(v) RNA polymerase during trophophase can initiate transcription of trophophase genes only, because it cannot attach to promotor sites of idiophase operons. After growth ceases, the structure of the enzyme changes, allowing it to initiate transcription of idiophase genes. The basis for this hypothesis is the finding that in *B. subitilis*, an organism known to make antibiotics after growth and prior to the appearance of spores, the template specificity of RNA polymerase does change after growth, i.e. the enzyme can no longer transcribe DNA from phage φe.[60] The structural change apparently involves cleavage of one of the polypeptide subunits of the core enzyme from a molecular weight of 155 000 to one of 110 000 daltons.

Repression of secondary biosynthetic enzymes during trophophase can apparently be disturbed by certain nutritional and genetic modifications. Although other explanations are possible, the observations listed below can best be understood if we assume the presence of a regulatory gene which is transcribed into regulatory messenger RNA which in turn is translated into a regulatory protein (repressor) that represses the structural gene for an idiophase enzyme (Figure 2). Any manipulation which

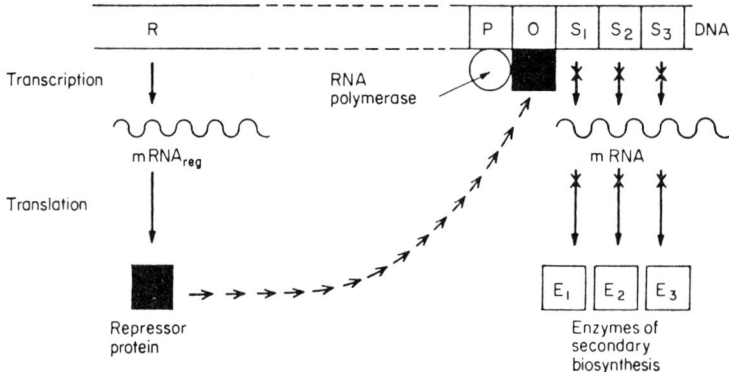

Figure 2. Possible scheme of trophophase repression of idiophase enzymes. R, repressor gene; P, promotor gene; O, operator gene; S, structural gene; E, enzyme; X, inhibited reactions.

interferes with this type of regulation without disturbing transcription and translation of the genes directing secondary product synthesis would be expected to derepress the enzymes forming secondary products. The observations are as follows.

(i) A mutant strain of *Penicillium patulum*, isolated by Light,[61] produced 6-methylsalicylic acid synthase and 6-methylsalicylic acid during growth. The addition of low levels of protein-synthesis inhibitors (cycloheximide, DL-*p*-fluorophenylalanine or DL-4-methyltryptophan) stimulated formation of 6-methylsalicylic acid synthase while partially inhibiting bulk protein synthesis.[62] Higher levels completely inhibited synthesis of both bulk protein and 6-methylsalicylic acid synthase. Apparently, translation of regulatory mRNA is more sensitive to inhibition by these agents than is translation of mRNA for 6-methylsalicylic acid synthase. The mutation described above probably occurred in the regulatory gene, producing an inactive repressor protein.

(ii) Jones and Weissbach[63] found that addition of 5-fluorouracil, an inhibitor of RNA synthesis, to trophophase mycelia of *Streptomyces antibioticus* caused increased and "premature" formation of phenoxazinone synthetase, a key enzyme of actinomycin synthesis. In this case, it appears that transcription of the regulatory gene into regulatory messenger RNA is more sensitive to inhibition than is transcription of the gene coding for phenoxazinone synthetase.

(iii) In the cases of penicillin,[64] chloramphenicol[27] and colistin,[65] fermentations in complex media follow typical trophophase-idiophase kinetics. The use of chemically defined media which support slower growth, however, results in antibiotic production during growth. Either the low rate of growth or the absence of some unknown medium ingredient or both probably interfere with production of the repressor protein.

9. A role for continuous culture

In a recent thought-provoking review, Tempest[66] pleads for the use of continuous culture as a tool for advancing microbiological research. In my opinion, the open system of continuous-flow can best prove itself in the area of metabolite overproduction, especially overproduction of secondary metabolites. Thus far, I have described how results of batch fermentations indicate that feedback regulation, catabolite regulation, induction and energy charge regulation play roles in secondary biosynthesis. I am afraid that these results will remain as mere indications unless the emphasis in our research is shifted from cells which are continuously undergoing physiological change to cells growing in unique and controllable steady-state environments.

Being a non-mathematical person and never having conducted a continuous flow experiment, I properly should end my presentation here, deriving a measure of security from the vagueness of the above paragraph. However, not being conservative at heart, I have succumbed to the temptation of making some suggestions for future research designed to gain a better understanding of secondary biosynthesis.

9.1. Feedback regulation

The indication that chloramphenicol inhibits its own synthesis was described almost 20 years ago,[67] but little progress has been made, due to the complication that the antibiotic is degraded by the producing organism.[27] The problems concerned with the long residence time of added chloramphenicol in the batch fermentation and the different rates of chloramphenicol breakdown at the different growth rates of the batch cycle might be by-passed easily in the chemostat.

In branched pathways leading to both a primary product and a secondary product, the true physiological importance of feedback regulation by the primary end product on production of the secondary metabolite could be ascertained in the chemostat. Pirt[1] has already suggested lysine limitation of a lysine auxotroph (blocked after aminoadipic acid) of *P. chrysogenum* to determine whether endogenous lysine limits penicillin production.

9.2. Catabolite regulation

Catabolite regulation of secondary biosynthesis is difficult to study in batch cultures because of the widely different growth rates obtained on different carbon sources. A further complication, revealed by the elegant experiments of Pirt and Righelato[63] on penicillin formation, is that the rate of decay of the specific rate of product formation (q_{pen}) after growth ceases is inversely related to the previous growth rate. The use of the chemostat would allow various carbon sources to be examined at the same growth rate. Since secondary metabolites can be produced during growth in a single-stage chemostat, repressive carbon sources should give low specific rates of product formation while nonrepressive carbon compounds should yield high rates. In a related area, Dawes and Mandelstam[68] found that glucose or nitrogen limitation of *B. subtilis* resulted in a high degree of sporulation whereas limitation of tryptophan, citrate, Mg or phosphate did not, thereby supporting an earlier suggestion[47] that a metabolite containing nitrogen and carbon was responsible for repression of spore formation. Even more revealing than rates of product formation would be specific activities of key enzymes such as phenoxazinone synthetase of actinomycin biosynthesis. The effect of mixtures of carbon sources at varying concentration ratios on the formation of these enzymes could be studied in a manner similar to studies on enzymes of primary metabolism.[66, 69, 70] Also of interest will be the effect of growth rate on the repressive effect of particular carbon sources. In *Escherichia coli*, Silver and Matales[70] found that the degree of glucose repression of β-galactosidase and aspartase increased as growth rate increased. Similarly, sporulation of carbon- or nitrogen-limited *B. subtilis* is repressed to a greater extent as growth rate increases.[68]

The studies outlined above would reveal whether the usefulness of carbon sources previously known to be "good" or "poor" for product formation results from their inability or ability to repress secondary biosynthetic enzymes. Even more important is the determination of whether catabolite repression causes the typical trophophase-idiophase kinetics of any batch process. If secondary biosynthetic enzymes are indeed found to be more repressed at higher growth rates, a link between catabolite regulation and batch culture kinetics will have been established. Although Pirt and Righelato[64] found q_{pen} to be independent of growth rate over the range of 0.014 to 0.086 h^{-1} (μ_{max} near 0.095 h^{-1}) in glucose-limited chemostat cultures of *P. chrysogenum*, the defined medium used did not exhibit trophophase–idiophase kinetics in batch culture. The addition of corn steep liquor to this medium resulted in typical batch culture kinetics, but only a limited amount of work was done with this complex medium in the chemostat. Only two growth rates were studied (0.033 and 0.077 h^{-1}) both yielding the same q_{pen} as in defined medium. Since the μ_{max} is probably considerably greater than 0.095 h^{-1} in the corn steep medium, it would be interesting to determine q_{pen} at growth rates near μ_{max} in this medium. Furthermore, other types of limitation should be examined to determine the effect of growth rate in excess glucose on q_{pen}. One might profitably employ the system of Řičica et al.[71] (two-stage system with flow of fresh medium into the second stage) to provide steady-state conditions at high growth rates in excess glucose.

9.3. Energy charge regulation

The studies of Sikyta, Slezák and Herold[72] on the continuous chlortetracycline fermentation revealed a finding which may aid assessment of the role of energy charge in secondary biosynthesis. These workers used a two-stage system and a sucrose-containing complex fermentation medium. In the batch process, chlortetracycline production begins at the point of phosphate exhaustion. Continuous culture at dilution rates of 0.151, 0.138 and 0.092 h^{-1} showed steady levels of antibiotic in both stages throughout the fermentations. The incoming medium contained 50 to 60 µg/ml phosphorus, about half from potassium phosphate and half contained in corn-steep liquor. Soon after the medium flow started, the phosphorus level in the chemostat was undetectable. When fermentation was conducted at 0.049 h^{-1}, a curious observation was made: the chlortetracycline levels decreased throughout the course of the experiment in both vessels. Although at this low dilution rate, as in the others, phosphorus was almost zero, elimination of the potassium phosphate component of the medium prevented the decrease in antibiotic titer. This suggests that only at the low dilution rate is the intracellular phosphate available to repress chlortetracycline biosynthesis. It would be particularly instructive to examine the pools of cultures growing at 0.049 and 0.092 h^{-1} for their content of AMP, ADP and ATP.

This last example suggests an extremely important point. To gain truly significant information about regulatory controls we cannot be satisfied with merely assaying the content of a compound in the incoming and outgoing broth. The intracellular concentration provides the most important data. For example, production of glutamine synthetase has been reported, on the basis of results obtained in batch culture, to be repressed by glutamine, asparagine, arginine, methionine, tryptophan and ammonia and to be induced by glutamate. However, when Meers and Tempest[73] examined it it in the chemostat, they found that the sole controlling factor was the intracellular ammonia concentration.

10. Summary

An effective fermentation organism is a wasteful creature which overproduces and excretes its metabolic intermediates and end products. Cultures obtained from screening programmes usually possess subnormal regulatory controls. Development programmes to increase product formation modify the residual control mechanisms so that the culture's "inefficiency" is increased.

For production of primary metabolites, feedback inhibition and repression must be by-passed. This is usually accomplished by decreasing the intracellular concentration of feedback inhibitors and repressors. Auxotrophic mutants and analogue-resistant mutants are most often used for this purpose. Development of fermentations for secondary metabolites such as antibiotics are less rational, due to our ignorance of the biosynthetic pathways involved. However, evidence is accumulating that such fermentations are subject to (i) feedback regulation by the final product, (ii) feedback regulation by primary metabolites which are precursors of the secondary metabolite, (iii) feedback regulation by primary metabolites which share a branched biosynthetic pathway with the secondary metabolite, (iv) feedback regulation by inorganic phosphate, (v) catabolite regulation by glucose and other rapidly utilised sugars, (vi) induction by primary metabolites which are precursors of the secondary metabolite and (vii) energy charge regulation.

Secondary metabolites are usually formed during idiophase. Enzymes of secondary metabolism appear to be repressed during trophophase. We have no clear idea concerning the type of repression control involved but the possibilities include inducer production, feedback repression, catabolite repression, energy charge regulation and modification of RNA polymerase. Repression of secondary biosynthetic enzymes during trophophase can apparently be disturbed by certain nutritional and genetic manipulations.

Continuous culture can profitably be used to answer many questions concerning metabolite overproduction and regulation, because continuous culture can provide an inexhaustible number of unique and controllable steady-state environments.

Acknowledgment

The preparation of this manuscript was supported by U.S. Public Health Service Research Grant AI-09345 from the National Institute of Allergy and Infectious Diseases.

References

1. Pirt, S. J. *Symp. Soc. gen. Microbiol.* **1969**, 19, 199.
2. Clarke, P. H.; Lilly, M. D. *Symp. Soc. gen. Microbiol.* **1969**, 19, 113.
3. Stadtman, E. R. *Adv. Enzymol.* **1966**, 28, 41.
4. Stadtman, E. R. *Ann. N.Y. Acad. Sci.* **1968**, 151, 516.
5. Datta, P. *Science, N.Y.* **1969**, 165.
6. Paigen, K.; Williams, B. *Adv. microbiol. Physiol.* **1970**, 4, 251.
7. Rebello, J. L.; Jensen, R. A. *J. biol. Chem.* **1970**, 245, 3738.
8. Atkinson, D. E. *A. Rev. Microbiol.* **1969**, 23, 47.
9. Edlin, G.; Broda, P. *Bact. Rev.* **1968**, 32, 206.
10. Nakayama, K.; Tanaka, H.; Hagino, H.; Kinoshita, S. *Agric. biol. Chem.* **1966**, 30, 611.
11. Shiio, I.; Ishii, K. *J. Biochem.* **1969**, 66, 175.
12. Lingens, F.; Goebel, W.; Uesseler, H. *Eur. J. Biochem.* **1966**, 2, 442.
13. Westlake, D. W. S.; Vining, L. C. *Biotechnol. Bioengng* **1969**, 11, 1125.
14. Jensen, R. A.; Rebello, J. L. *Devs ind. Microbiol.* **1970**, 11, 105.
15. Demain, A. L.; Birnbaum, J. *Curr. Topics Microbiol. Immunol.* **1968**, 46, 1.
16. Matsuo, T.; Oyama, Y.; Tanimoto, H.; Hashida, W.; Terramoto, S. *Hakko to Taishi* **1966**, 12, 915.
17. Demain, A. L. *Prog. ind. Microbiol.* **1968**, 8, 35.
18. Shiio, I.; Nakamori, S. *Agric. biol. Chem.* **1970**, 34, 448.
19. Reh, M.; Schlegel, H. G. *Arch. Mikrobiol.* **1969**, 67, 110.
20. Demain, A. L.; Jackson, M.; Vitali, R. A.; Hendlin, D.; Jacob, T. A. *Appl. Microbiol.* **1966**, 14, 821.
21. Kleinkauf, H.; Gevers, W. *Cold Spring Harb. Symp. quant. Biol.* **1969**, 34, 805.

22. McCormick, J. R. D. In *Antibiotics. II. Biosynthesis*. (D. Gottlieb and P. D. Shaw, eds.) Springer-Verlag, New York. **1967**, p. 113.
23. Kominek, L. A. In *Antibiotics. II. Biosynthesis*. (D. Gottlieb and P. D. Shaw, eds.) Springer-Verlag, New York. **1967**, p. 231.
14. Paulus, H. In *Antibiotics. II. Biosynthesis*. (D. Gottlieb and P. D. Shaw, eds.) Springer-Verlag, New York. **1967**, p. 254.
25. Williams, P. R.; Hearn, W. R. In *Antibiotics. II. Biosynthesis*. (D. Gottlieb and P. D. Shaw, eds.) Springer-Verlag, New York. **1967**, pp. 410, 449.
26. Martin, J. R.; Rosenbrook, W. *Biochemistry* **1967**, 6, 435.
27. Malik, V. S.; Vining, L. C. *Can. J. Microbiol.* **1970**, 16, 173.
28. Bu'Lock, J. D.; Hulme, M. A.; Shepherd, D. *Nature, Lond.* **1966**, 211, 1090.
29. Woodruff, H. B. *Symp. Soc. gen. Microbiol.* **1966**, 16, 22.
30. Trenina, G. A.; Trutneva, E. M. *Antibiotiki* **1966**, 11, 770.
31. Dulaney, E. L.; Dulaney, D. D *Trans. N.Y. Acad. Sci.* **1967**, 29, 782.
32. Goulden, S. A.; Chattaway, F. W. *J. gen. Microbiol.* **1969**, 59, 111.
33. Elander, R. P.; Mabe, J. A.; Hamill, R. L.; Gorman, M. Abstracts. 1st int. Symp. Genetics of Industrial Micro-organisms. **1970**, p. 166.
34. Demain, A. L. In *Biosynthesis of Antibiotics*. (J. Snell, ed.) Academic Press, New York, **1967**, p. 29.
35. Goulden, S. A.; Chattaway, F. W. *Biochem. J.* **1968**, 110, 55P.
36. Masurekar, P. S.; Demain, A. L. *Bact. Proc.* **1971**, 8.
37. Demain, A. L.; Jackson, M.; Trenner, N. R. *J. Bact.* **1967**, 94, 323.
38. Gross, S. R. *J. biol. Chem.* **1958**, 233, 1146.
39. Walker, J. B. *Devs ind. Microbiol.* **1967**, 8, 109.
40. Demain, A. L.; Inamine, E. *Bact. Rev.* **1970**, 34, 1.
41. Nomi, R.; Nimi, O. *Agric. biol. Chem.* **1969**, 33, 1459.
42. Miller, A. L.; Walker, J. B. *J. Bact.* **1970**, 104, 8.
43. Demain, A. L. *Lloydia* **1968**, 31, 395.
44. Marshall, R.; Redfield, B.; Katz, E.; Weissbach, H. *Archs Biochem. Biophys.* **1968**, 123, 317.
45. Katz, E. Personal communication. 1970.
46. Inamine, E.; Lago, B. D.; Demain, A. L. In *Fermentation Advances*. (D. Perlman, ed.) Academic Press, New York. **1969**, p. 199.
47. Shaeffer, P. *Bact. Rev.* **1969**, 33, 48.
48. Pastan, I.; Perlman, R. *Science, N.Y.* **1970**, 169, 339.
49. Robbers, J. E.; Floss, H. G. *J. pharm. Sci* **1970**, 59, 702.
50. Bu'Lock, J. D.; Barr, J. G. *Lloydia* **1968**, 31, 342.
51. Vining, L. C. *Can. J. Microbiol.* **1970**, 16, 473.
52. Floss, H. G. In *Internationalishe Symposium am Biochemie und Physiologie der Alkaloide*. Akademie-Verlag, Berlin. **1969**, p. 21.
53. Hošťálek, Z. *Folia microbiol., Praha* **1964**, 9, 78.
54. Janglová, Z.; Suchý, J.; Vaněk, Z. *Folia microbiol., Praha* **1969**, 14, 208.
55. Bu'Lock, J. D. *Essays in Biosynthesis and Microbial Development*. John Wiley, New York. **1967**.
56. Pruess, D. L.; Johnson, M. J. *J. Bact.* **1967**, 94, 1502.
57. Bunner, R.; Roehr, M.; Zinner, M. *Hoppe-Seylers Z. physiol. Chem.* **1968**, 349, 25.
58. Kurahashi, K.; Yamada, M.; Mori, K.; Fujikawa, K.; Kambe, M.; Imae, Y.; Sato, E.; Takahashi, H.; Sakamoto, Y. *Cold Spring Harb. Symp. quant. Biol.* **1969**, 34, 815.
59. Katz, E. In *Antibiotics. II. Biosynthesis*. (D. Gottlieb and P. D. Shaw, eds.) Springer-Verlag, New York. **1967**, p. 276.
60. Losick, R.; Shorenstein, R. G.; Sonenshein, A. L. *Nature, Lond.* **1970**, 227, 910.
61. Light, R. J. *J. biol. Chem.* **1967**, 242, 1880.
62. Light, R. J. *Archs Biochem. Biophys.* **1967**, 122, 494.
63. Jones, G. H.; Weissbach, H. *Archs Biochem. Biophys.* **1970**, 137, 558.
64. Pirt, S. J.; Righelato, R. C. *Appl. Microbiol.* **1967**, 15, 1284.
65. Ito, M.; Aida, K.; Uemura, T. *Agric. biol. Chem.* **1969**, 33, 262.
66. Tempest, D. *Adv. microbiol Physiol.* **1970**, 4, 223.
67. Legator, M.; Gottlieb, D. *Antibiotics Chemother.* **1953**, 3, 809.

68. Dawes, I. W.; Mandelstam, J. In *Continuous Culture of Micro-organisms*. Proceedings of the 4th International Symposium on the Continuous Cultivation of Micro-organisms. (I. Málek *et al.*, eds.) Academia, Prague. **1969**, p. 157.
69. Hamlin, B. T.; Ng, F. M. W.; Dawes, E. A. In *Microbial Physiology and Cell Culture*. Proceedings of the 3rd International Symposium on the Continuous Cultivation of Micro-organisms. (E. O. Powell *et al.*, eds.) H.M.S.O., London. **1967**, p. 211.
70. Silver, R. S.; Matales, R. I. *J. Bact.* **1969**, 97, 535.
71. Řičica, J.; Nečinová, S.; Stejskalová, E.; Fencl, Z. In *Microbial Physiology and Cell Culture*. Proceedings of the 3rd International Symposium on the Continuous Cultivation of Micro-organisms. (E. O. Powell *et al.*, eds.) H.M.S.O., London. **1969**, p. 196.
72. Sikyta, B.; Slezák, J.; Herold, M. In *Continuous Culture of Micro-organisms*. Proceedings of the 2nd International Symposium on the Continuous Cultivation of Micro-organisms. (I. Málek *et al.*, eds.) Publishing House of the Czechoslovak Academy of Sciences, Prague. **1964**, p. 173.
73. Meers, J. L.; Tempest, D. W. *Biochem. J.* **1970**, 119, 603.

Part IX

KINETICS AND MODELING

Editor's Comments
on Papers 19 and 20

19 GADEN
Fermentation Process Kinetics

20 CALAM, ELLIS, and McCANN
Mathematical Models of Fermentations and a Simulation of the Griseofulvin Fermentation

Gaden's paper on fermentation process kinetics appeared when there was a growing realization among leading fermentation technologists that the principles of chemical kinetics could be applied to biochemical reaction systems. Gaden foresaw and articulated the importance of analyzing chemical changes occurring in fermentation processes, calculating volumetric and cell-specific rates, displaying rate processes graphically, and on this basis characterizing and typing batch fermentation processes. Detailed kinetic characterization of the physical and chemical rate processes occurring necessarily under strict control of process variables was seen to be essential for designing continuous processes, and it was predicted to become invaluable for improving batch processes. In Paper 19 Gaden reviewed the development of the theory of fermentation kinetics, described what he saw as fundamental rate processes in fermentations, and offered a pragmatic classification scheme for a few simple and complex batch fermentation processes concerning which many published data were available.

Mathematical modeling or simulation is the art of developing a set of algebraic or differential equations based on careful characterization of chemical and physical rate processes in fermentations. Although mathematical models are being used increasingly in attempts to enhance knowledge of fermentation kinetics in batch and continuous processes, and to improve the design of batch, continuous, or quasi-continuous processes, it seems that with all but the simplest mathematical models the use of advanced computational methods, usually employing digital computers, is required. In Paper 20 Calam et al. have reviewed the principal models that have been employed to simulate fermentation processes and have described the development of a mechanistic model expressed by 17 differential equations, based on experimental data, time-course graphical display and analysis of fermentation patterns, and conception of a schematic graphical model of the key biochemical events presumed to govern the fermentation.

Mathematical solution of the equations was not necessary because a program employing a simulation language was devised and a computer of advanced design was used to produce graphs that, after adjustment of rate constants and reconsideration of stoichiometric relationships, were made to simulate the griseofulvin fermentation very well. Calam and coworkers argue convincingly for the general utility of their methods for simulation and optimization of fermentation processes.

19

Copyright © 1959 by John Wiley and Sons, Inc.

Reprinted from *J. Biochem. Microbiol. Tech. Eng.*, 1(4), 413–429 (1959)

Fermentation Process Kinetics*

ELMER L. GADEN, JR.
*Department of Chemical Engineering, Columbia University,
New York 27, N.Y.*

Summary. Information on fermentation process kinetics is potentially valuable for the improvement of batch process performance; it is essential for continuous process design. An empirical examination of rate patterns in various fermentations discloses three basic types: (1) 'growth associated' products arising directly from the energy metabolism of carbohydrates supplied, (2) indirect products of carbohydrate metabolism and (3) products apparently unrelated to carbohydrate oxidation. Effects of operating variables on the primary kinetic processes, growth, sugar utilization and antibiotic formation, in the penicillin process, illustrate the special nature of this type.

Introduction

In the design of any chemical, or biochemical, process one must consider two more or less distinct aspects. First, there are the chemical reactions themselves and secondly, the numerous physical processes which precede, accompany and follow them. Some of these physical processes are quite clearly separate, like the purification of raw materials and products. Others, like the transport of materials to and from the surface of a solid catalyst, are intimately bound up with the reactions themselves.

For a long time, methods available for dealing with the physical aspects of chemical processes were better developed than those for handling the chemical changes themselves. This was largely the result of empirical simplifications offered by the 'unit operations' concept in chemical engineering. With the rapid development of chemical kinetics and, equally important, methods for applying kinetic relationships to process design, this disparity has been overcome.

* Presented at the 134th National Meeting of the American Chemical Society, Chicago, September 1958.

Kinetics is concerned with reaction rates in general; 'process kinetics' simply suggests a primary concern with the rates of commercially practised reactions and, particularly, with the effects of process variables on them.

Since fermentation is only another type of chemical process, albeit a special and complex one, possibilities for applying ideas and techniques developed for more conventional chemical systems should always be sought. This is especially true for kinetics. Although the study of fermentation rates is relatively new, it promises much for the fuller and more efficient exploitation of biochemical reaction systems.

Development of Fermentation Kinetics

Final product yields and substrate conversions were the only criteria of performance in early commercial fermentations. As the technology developed, however, greater attention was paid to time factors; 'productivity', the average rate of product formation (Fig. 1), soon became popular as a basis for comparison. On the other hand, instantaneous rates were largely ignored until the studies of gluconic acid production by Wells, Moyer, Gastrock et al.[12, 19, 20, 26] in the late 1930's. They were among the first to report rates of sugar utilization and acid formation in detail.

The introduction of antibiotic fermentations greatly stimulated interest in fermentation rates. It was recognized from the first that these processes were markedly different from most earlier fermentations. Studies of the chemical changes in penicillin biosynthesis required frequent analysis of carbohydrate and nitrogen levels, cell weight and antibiotic titre. From these, general rate patterns could be discerned and it was soon noted that the process comprised two more or less distinct phases; growth and antibiotic production.

Dulaney et al.,[8] noticed the same general behaviour in streptomycin fermentations. They defined an initial 'growth phase' in which mycelium was rapidly generated, accompanied by a reduction in soluble medium constituents (carbon, nitrogen, phosphorous), rapid sugar utilization and high oxygen demand. Virtually no streptomycin was produced. Following this was an 'autolytic phase', characterized by a marked drop in mycelial weight, release of nitrogen and inorganic phosphorous to the medium, low oxygen demand and rapid antibiotic synthesis. All strains examined exhibited the same basic pattern and gross medium changes had little effect on it.

Calam, Driver and Bowers[6] were among the first to support these general observations with specific experiments. Penicillin fermentations were carried out at several temperatures between 12° and 32 °C and average rates of growth, respiration and penicillin synthesis noted. By plotting the observed rates in the Arrhenius manner (logarithm of rate versus reciprocal absolute temperature) it was possible to characterize each process by the slope of the line obtained, the 'thermal increment'. Since these three rates all exhibited significantly different thermal increments, the authors concluded that the 'pace-setting enzyme-systems' involved are different.

Any survey of the literature on fermentation rates underscores the dearth of direct kinetic studies of this type. One cardinal reason for this is the matter of experimental procedure itself. Rate information can best be obtained in steady-state (continuous) systems with automatic control of process variables.

In an excellent example of this approach, Kempe, Gillies and West[15] studied rates of acid production by *Lactobacillus delbrueckii* at controlled pH. Rates were determined by differentiating the automatically recorded curve of alkali addition. Steady-state operation at various temperatures provided values for an Arrhenius-type plot which gave an activation energy of 17 kcal/g mole, a value in the range characteristic of many chemical reactions.

For one reason or another satisfactory methods for automatic regulation and control in fermentation studies have only recently been introduced and most experiments so far reported involve the classical batch technique. Data which permit the computation of rates are rare—and often inadequate because of the absence of key values. Of course, the aim of these experiments was yield improvement in batch processes, not the gathering of kinetic data. Still, despite the inherent limitations of the unsteady-state, batch technique, a surprising amount of information has been accumulated and a great deal has been learned about the general kinetic aspects of various fermentation processes.

From an analysis of the rate patterns in batch alcohol, citric acid and penicillin fermentations, for example, Gaden[9] distinguished between three broad kinetic groups.

(1) Processes in which the desired products (ethanol, gluconic and lactic acids, for example) arise directly from oxidation of the primary carbohydrate.

(2) Processes in which the products (citric acid, for example), though also resulting from carbohydrate dissimilation, do so indirectly and accumulate only under conditions of restricted or abnormal metabolism.

(3) Processes in which product formation has no apparent association with carbohydrate oxidation (penicillin and many other antibiotics are examples of this type).

It must be recognized that a classification of this sort is based on purely empirical examination of batch fermentation results, not on a full and complete understanding of the individual mechanisms involved and their relationships to one another. Still, until such understanding has been achieved, empirical analysis is a powerful and useful tool—so long as its limitations are kept constantly in mind.

More recently Luedeking[16] investigated the kinetics of the lactic acid fermentation using a batch process at controlled pH. He showed that rate of product formation is indeed proportional to the rate of substrate utilization as expected. Furthermore, rates of acid production could be related to rates of growth by a simple expression involving two constants dependent on the pH of the fermentation.

Subsequently, the performance of single or multi-stage continuous lactic acid processes were predicted from these batch results by analytical and graphical methods.[17] Equations for both transient and steady-state operations of the continuous system have been developed.

Kinetic Phenomena in Fermentation

The first problems in studying fermentation kinetics are (1) the establishment of consistent rate expressions, and (2) the selection of meaningful rate processes to be measured.

Rates and Productivity

To avoid confusion, the term 'productivity' has been recommended for the time-average output of a process.[9] The expression 'fermentation rate' can then be reserved for the instantaneous rate of change of any concentration factor—sugar, product, cell weight, etc. These distinctions are shown graphically in Fig. 1.

Productivity is defined as the final product concentration divided by the time from inoculation to delivery of the batch. It might seem more reasonable to divide by the total process time from delivery of one batch to delivery of the next. This would include many operational factors involved in turnover of a tank,

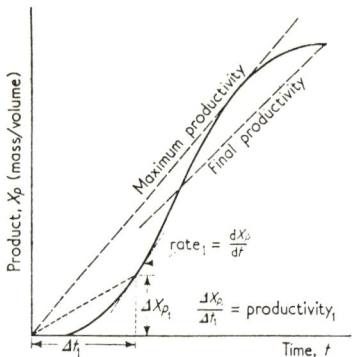

Fig. 1. Fermentation rates and productivity

like cleaning, batching and filling, which have little or nothing to do with the actual fermentation system. While it is essential for proper economic analysis of the plant, such an overall productivity has little use in analyzing the fermentation process itself.

Two bases for expressing fermentation rates have been proposed:[9]

(1) The *volumetric* rate, or the rate of change of concentration with time; its units are mass/unit time (unit volume).

(2) The *specific* rate, or the volumetric rate divided by cell concentration; its units are mass/unit time (unit cell mass).

The first is the preferred form for process design, especially for continuous systems, because it includes a volume term. The second is best for kinetic analysis because it puts everything on a comparable basis—unit mass of tissue. It does not follow, of course, that this unit tissue mass is physiologically identical throughout the fermentation process.

Rate process in fermentation

Rate measurements may be applied to an almost infinite number of factors in a fermentation system. Three of these, however, have been consistently singled out for study—growth, sugar utilization and product formation.

Growth is taken as a rough expression of the total catalytic activity in the system. Admittedly, tissue accumulation is only the crudest expression of the true levels of activity of the various enzyme systems involved. Until these can actually be determined, however, it is the best measure we have.

Synthetic processes require the metabolic energy released by oxidation of primary carbon sources and sugar utilization is generally taken as an indication of the rate of energy release to the system. While it is true that proteins and fats are similarly degraded, with accompanying energy release, carbohydrate sources are ordinarily the major energy suppliers. At the same time these materials are frequently the substrates from which specific products are formed. The key rate, product formation, needs no further elaboration.

Perhaps the greatest difficulty encountered in the examination of any complex fermentation process is the lack of any stoichiometric relationship between reactants and products. Lacking this, measurements of the three basic rates defined above may still be made. They offer the singular advantage of being determined directly from the measurements most commonly made in fermentation process studies, tissue mass, sugar and product concentrations.

Complete rate patterns, on both volumetric and specific bases, for a typical complex fermentation (streptomycin biosynthesis) are shown in Fig. 2. They were calculated from data of Sikyta et al., in the manner previously described.[9]

Fermentation process variables

Primary process variables in fermentation are: (1) temperature, (2) pH, and (3) nutrient (or reactant) concentration (including oxygen). In addition certain conditions of the physical environment, like fluid turbulence and equipment design features which affect mass transfer in the reaction zone, must be considered.

Note that the fundamental composition of the nutrient environment, as opposed to the concentration of specific components (sugar, nitrogen sources, etc.), is not included. This is considered an inherent feature of the process system and not a 'process variable' in the usual sense. While this view is reasonable for most other chemical reaction systems, it may not be so for fermentation. One cannot synthesize ammonia unless the reaction mixture contains both nitrogen and hydrogen (the mole ratio of these reactants is the 'process variable') but tetracycline can be produced in a wide variety of nutrient media.

Fermentation Process Types

Fermentation processes may be classified in a number of different ways. The first systematic approach was proposed by Gale[11] who grouped microbiological processes in a series of type groups, oxidation, reduction, hydrolysis, etc. Such an arrangement though fundamentally attractive, is only suitable for specific reactions operating on specific substrates to yield specific products. Unfortunately, many commercially important fermentation processes cannot be so neatly described.

Gale's classification scheme has recently been extended by Stodola[24] and others,[25] who have proposed a more detailed breakdown of 'type reactions'. In this scheme, micro-organisms, or more specifically their enzyme complements, are looked at as added means for controlled organic synthesis. Again, this concept is not applicable to most of the fermentation processes now practised commercially—at least at the present level of knowledge regarding mechanisms.

A different approach was proposed by Gaden.[10] It is summarized in modified form in Table I. Here fermentation processes rather than specific reactions are grouped together and the overall free energy change involved is the basis for classification.

The primary advantage of this scheme is technological; it coincides with the general classification of fermentation rate patterns suggested earlier.[9] Experience has shown that fermentation processes fall more or less into three kinetic groups, which may be designated 'types I to III' for convenience. Their relationship to the general reaction types is shown in Table I and summarized below:

Type I: processes in which the main product appears as a result of primary energy metabolism. Examples of this type of system are most common in the older branches of fermentation technology, for instance: (1) aerobic yeast propagation (mass propagation of cells in general), (2) alcoholic fermentation, (3) oxidation of glucose to gluconic acid, and (4) dissimilation of sugar to lactic acid.

Type II: processes in which the main product arises indirectly from reactions of energy metabolism. In systems of this type the product is not a direct residue of oxidation of the carbon source but the result of some side-reaction or subsequent interaction between these direct metabolic products. Examples are: (1) formation of citric and itaconic acids, and (2) formation of certain amino acids.

Type III: processes in which the main product does not arise from energy metabolism at all but is independently elaborated or accumulated by the cells. It is perfectly true that carbon, nitrogen, etc., provided in essential metabolites appear in product molecules but the major products of energy metabolism are CO_2 and water. Antibiotic synthesis (Fig. 2) is a prime example of this type.

Table I. Fermentation Process Types

Dissimilation Reactions $\Delta F = -$	Biosynthesis $\Delta F = +$
Type I. Simple: $A \to$ products $A \to B \to C \to$ products Type II. Complex: $A \to B \to C \to$ $\quad\quad\;\; \downarrow \quad \downarrow$ $\quad\quad\;\; D \quad E$ $\quad\quad\;\; \searrow \swarrow$ $\quad\quad\quad \downarrow$ $\quad\quad\;\;$ products	Type III. Biosynthesis of complex molecules: Polymerizations—carbohydrates, proteins Antibiotics, vitamins, etc. Fats

Each of these types demonstrates a fairly distinctive rate pattern. These are shown schematically in Fig. 3. The Type I processes show only one maximum for each of the rate processes and these are virtually coincident, hence the term 'growth-associated' often used for products of this process type.

In the Type II process two rate maxima are distinguishable. In the first phase tissue is produced with little product formation;

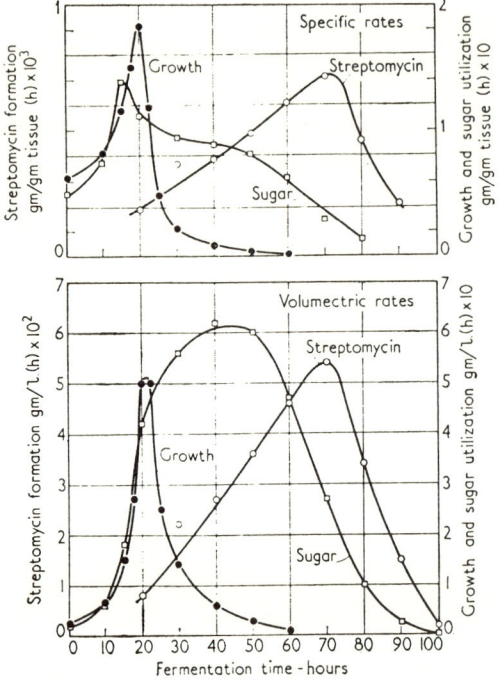

Fig. 2. Streptomycin fermentation rates

in the second product formation rate is maximized. Rapid carbohydrate utilization is common to both. Unfortunately, very few kinetic data are available for this group. In fact, until the recent development of microbiological processes for amino acids (probably Type II), the citric acid fermentation was the only example for which rate information had been published.

Type III processes again show two distinct phases. In the first tissue accumulation and all aspects of energy metabolism are maximized with virtually no accumulation of the desired product; in the second oxidative metabolism is practically over and product accumulation is maximum. Both penicillin and streptomycin (Fig. 2) fermentations are excellent examples of the Type III kinetic pattern.

It must be emphasized that these are only generalizations for technological convenience. They are neither perfect nor comprehensive and great variations may occur. A particular fermenta-

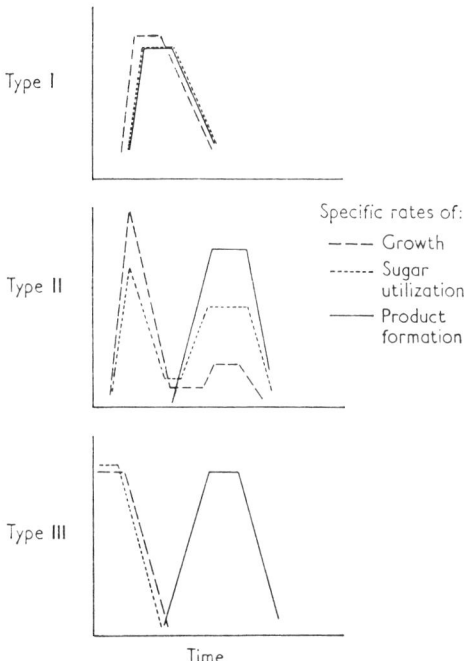

Fig. 3. Fermentation rate patterns

tion type may exhibit widely different behaviour with major changes in medium composition and process conditions. Strain variations, on the other hand, seem to have little effect on the general rate patterns.

Exceptions are found in all groups, especially Type III. In fact it may prove necessary to subdivide this group further as more kinetic information on complex processes becomes available.

One apparent exception is the production of oxytetracycline. Doskočil et al.[7] have presented a very complete study of metabolic changes observed during this fermentation. Rate curves calculated from these are shown in Fig. 4.

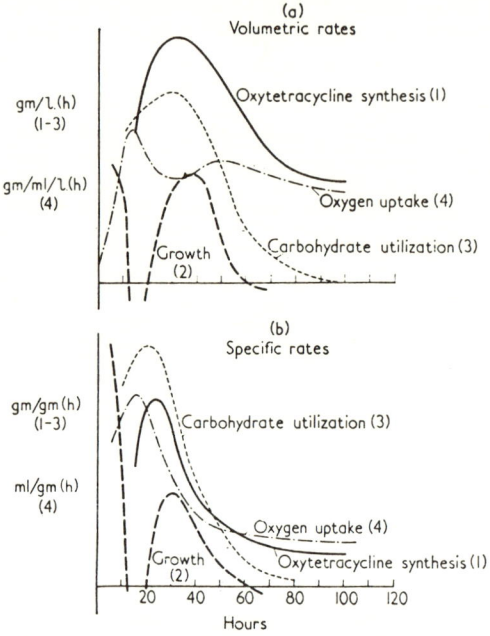

Fig. 4. Oxytetracycline fermentation rates

These authors[7] did not attempt any detailed analysis of rates, but they did suggest a multiphase nature for this process. Specifically they proposed five periods as follows:

(1) Lag: virtually no metabolic activity.
(2) Growth of primary mycelium: very high level of metabolism (respiration, nucleic acid synthesis, etc.), no antibiotic formation.
(3) Fragmentation of primary mycelium: respiration and nucleic acid synthesis fall, antibiotic synthesis is just starting.

(4) Growth of secondary mycelium; rapid antibiotic production, renewal of nucleic acid synthesis, further decrease in respiration.

(5) Stationary phase: no further growth, metabolic activity low but antibiotic synthesis continues.

Another process which one would expect to fall in Type III is the chloramphenicol (chloromycetin) fermentation. On the basis of very scanty data, however, it too appears to be an exception to the general pattern. If it is in fact, then the two processes which give a typical behaviour both involve organisms which normally fragment during growth. This may well lead to a characteristic kinetic behaviour different from that for streptomycin; unfortunately, the information available is not sufficiently complete to permit any firm conclusion.

A generous amount of sub-classification would undoubtedly remove most discrepancies. At the same time, however, it would make void the primary purpose of this approach—the establishment of certain reasonably reliable generalizations about fermentation rate patterns which can serve as a basis for further kinetic studies.

Fermentation Kinetics and Continuous Processes

Many reasons, both practically useful and intellectually satisfying, can be offered to justify more intensive study of fermentation kinetics but one outweighs all others: we cannot hope to operate continuous processes at a predictable steady state unless the relationships between major rate processes and the effects of process variables on them are known.

The reactor system which is apparently best adapted to continuous fermentation is the homogeneous, overflow type, with virtually complete backmixing. To establish an overflow reactor at steady state all rate processes must be in balance. It is possible to achieve this by simply letting the system hunt for such a point, but no one can predict in advance where this point will be. Such a procedure is hardly an adequate basis for plant operation.

For the 'kinetically simple' Type I fermentations, prediction of continuous steady state operating conditions from batch data is theoretically possible.[13, 17, 18] Furthermore, this type of process can be operated satisfactorily in a single stage system, although additional stages may be added to ensure economical utilization of nutrients supplied. Both these points have been demonstrated experimentally in a number of cases.[5, 13, 17]

On the other hand, kinetic considerations alone demand at least two stages for satisfactory operation of the more complex process types (II and III). In the first, conditions will be adjusted to provide maximum rates of growth, and energy metabolism, in the second, for maximum product formation. Kinetic studies for continuous process design should therefore be aimed primarily at elucidating the relationships between these various rates and the major, controllable process variables.

The only complex fermentation process for which studies of this sort have been made is the biosynthesis of penicillin. In the final section of this paper, that information will be collected and related to illustrate the kinetic nature of the Type III process.

Penicillin Process Kinetics

Early attempts to clarify the effects of process variables on the two phases of the penicillin fermentation were seriously handicapped by the inadequacies of available experimental techniques. Even so a general picture was obtained. With improved procedures this has been greatly amplified over the last decade until the effects of major process variables on growth and antibiotic formation are reasonably understood. Temperature and pH are the best examples.

Temperature

Stefaniak *et al.*[23] found no effect on overall penicillin yields between 20° and 29°C with an early culture (X–1612). At 32°C, however, antibiotic yields fell while oxidative metabolic processes (sugar utilization, etc.) were more rapid.

In the work previously cited, Calam, Driver and Bowers[6] set the optimum temperatures for growth and penicillin formation at 30° and 25°C, respectively. These conclusions were arrived at rather indirectly because they did not, in fact, separate the two phases of the process experimentally.

This was done by Owens and Johnson[21] who showed that growth rates were highest around 30°C while penicillin synthesis proceeded most rapidly near 20°C. A two-stage fermentation with the temperature reduced from 30° to 20°C after 40 h gave the highest penicillin titre.

pH

The importance of pH in the penicillin fermentation was early recognized. Lacking reliable means for external control, most

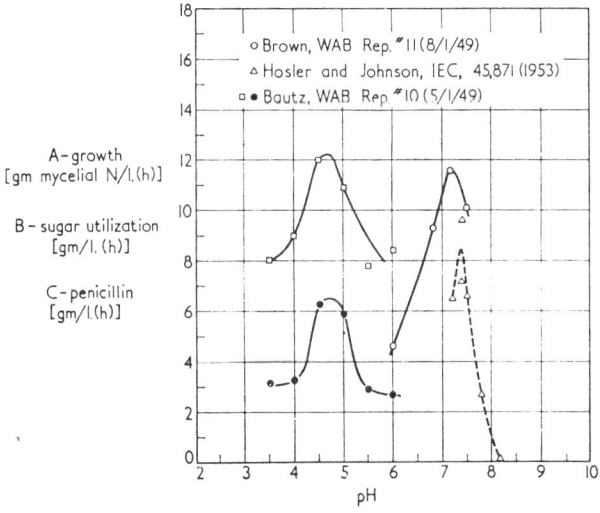

Fig. 5. pH effects in penicillin biosynthesis

processes employed medium formulations which provided a degree of internal buffering. A number of laboratory studies with externally controlled pH have been reported,[2, 3, 14] however, and the results are plotted on a common basis in Fig. 5. Note that the rates indicated are average rather than instantaneous. This does not alter the fundamental relationships shown.

From these experiments it is clear that the growth phase of the penicillin fermentation should be operated at a pH value around 4·5–5 while antibiotic formation will be maximized around 7–7·5.

It is also interesting to note the effect of external pH control on rate patterns in a penicillin fermentation. Brown and Peterson[4] have reported batch fermentations employing a medium which tended to become alkaline. After 30 h, the pH was adjusted to 7·0 with acid and held there (approximately) by controlled acid addition. Volumetric and specific rate patterns calculated from their results are shown in Fig. 6. Since no determinations of

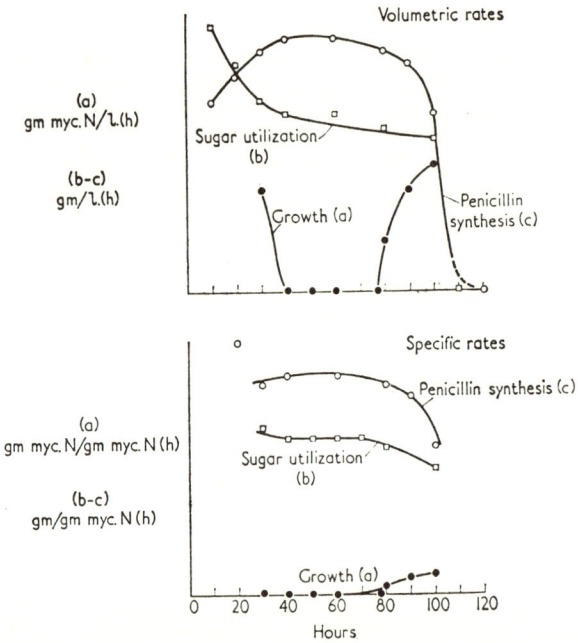

Fig. 6. Pencillin fermentation rates with pH control

mycelial nitrogen were made before the 30-h point, specific rates (based on mycelial nitrogen, not dry tissue in this case) cannot be computed for the early hours.

With pH control, constant rates of metabolism and product formation may be sustained for a long time, even in the unsteady-state batch process. Extended batch processes of this type may very well be practical competitors of continuous operations, particularly if the operating problems which have often been encountered in continuous systems prove difficult to overcome. The limit on such a process, assuming continuous nutrient addition as well as pH control, will presumably be imposed by the accumulation of products toxic to the organism or inhibitory to its enzyme systems.

Acknowledgement. This study was aided by a grant from the National Science Foundation, whose support is gratefully acknowledged.

References

[1] Adams, S. L. and Hungate, R. E.   *Industr. Engng. Chem. (Industr.),* 42 (1950), 1815
[2] Bautz, M.   *Antibiotics Research,* Report No. 10, University of Wisconsin (May 1, 1949)
[3] Brown, W. E.   *Antibiotics Research,* Report No. 11, University of Wisconsin (August 1, 1949)
[4] Brown, W. E. and Peterson, W. H.   *Industr. Engng. Chem. (Industr.),* 42 (1950), 1769
[5] Butlin, K. R.   *Continuous Culture of Microorganisms,* Czechoslovak Academy of Science, Prague (1958)
[6] Calam, C. T., Driver, N. and Bowers, R. H.   *J. Appl. Chem.,* 1 (1959), 209
[7] Doskočil, J., Sikyta, B., Kašparová, J., Doskočilová, D. and Zajíček, J.   *J. gen. Microbiol.,* 18 (1958), 302
[8] Dulaney, E. L., Hodges, A. B. and Perlman, D.   *J. Bact.,* 54 (1947), 1
[9] Gaden, E. L., Jr.   *Chem. & Ind. (Rev.)* (1955), 154
[10] Gaden, E. L., Jr.   *Chem. Engng.* (April, 1956), 159
[11] Gale, E. F.   *Chemical Activities of the Bacteria.* (1947). New York: Academic Press
[12] Gastrock, E. A., Porges, N., Wells, P. A. and Moyer, A. J.   *Industr. Engng. Chem. (Industr.),* 30 (1938), 782
[13] Herbert, D., Elsworth, R. and Telling, R. C.   *J. gen. Microbiol.,* 14 (1956), 601
[14] Hosler, P. and Johnson, M. J.   *Industr. Engng. Chem. (Industr.),* 45 (1953), 871
[15] Kempe, L. L., Gillies, R. A. and West, R. E.   *Appl. Microbiol.,* 4 (1956), 175
[16] Luedeking, R., Ph.D. Thesis, Dept. of Chemical Engineering, University of Minnesota (1956)
[17] Luedeking, R. and Pivet, E. L.   *This Journal,* 1 (1959), 393
[18] Maxon, W. D.   *Appl. Microbiol.,* 3 (1955), 110
[19] Moyer, A. J., Wells, P. A., Stubbs, J. J., Herrick, H. T. and May, O. E.   *Industr. Engng. Chem. (Industr.),* 29 (1937), 777
[20] Moyer, A. J., Umberger, A. J. and Stubbs, J. J.   *Industr. Engng. Chem. (Industr.),* 32 (1940), 1379
[21] Owens, S. P. and Johnson, M. J.   *Appl. Microbiol.,* 3 (1955), 375
[22] Sikyta, B., Doskočil, J. and Kašparová, J.   *This Journal,* 1 (1959), 379

[23] Stefaniak, J. J., Gailey, F. B., Jarvis, F. G. and Johnson, M. J. *J. Bact.*, 52 (1946), 119
[24] Stodola, F. H. *Chemical Transformations by Microorganisms.* (1958). New York; Wiley
[25] Wallen, L. L., Stodola, F. H. and Jackson, R. W. *Type Reactions in Fermentation Chemistry.* (1959). U.S. Dept. of Agriculture
[26] Wells, P. A., Moyer, A. J., Stubbs, J. J., Herrick, H. T. and May, O. E. *Industr. Engng. Chem. (Industr.)*, 29 (1937), 653

20

Copyright © 1971 by C. T. Calam, S. H. Ellis, and M. J. McCann

Reprinted from *J. Appl. Chem. Biotechnol.*, **21**, 181–189 (July 1971)

Mathematical Models of Fermentations and a Simulation of the Griseofulvin Fermentation*

by C. T. Calam**, S. H. Ellis and M. J. McCann†

I.C.I. Ltd., Pharmaceuticals Division, Alderley Park, Macclesfield, Cheshire

(*Manuscript received 18 January, 1971*)

Introduction

A FEATURE of the microbiological literature, in recent years, has been the appearance of papers describing the construction of mathematical models of fermentation. It is not intended to attempt to review every model so far described, but rather to indicate the main trends developing and the problems of different types of model. An account is also given of an attempt to simulate the griseofulvin fermentation. The present paper deals with batch cultures, particularly those of the industrial type, using complex media.

Models have long been used in engineering as means of investigating the behaviour of large or complex structures or machines and are useful when the behaviour of the structure cannot be predicted by direct analysis on account of the large number of factors involved. On account of the reduction in scale the model has to be a simplified version of the main structure, and the selection of the correct features for incorporation in the model is therefore of great importance. The model thus provides only a limited amount of information and is usually intended to deal with certain specific problems. When these have been solved the model is usually discarded. Models are also often made for pleasure or for instruction. All these points can be observed in connexion with models of fermentation.

The use of physical models in engineering is tending to be replaced by mathematical models in which the behaviour of the object being investigated is expressed by suitable mathematical equations. In chemical engineering and fermentation work, mathematical models are the only ones that can be applied. Fermentations are often represented by two or three differential equations. Many fermentation workers find equations of this type difficult to understand and to manipulate, even when the differential equations are solved and expressed algebraically. For these reasons models have been of limited interest hitherto, but several factors have tended to increase interest in modelling, e.g. the availability of computers which make the mathematical part of modelling much easier. Of more practical importance has been the introduction of simulation languages which make the writing of programmes easier and render unnecessary the solution of the equations. The ever-increasing use of computer terminals (operating via the telephone system) is also making computers available on a rapidly increasing scale. Although mathematical assistance is needed to construct the model, it is now possible for microbiologists to use and modify them with a reasonable degree of ease. The expense of computation is of course important. Fermentation experiments are, however, extremely expensive, and it can reasonably be argued that the relatively small extra cost of a mathematical analysis is justified if it increases the amount of information recovered from the experimental data.

The fact that a model deals with only a few of the factors involved in the real situation stresses the need for a careful choice in its construction. This is in many ways more important than the purely mathematical side. Since models are made and used to solve particular problems, their interest to others is usually limited. This presents a serious problem in description, as it is not always easy to see how a model intended for one purpose can be adapted to another.

The metabolic pattern in a fermentation process involves the interaction of a number of simultaneous and partly independent processes. These are usually represented graphically. Such behaviour is too complex to visualise and express simply. A model enables the metabolic pattern to be described in a few easily understood parameters. The effect of changes in fermentation conditions can be expressed in terms of these parameters.

Locker[1] has pointed out the existence of a gap between the fundamental theories of science and factual data, and that models are an attempt to bridge this gap. The use of modelling must thus be distinguished from a complete scientific study of a situation. The model in fact offers a short cut to a workable solution as opposed to a fully scientific answer to a problem. It is important to realise this, since models can appear to be very convincing and the effects of the assumptions made in their construction and the errors in the experimental data used to determine their parameters can easily be overlooked; attempts may be made to reach conclusions which the experimental evidence is not really able to support.

It should be added that models and results obtained from modelling are only as good as the data employed. In industrial work, fermentations are often assessed on the basis of a few figures giving the maximum yield, perhaps accompanied by pH measurements and a few analyses. Data relating to the early stages of growth and production are often lacking. Modelling, at least during the construction of

---

* Based partly on a lecture given to the S.C.I. Microbiology Group, 23 June, 1970.
** Address from Sep. 1971: Biological Dept., Liverpool Polytechnic, Liverpool 3.
† I.C.I. Ltd., Central Management Services Division. Fulshaw Hall, Wilmslow, Cheshire.

the model, often requires a great deal of accurate data so that curves and equations can be adequately defined. When complex media are used it may be difficult to define growth curves. This problem has to be borne in mind when modelling work is planned, otherwise repetition may be necessary. Some of the problems of evaluating growth when using complex media have been discussed.[2]

A model is a tool which can be used to help in the understanding and improvement of fermentation processes. As such it is necessary, at the present early stage of development, to spend time on the design of models so as to produce the most suitable types for use in different situations, and also to determine the best ways in which they should be used.

Mathematical models of fermentation

Basic features of modelling

At a minimum level a fermentation can be regarded as involving three main processes—growth, product formation and utilisation of substrate. On this basis models will involve one, two or three differential equations, each corresponding to the processes indicated. In more elaborate models more processes may be introduced so as to give a more satisfactory representation of the fermentation, and in these cases a correspondingly large number of equations will be needed. While some models are based on reaction mechanisms, in most cases differential equations are used which represent curves of metabolism adequately, without implying that any particular type of reaction mechanism is involved.

The word 'structure' has been used in several ways in connexion with models. Ramkrishna et al.[3] referred to unstructured and structured models. By unstructured models they meant curve-fitting models in which arbitrarily chosen equations were used to produce curves which fitted the experimental data. Structured models involved stoicheiometric relationships between cell-mass, substrate consumed and other components of the system. The word 'structured' can also be used with another meaning, that the equations not only express the types of reaction involved in the fermentation, but that they are closely linked together, e.g. by feedback loops or other control mechanisms. In this way the model resembles more closely the microbial cell, in which the network of reactions is also closely regulated by a number of control mechanisms. With a modern computer language it is possible to deal with models containing complex systems of this kind. However, such systems may become 'stiff' mathematically, and as a result of this computation can prove excessively slow.

Ideally models are autonomous, i.e. the rates of change of the variables depend explicitly only on the levels of the variables, and not on time; in other words, any changes in metabolic pattern arise spontaneously in the model and are not introduced by timing devices. Making the rates of change depend on time can however be of considerable value in the development stages of modelling, as it may be difficult to visualise the complete control system until some experience has been gained as to how the rest of the model behaves.

Representation of growth and other features

Four main phases of growth have long been known to occur with micro-organisms: an initial lag phase when growth is very slow; a phase of exponential growth, with a fixed doubling time; a stationary phase when the cell concentration achieved by exponential growth remains constant; and a decline phase when the cells die off or lyse.

In industrial fermentations the lag phase is usually avoided by the use of large rapidly growing inocula, and fermentations normally end before the decline phase really begins. Attention is therefore usually concentrated on the exponential (or logarithmic) phase and the stationary phase.

Although growth can be expressed in this simple way, it is well known that growth is a complex process, and when detailed study is made of growth curves it is usually found that they possess a variety of irregularities. For instance if complex media are used, during the rapid growth phase there are often changes in growth rate due to different medium constituents being used at different times. Even during true exponential growth a proportion of cells may die. In other cases an obvious break may occur because of exhaustion of a major component followed by further rapid growth (diauxie). The accumulation of toxins may interfere with growth. However, basically, growth can usually be described in the simple manner quoted above.

For a general discussion of growth equations, reference may be made to standard textbooks.[4,5] Although most of the work was carried out with bacteria, it is found that moulds and yeasts follow a similar pattern.

Growth is usually expressed by the equation:

$$\frac{dx}{dt} = K\left(\frac{S}{S + K_M}\right)x \quad \ldots\ldots\ldots\ldots\ldots\ldots(1)$$

where x = concentration of cell-mass, S = concentration of substrate, K = reaction rate coefficient and K_M = Michaelis constant.

This is essentially the Michaelis–Menten equation. The use of this equation for this purpose is attributed to Monod.[6] The value of the Michaelis constant, K_M, is usually very low; Monod found a value of 4 mg glucose/litre for *Escherichia coli* when glucose was the main carbon source. As a rule, substrate concentrations exceed the value of this Michaelis constant, and Equation (1) reverts to the simple form $dx/dt = Kx$.

Assuming that the substrate(s) is used up in proportion to the weight of the cells produced, its disappearance can be written:

$$-\frac{dS}{dt} = a\frac{dx}{dt} \quad \ldots\ldots\ldots\ldots\ldots\ldots(2)$$

where a is a constant.

Equations (1) and (2) can be combined and integrated, enabling cell mass and substrate utilisation to be calculated at different times.[4,5]

Characteristically, growth equations of this type give a rapidly rising growth curve with a rather abrupt change to the stationary condition when the substrate is exhausted. It is evident that in real fermentations growth often slows down for other reasons before the substrate is completely exhausted. In a number of models Equations (1) and (2) are therefore modified. For instance, Ramkrishna et al.[3] added terms to simulate the formation of toxins which may either stop growth or dissolve the cells.

The other type of equation used to represent the growth curve is the so called 'logistic law':

$$\frac{dx}{dt} = b_1 x\left(1 - \frac{x}{b_2}\right) \quad \ldots\ldots\ldots\ldots(3)$$

where b_1 is a rate coefficient and b_2 is the maximum cell-mass that can be achieved. It will be seen that when x is small x/b_2 is negligible and the equation reverts to $\frac{dx}{dt} = b_1 x$ or the exponential growth equation. As x increases, x/b_2 becomes larger and dx/dt declines. An S-shaped growth curve is thus produced, very similar to growth curves observed in practice.

The rate of product formation is represented by equations which are basically similar to growth equations:

$$\frac{dp}{dt} = K_p\left(\frac{S}{S + K_{M_p}}\right)x \quad \ldots\ldots\ldots\ldots(4)$$

where p is product concentration, K_p is the reaction rate coefficient and K_{Mp} the Michaelis constant for production. Usually other terms are included to allow for competing processes; these are illustrated in the examples given in the next section. The use of substrate for product formation is also allowed for in a manner similar to that with growth.

Equations (1), (2) and (4) form a system capable of modelling a variety of fermentations. The letter S is used to represent 'substrate' for both growth and product formation, but in fact the limiting substrates for these two processes may not be the same. In this case S will refer to different substrates, e.g. nitrogen-source for substrate formation, sugar for product. In one of the examples given below,[11] Equation (1) has been modified, and the limiting substrate for growth in biotin. This example illustrates well the way in which this group of equations can be used for modelling, and the sort of modifications that are possible.

An alternative is the equation:[7]

$$\frac{dp}{dt} = b_3 \frac{dx}{dt} + b_4 x \quad \ldots\ldots\ldots\ldots\ldots(5)$$

The values of b_3 and b_4 can be obtained by plotting the specific production rate $\left(\frac{1}{x} \cdot \frac{dp}{dx}\right)$ against specific growth rate $\left(\frac{1}{x} \cdot \frac{dx}{dt}\right)$. The resulting straight line has a slope b_3 and cuts the vertical axis at a value equal to b_4. b_3 is thus 'growth associated' and b_4 'cell-mass associated'.[7] The equation may be modified to allow for cessation of production at specific times and in other ways. Equations (3) and (5) form the simplest way to represent a fermentation. A good example of the exploitation of this type of model is given in Ref. 10.

Some typical mathematical models

A large number of models have been described in the literature. A few of these have been selected here to illustrate the types of model being developed and the objectives of their authors. For the sake of convenience, the notation employed by the original authors has been retained.

Table 1 lists seven models, giving the type of model concerned, the number of different equations, and tools required to use the model and the subject being studied. The different degree of complexity is illustrated both by the number of equations involved and the tools required to use the models.

Objectives of modelling

The work of Leudeking & Piret[7] was intended to provide instantaneous rate measurements so as to improve knowledge of fermentation kinetics in both batch and continuous cultures.

Hockenhull & MacKenzie[8] used a model to determine optimum feed systems for the penicillin fermentation.

Maxon and his collaborators[9] were interested in mechanistic models which were simple enough to be usable and which related operating variables to kinetics. In steroid conversions, factors such as solubility and product inhibition make the process extremely complex, especially as the substrate may be added incrementally or as a feed. It was hoped the model would reduce the amount of experimental work involved. It was also hoped the model would help in designing optimum instrumentation and control systems. The usefulness of the models actually developed was limited — although they threw much light on the bioconversions, the effects of aeration, pH, type of culture, etc. were not even hinted at. None the less they felt that models, if used with discretion, are powerful tools in the study of fermentation.

Constantinides et al.[10] were attempting a new venture in modelling by introducing an optimisation stage, and demonstrating the practical advantages of using modelling techniques for optimisation purposes and illustrated these by a study of the effect of temperature on the penicillin fermentation.

Yamashita et al.[11] developed models of the glutamic acid fermentation with a view to on-line optimisation of the plant process. This could be done by using the model to determine the future course of a fermentation while it was running; control could be optimised or the fermentation stopped at the most profitable time.

Koga et al.[12] were interested in the mathematical technique used in modelling and demonstrating the utility of this kind of work. Lactic acid production could be defined in terms of four parameters, which, they suggest, could be used as a basis for optimisation work.

With the earlier models, it is possible to determine the parameters involved graphically; the equations can be solved analytically and the model curves defined by laboriously calculating individual points. It is easier and more accurate to use computational methods for these purposes and with the more complex models digital computers are normally employed. If the equation can in fact be solved, the algebraic form is often cumbersome and less meaningful

TABLE 1
Examples of mathematical models
All models relate to experimental data

Reference	Type of model	Number of equations	Tools required or method of calculation	Purpose of model
7	Curve-generating, autonomous	1	graphical and analytical	Relates product formation to cell growth
8	Curve-generating, autonomous	1	graphical and analytical	Defines a growth curve in mathematical terms
9	Stoicheiometric	1	graphical and analytical	Study of steroid conversion
10	Curve-generating, autonomous	2	analytical or graphical solution possible, but used digital computer	Describes growth and production, leading to mathematical optimisation
11	Curve-fitting, stoicheiometric, autonomous	3	analytical solution possible but used digital computer	To represent and control glutamic acid production
12	Partly mechanistic, stoicheiometric	4	digital computer (simulation language)	Relates growth and product formation
This paper	Mechanistic, stoicheiometric, autonomous conservative. Capable of extension	17	digital computer (simulation language)	Describes growth, product formation and internal reactions, griseofulvin fermentation

than the original differential equation. It is obvious from Table I that unless the model contains a large number of equations, the information it can provide is strictly limited.

Three models

Yamashita et al.[11] have produced a model of the glutamic acid fermentation. This involved three equations:

growth
$$\frac{dx}{dt} = K\frac{(B_0 - Z_m)x}{B_0 + (J - Z_m)X}$$

glutamic acid production
$$\frac{dp}{dt} = b\left(\frac{S}{K_M + S}\right)X - a\frac{dx}{dt}$$

sugar consumption
$$\frac{ds}{dt} = -\frac{1}{Y_G} \cdot \frac{dx}{dt} - \frac{1}{Y_M} \cdot \frac{dp}{dt} - mx$$

where B_0 and Z_m are the concentrations of biotin in the medium and in the cells, J is a constant and Y_G and Y_M are yield constants for growth and production, whilst mx makes an allowance for maintenance energy. The growth equation has been modified from Equation (1) above, and the use of different limiting substrates for growth and production should be noted. Details are given[11] of methods used to measure the various parameters involved and the good degree of fit between model curves and actual data.

Constantinides et al.[10] have developed a model used to describe the penicillin fermentation. The two basic equations were:

growth
$$\frac{dx}{dt} = b_1 x \left(1 - \frac{x}{b_2}\right) \text{ (logistic law)}$$

production
$$\frac{dp}{dt} = b_3 x - b_4 p$$

where b_3 related cell mass to penicillin production and b_4 to decay of penicillin. (Note that b_3 and b_4 have different meanings in Equation (5) above.) A delay of 20 hours was allowed between cell formation and penicillin production.

The values for all the parameters, including x_0 (cell-mass at zero time) were obtained from a multi-variate, non-linear regression programme. Values for the parameters b_1, b_2, b_3 and b_4 were obtained for the different fermentation temperatures from which the model curves of growth and penicillin production could be constructed.

It can be seen from the published information that very good fits were obtained between the model curves and experimental curves and experimental data. The values of the four parameters, at different temperatures, were now applied to an optimisation programme. Regression equations were calculated for each parameter against temperature, so that fermentation performance could be calculated accurately for any temperature. The optimisation programme involved the use of Hamiltonians and a Fibonacci search technique for optimum temperatures throughout the run. A considerable advantage was indicated from the use of the best temperature profile. Typically, best results would be expected in a fermentation begun at 30°C the temperature being lowered to 18·7° after 80 h. It has been known for some years[13] that the use of a high temperature followed by a lower temperature can give improved results in the penicillin fermentation, but the general method described by the Columbia workers is obviously a fundamental step towards a new technique for fermentation development.

As a matter of interest the present writers investigated the possibility of using the original Equation (4), above, to represent penicillin production, instead of the one originally used.[10] Provided that an extra term was added to set b_3 to zero when the specific growth rate fell to a very low level, equally good fits to the experimental data could be obtained, though rather surprisingly penicillin production was found to be partly growth-associated. An optimisation procedure gave results similar to those previously predicted.

It was noticed at this time that the value for initial cell mass had been taken as 5 g/kg by Constantinides et al.[10] whereas in reality the value was only about 1 g/kg. With such a low starting weight it is not possible to get the logistic law curve to fit the data satisfactorily. This is due to growth occurring in two waves, an initial fast stage, followed by a later period of slower growth. The general model also failed to deal with penicillin fermentations in which experimental growth had been produced by calculated sugar feed rates.[14]

These and other criticisms might lead to the view that the Columbia model is inadequate. This, however, is to misunderstand the purpose of a model of this type. The Columbia model was intended to be as simple as possible and easy to use, and to allow for the application of optimisation procedures. It illustrates that provided a model can be used for its intended purpose, the simpler it is the better. On the other hand, if it was desired to throw light on the mechanisms of penicillin production, a different model would be more suitable.

Hockenhull & MacKenzie[8] used the logistic law to obtain the growth curve of *Penicillium chrysogenum* in the penicillin fermentation, and used this as a basis to calculate optimum rates of feeds of sugar.

Two theoretical papers should be mentioned which report investigations into types of equations which can be used to simulate fermentation curves. Ramkrishna et al.[3] described several equations, giving particular weight to the effect of inhibition of growth due to toxins. Kono & Asai[14] have described complex equations which can be used to simulate growth and production curves, particularly where the growth-rate varies during the fermentation. Excellent fits have been obtained between model curves and fermentation data.

Simulation of the griseofulvin fermentation

The stirred culture process for the production of griseofulvin was originally developed by Glaxo Laboratories.[15] The fermentation process was reviewed by Rhodes[16] and this paper also gave an account of the biosynthetic route to griseofulvin, which has been worked out in some detail, though not completely. In the Glaxo process, an inoculum was added to a corn-steep medium and after about 8 h, a glucose feed was started at such a rate that it held the pH at 7·0. It was necessary to adjust the pH at frequent intervals. Data in the patent specification showed that griseofulvin production reached over 7–8 g/kg culture. Table II gives for this fermentation, the balance sheet based on information given in the patent example and a certain degree of speculation. It shows that the main raw material was glucose and that the main product was carbon dioxide.

The decision to attempt to model this fermentation arose from attempts to investigate the relation between the change in pH during culture, rates of feed and griseofulvin pro-

TABLE II
220-h griseofulvin fermentation

Input, lb		Production, lb	
(4400 lb initial volume)		(5000 lb final volume)	
Corn steep solids	62	Mycelium	200
Glucose	520	Griseofulvin	35
Oxygen	368	Carbon dioxide	505
		Water	206
Total	950		946

Assumes 66% conversion of glucose to CO_2 and water, and production of 40 g/kg of mycelium.

duction. It was found that it was extemely difficult to establish a correlation between these important factors, and it was thought that this was because of complexities of the process and lack of knowledge of the mechanisms involved. It was considered that if a mechanistic model could be developed it would throw a light on these points and also make it possible to correlate fermentations carried out on different scales.

It was also hoped that by attempting to develop a mechanistic model, based on biochemical or 'pseudobiochemical' reasoning, it would be possible to make use of biochemical theory more directly in the improvement of industrial fermentations. Basically, however, the model was intended to be of practical rather than theoretical interest. Another possibility envisaged was the more intensive use of limited experimental data, especially as griseofulvin fermentations are difficult to perform in the laboratory.

A model of the type being discussed involves three parts — first the provision of experimental data; secondly, the development of a biochemical image from these data, which is intended to represent the main processes taking part; lastly the expression of this biochemical image mathematically. Each of these stages reacts on the others. The choice of a biochemical image, for example, is limited to ideas which can be expressed precisely in mathematics. Where experimental work is difficult, the biochemical image which must be chosen will make the best use of existing data, while at the same time adopting flexible but reliable biochemical theory.

Although the model described here has fulfilled some of the purpose for which it was intended, it represents only a stage in the development of modelling. It is hoped that a description of the techniques employed will be of value to other workers in the fermentation field.

Fermentation pattern

Fig. 1 summarises the general metabolic pattern of the batch which was used for modelling work. The pH reached 7 after ~ 15 h. The sugar feed, which started at 8 h, was rapid for a time but at ~ 100 h slowed down considerably, later it was increased and continued fairly steadily until the end of the run. Growth was rapid at first, giving a rather convex curve instead of the type of curve which is usually obtained when exponential growth is involved; it is thought that the process must reflect the presence in the medium of a variety of nutrients which are used up in succession, each corresponding to the different rate of growth. Rapid growth continued for ~ 100 h after which it halted for a time and was then renewed at a slower rate. The end of rapid growth occurred not long after the ammonia was used up. The slow growth which occurs at the end of the fermentation appears to occur without the utilisation of any further nitrogen from the medium.

Griseofulvin production starts at ~ 60 h and continues steadily until the end of the fermentation. Oxygen consumption shows a general relationship to the sugar used but is not so clearly related to growth. Oxygen consumption was measured via the carbon balance, the ratio CO_2/O_2 being obtained from analyses performed at different times during a number of fermentations. Although this method is probably somewhat unreliable it was considered adequate for the purpose.

The fermentation shows a number of patterns of behaviour, e.g. the growth curve, oxygen consumption, pH, griseofulvin production and changes in mycelial form. While the curves in Fig. 1 suggest the existence of relationships between these patterns of behaviour, the curves are obviously complex, and the mechanisms involved in the metabolic process suggest a complex reaction structure.

A feature of the fermentation is the series of changes which occur in the thickness of the culture. During the initial growth phase, up to ~ 100 h, the culture becomes extremely thick. After this it gradually becomes thinner and more and more mobile, owing to fragmentation of the mycelium. During the initial period of rapid growth the cells appear under the microscope as young, healthy, threads. After this much of the mycelium gradually dies off. Short threads grow out from the old mycelium and form the basis of a fragmented type of culture which gradually increases in weight.

Biochemistry of the fermentation

Fig. 2 shows the general transfer of materials in the process, in a very simplified form. Raw materials are seen as passing through intermediates and then being converted to cells and griseofulvin. An important part of this system is the conversion of intermediates via the respiratory system into energy (ATP) and carbon dioxide. This energy is passed to the synthesis of cells and griseofulvin and is closely coupled to it. Some of the glucose however, is converted directly to cells where it is used to form carbohydrates and cell walls. Half to two-thirds of the glucose is converted to energy which is used for biosynthesis or lost as heat, or in other ways.

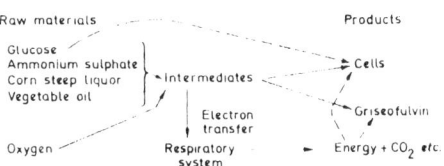

Fig. 2. *Transfer of materials*

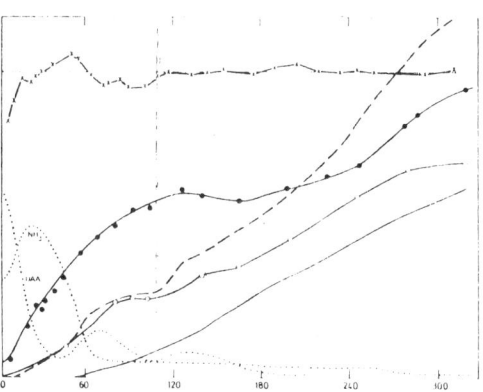

Fig. 1. *Griseofulvin batch metabolism*
× pH; ● cell mass; ○ oxygen; - - - sugar added; —··— griseofulvin formed; UAA utilisable amino acid; NH_3 ammonia

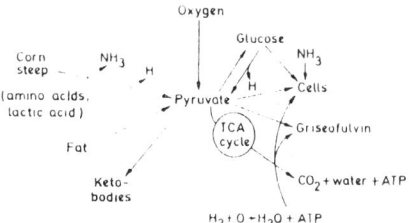

Fig. 3. *Model showing biochemical pathways*

Biochemical model of the fermentation

Biochemical modelling required the development of a suitable framework that can deal with the reactions involved in the processes shown in Fig. 2. Such a scheme must be able to take account of all the materials involved, the pathways being appropriately linked together. It will be noted that in order to account fully for all the materials used it is necessary to incorporate a respiratory energy-transfer system. It is evident that such a scheme must involve intermediate substances, so as to allow for all types of synthetic activity.

After several systems had been considered, the model shown in Fig. 3 was decided upon. It enables all the raw materials and products to be conveniently linked. The intermediates are pyruvate, ammonia, hydrogen and ATP. The expression ATP is used to indicate units of high-energy phosphate and not adenosine triphosphate in particular, and the energy-transfer system is not defined in detail. It is assumed that the medium (referred to as 'water') contains adequate metals, phosphate and growth factors.

The reactions involved in the model are: conversion of corn-steep liquor (amino acids and lactate) to pyruvate plus ammonia; glycolysis; cell production (formation of protein, walls, carbohydrate, based on analyses); griseofulvin biosynthesis; production of ATP; formation of glucose from pyruvate; and fat metabolism.

Notes on reactions

The formation of pyruvate from corn-steep liquor occurs in two ways, by conversion of lactic acid by removal of two hydrogen atoms, and by oxidative deamination of the amino acids present to give pyruvate and ammonia. About two-thirds of the amino acids are used, these being referred to as utilisable amino acids (UAA). Assuming that the amino acids of corn-steep liquor are similar to those in the protein of maize, an average analysis corresponding to $C_4H_9O_2N$ would be expected; although this amino acid does not exist in nature it was adopted in the model as the standard amino acid both for breakdown and for biosynthesis. These reactions and all the others except the cell production and the synthesis of griseofulvin, can be found in the standard textbooks. Enough is known of the biosynthesis of griseofulvin to show that it is formed from seven acetyl groups plus three methylene groups. A biosynthesis based on this can be readily devised and chlorination is assumed to take place by a reaction similar to the chloro-peroxidase system.[16] The following are typical examples which illustrate the manner in which the reactions are presented.

180GLU → 176PYR + 4HYD + 2ATP
88PYR + 64OXY = 132CO2 + 18WAT + 2HYD + 12ATP
38PYR + 73GLU + 11·5NH3 + 2HYD + 5ATP = 100CMS + 26·5WAT (Reaction 4A)
131UAA + 80OXY = 176PYR + 17NH3 + 18WAT
2HYD + 16OXY = 18WAT + 3ATP

The production of cells involves the formation of protein (PRO), cell-walls and carbohydrate (CBH). The formation of protein occurs by the condensation of amino acids, and the condensation of glucose gives carbohydrates. Similarly, walls are seen as being formed from glucose and ammonia giving a combination of glucose and glucoseamine units.

In formulating cell production (CMS), weights of raw materials for the production of 100 units of cell mass were calculated. The analysis of mycelium varies with time (Table III). Three equations were used for cell formation — one from 0 to 20 h and requiring a high proportion of ammonia, another from 20 to 70 h which also involved ammonia, and a third in which cells were produced without further ammonia being required. All the cell material can be derived from pyruvate and glucose plus certain amounts of ammonia. It

TABLE III

Analysis of mycelium
(Approximate only)

Age	N, %	PRO, %	CBH, %	Cell walls, %		
				Keratin	Cellulose	Total
0	11·3	30	5	60	0	60
20	8·2	22	13	45	12	57
70	6·5	20	23	39	12	51
100	6·2	20	27	36	12	48
200	4·8	17	30	28	15	43
300	3·6	12	29	20	23	43

follows that whatever mixture of protein, keratin, and cellulose is involved, the actual material requirements are similar for cells of different compositions except for the amounts of ammonia needed. It is known from analyses that the quantity of nucleic acid in the cells amounts to a very small percentage and this has been regarded as insignificant.

As already mentioned, during the first 70 h the growth curve shows a convex form, cell accumulation being particularly rapid at first. It is considered that this is due to the corn-steep liquor containing nutrients which are used at differing rates. To cover this, the usable amino acid fraction was divided into three portions, each of which was broken down via the same stoicheiometry but with different reaction rate constants (reactions R1A, R1B, R1C: rate constants K1A, K1B, K1C).

Not only have these reactions been extremely simplified, but in some cases reactions which may not exist in nature have been employed. The object has been to use the simplest possible system in the formation of the model. It has been found by experience that if a complicated reaction system is used, so as to improve the veracity of the model, great difficulties arise in controlling its behaviour, whereas considerable variations in the stoicheiometry can be withstood if the structure of the model is sound. Note that the model is conservative, i.e. the weight of material produced is equal to the weight of material put in.

Formation of the mathematical model

This task is rendered easier by the existence of simulation languages[17,18] such as 360/C.S.M.P. which are specially designed for this type of work and reduce the programme to a relatively simple but also intelligible form.

There is a total of 17 differential equations in the present model, one of which is a triplet describing the breakdown of amino acids to ammonia and pyruvic acid, at three different rates. A typical equation within this group of equations deals with ammonia concentration:

$$NH3 = INTGRL (NH3Z, NH3D)$$

which indicates to the machine that at any time the weight of the ammonia present can be calculated from the initial value (NH3Z) and the amount transferred into or out of the system by the various reactions. The value of NH3Z can be obtained from the initial conditions (INCON) which have to be listed in the programme. The value of NH3D (i.e. the rate of accumulation of NH3 at any time) can be obtained from the equation:

$$NH3D = 17*(R1A + R1B + R1C) - 11·5*R4A - 7*R4B$$

(Note the use of the FORTRAN notation, e.g. * to mean multiply).

Here, R1A, B, C are the three parallel reactions producing one molecule of ammonia (i.e. 17 units) from amino acids (at different times) while R4A and R4B are reactions which produce cells, and which utilise 11·5 and 7 units of ammonia, respectively. Reaction 4A given above shows how the 11·5 units of ammonia are required for the production of 100 units of cells (CMS).

At the next stage, the values for R must be defined for each equation. Referring again to ammonia, one of these is:

R4A = K4A∗PYRR∗GLUR∗NH3R∗HYDR∗ATPR∗CMS

The types of parameter in this equation are: the reaction rate coefficient K4A; the rate equation for each component, e.g. PYRR = PYRC/(PYRN + PYRC) where PYRC = concentration of pyruvic acid and PYRN the Michaelis constant (thus the reactions are of the Michaelis-Menten type); CMS = cell-mass, the weight of 'catalyst' present, which determines reaction rate. In many equations of this type, for instance the equation dealing with pyruvate, a 'limiter' is included, in the form PYRL = PYRN/(PYRC + PYRN) which prevents the concentration of pyruvate becoming too high.

As has been mentioned above, mould growth appears to involve three types of growth, starting at 0, 20 and 70 h, approximately. Ideally, in order to produce a fully autonomous model, the change from one equation to another should be produced by modifier expressions which would start and stop the reactions rather in the way the Limiters work. In the development stages steps can be used. The addition of ∗(1·0-STEP (20·0)) to the above reaction would instruct the computer to run the reaction for only 20 h. After this R4B would be started.

These three groups of equations define the model. The remainder of the programme consists of lists of the other terms needed. These are: definition of constants (e.g. K4A, PYRN); equations to calculate concentrations, etc. from the weights of materials used and the volume of the medium; a set of equations to calculate the quantities of glucose added, at any time, from the feed data; and instructions for tabulating or graphing the results (this includes means of checking the arithmetic during calculations).

The programme is then put onto punched cards and run on the computer. By using algebraic terms for some of the stoicheiometry (e.g. for quantities of ATP used), and giving the values of these quantities on separate cards, it is easier to adjust the equations if this is desired. In this model weights were used directly for calculating reaction rates, instead of molecular ratios.

Running and adjusting the mathematical model

A programme was written and the stoicheiometric values were entered. Where it was desired to change from one reaction to another (e.g. in the equations involving growth), steps were used. In later versions of the model it has proved possible to discard this device. The programme was then run on an IBM 360/40 or 360/65 computer. Computing time was 1–4 min, representing 50–300 h fermentation time. The model immediately produced graphs resembling fermentations, though many of the constants were obviously incorrect.

By a series of trials, running for only 50–100 h of simulated fermentation time (2–3 min of computation) the model was gradually corrected. The rate coefficients and Michaelis constants had to be obtained mainly by informed guesswork, but experience obtained in the preliminary trials was of great assistance in choosing suitable values. This process also involved a few changes in the stoicheiometry, the need for which became apparent at this time.

These trials gave results illustrated in Fig. 4. There was good agreement between calculated and real values for utilisation of ammonia lactic acid and oxygen and the production of cell-mass. The adjustment of the stoicheiometry of oxygen use, mediated by ATP units of energy, proved interesting. When the growth equations were first formulated it was thought that one or two units of ATP would be required per 100 units of cells, since the reactions involved were simple condensations. This led, in the early models to too low levels of oxygen uptake and the number of units of ATP had to be increased to 5 or more. In a recent review,[19] it has been shown that 100 units of bacterial cells require about 10 units of ATP. Similar results have been recorded for yeasts; precise values depend on the age and type of culture and on growth conditions. The quantities indicated by the model are in reasonable agreement with these conclusions. The model assures a P/O ratio of 3, on the basis of the behaviour of mammalian cells and yeasts.

It was now considered that adequate agreement had been obtained between the model (referred to as Model 5) and the actual batch, during the early stages. On this basis the computer was set to run the fermentation for a longer period, initially to 200 h, as shown in Fig. 5. At this time the cell growth equations only allowed for formation of cells when ammonia was available, and so the model, as expected, showed constant cell weight after 100 h; at the same time a large quantity of griseofulvin was produced and glucose accumulated to a large extent in the medium. It was necessary now to introduce another equation for the production of cells without uptake of ammonia, so as to allow growth to restart. This would require further quantities of ATP. These reactions would consume the surplus glucose and reduce griseofulvin formation by competition. A suitable equation for cell growth was worked out and put into the programme, to commence at 70 h; again a 'step' was provided to initiate

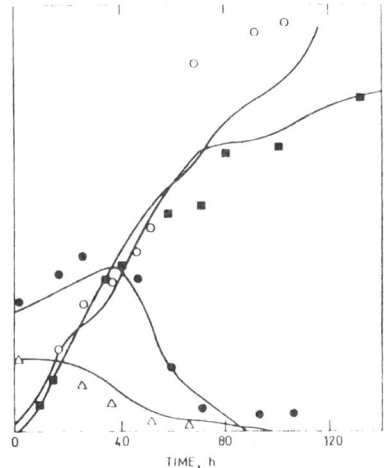

Fig. 4. Model and batch results, up to about 100 h
model
plant data: ■ cell-mass, ○ oxygen, ● ammonia, △ lactic acid

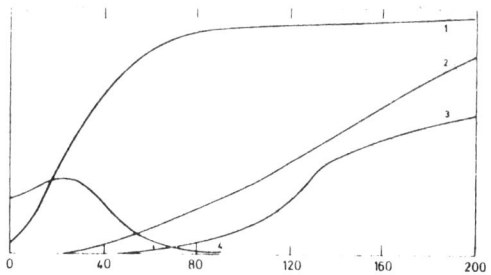

Fig. 5. Model 5, longer run
1. cell-mass, 2. griseofulvin, 3. glucose, 4. ammonia

the reaction. This equation had a very slow reaction rate and the formation of cells under these conditions required a large quantity of ATP; ten times as much as before. When the model was run on the computer the results obtained in Fig. 6 were obtained. This showed very good agreement between the model data and the actual batch results.

Examination of Fig. 6 shows there are some discrepancies but these are small and could be eliminated by altering some of the constants. It was not felt, however, that it would be worth the trouble of trying to get still closer agreement at this stage. It is emphasised that in modelling work it is the shape of the curves which is the main criterion of agreement, rather than precise agreement at the end of the run. The rather low predicted use of oxygen at 100–150 h probably indicates that the ATP requirements for the growth processes at this period have been set at too low a level.

Discussion

The model which has been described gives a reasonable representation of the behaviour of a griseofulvin fermentation and could be used in a preliminary way for testing the effects of different feed rates and for other purposes. This is not the only model that can be used, or necessarily the best model. The approach taken was, however, proved convenient because not only has it been possible to make it work but also it does follow lines of ordinary biochemical thinking. It is again stressed that in describing the model it was intended to give an example of a way in which models can be made, and that mathematical facilities are now available which are relatively easy to use. We have not attempted to describe a universal model which would be suitable for every kind of purpose. It is unlikely that a universal model would be of a great practical use because it would probably be far too cumbersome.

A good deal of thought has been given to weaknesses in this particular model. Basically probably too little attention has been given to the mechanisms which change the metabolic pattern during the growth cycle. It seems likely that growth stops at ~ 100 h because growth conditions and especially oxygen supply have become limiting; i.e. the biological space has been filled (cf. Hockenhull & McKenzie[8]). As a result of this the growth rate falls below that which is adequate to maintain the cells in good condition, and the main processes of secondary metabolism begin. This subject has been studied by Pirt and co-workers[20] who have shown that when the minimum growth rate is reached, significant changes take place in the metabolism of fungi. The extensive morphological changes which occur during the griseofulvin fermentation have been mentioned. From these it would seem apparent that during much of the time the mould is undergoing a series of transformations, presumably as a result of changes in the growth conditions involved. The very interesting effect of changes in growth rate and in dissolved oxygen tension on the composition of the enzymes in *Aspergillus nidulans* have recently been described.[21] Changes in enzymology are also well recognised in microbiology from studies on the effects of periods of starvation on sporulation. New enzyme systems are developed to deal with the changed growth conditions, and it is known that this is accompanied by requirements for large quantities of ATP. It seems probable that the high ATP demands which the model has shown to exist could well be associated with the extensive cell transformations which are continuously taking place.

As an example of the effects of oxygen uptake one can describe a batch which gave an unexpectedly high yield and also continued production longer than expected. The model was used to calculate production of griseofulvin and cell mass and utilisation of oxygen. The model gave a distinctly lower yield of griseofulvin and cells than were found in practice. When the equations in the model were altered to correspond to 70% of the standard model requirement of ATP it was found that agreement was greatly improved, although oxygen usage was still somewhat too high (Fig. 7). This draws attention to the importance of cell efficiency in the use of oxygen by cells as a factor in determining the yield of materials in the fermentation.

When attempts were being made to use the fall in the level of ammonia to a minimum value to start the final period of slow growth, as soon as the ammonia reached a low level, growth and griseofulvin production stopped and very large quantities of oxygen were found to be used. A careful study of the data in the model showed that this was due to a reduction of the glucose concentration to an extremely low level; this caused the reaction pyruvate to glucose to start working at a high rate, while subsequently the glucose was used for the formation of energy (the reaction pyruvate to glucose requires a considerable amount of ATP). In this way a cyclic system was set up in the model which converted the glucose to carbon dioxide with excessive use of oxygen, and at the same time cut off material for growth and griseofulvin production. It would seem likely that alternative modes can arise in fermentations under special conditions, triggered off by a sudden change in the concentration of important raw

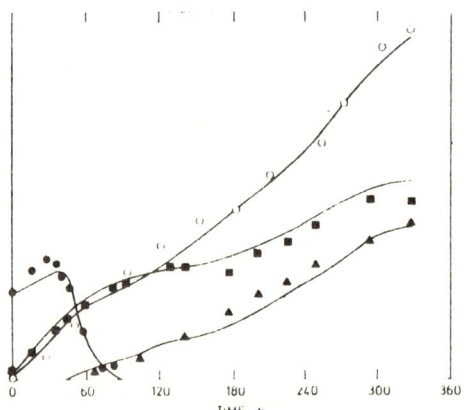

Fig. 6. *Griseofulvin fermentation dynamics*
—— model
experimental data: ■ cell-mass, ○ oxygen, ● ammonia, ▲ griseofulvin

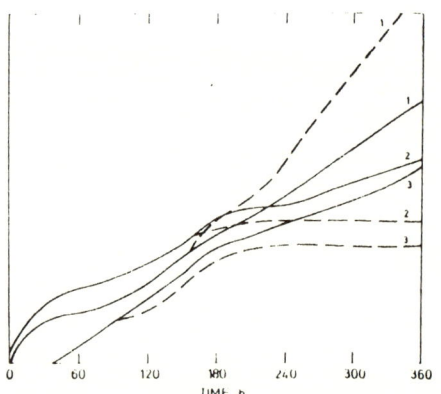

Fig. 7. *Efficiency of oxygen use*
—— at normal oxygen uptake, ----- at 70% oxygen uptake
1. oxygen, 2. cell-mass, 3. griseofulvin

materials in the medium. While it is unlikely that the particular alternative mode of metabolism adopted by the model would arise in real life, this particular incident draws attention to the possibilities in the fermentation which are not always foreseen in practical work.

When the model was operated with an inoculum only one fortieth of that usually used, it mimicked a batch run in this way, although the initial growth curves showed an entirely different pattern from the normal. The model thus showed its advantage over the curve-fitting type of model which always produces curves of the same type.

Considerable improvements and alterations to the model should be possible. The object of describing it has been to show the way in which new techniques can be applied, and it is hoped that these may be interesting to other workers in the fermentation field. It should be noticed that one aspect of the model was the estimation by informed guess work of the required reaction rate coefficients in other programmes totalling almost 20 in number. This was done successfully. This is a branch of the so called 'parameterless' mathematics, which is of increasing interest in certain fields at present. It shows that the new type of computer language enables work to be done where data are almost unavailable, and allows one to exploit minimum data in a way that would have been impossible a few years ago.

References

[1] Locker, A., 'Quantitative biology of metabolism', 1968, p. 1 (Berlin: Springer)
[2] Calam, C. T., *Meth. Microbiol.*, 1969, **1**, 567
[3] Ramkrishna, D., Fredrikson, A. G., & Tsuchiya, H. M., *Biotechnol. Bioengng*, 1967, **9**, 129
[4] Dean, A. C. R., & Hinschelwood, C., 'Growth function and regulation in bacterial cells', 1966 (Oxford: Oxford University Press)
[5] Clifton, C. E., 'Introduction to bacterial physiology', 1957 (New York: McGraw-Hill)
[6] Monod, J., 'Recherches sur la croissance des cultures bacterienne' 1958 (Paris: Hermann)
[7] Leudeking, R., & Piret, E. G., *J. biochem. microbiol. Technol. Engng*, 1959, **1**, 393
[8] Hockenhull, D. J. D., & MacKenzie, R. M., *Chemy Ind.*, 1968, p. 607
[9] Chen, J. W., Koepsall, H. J., & Maxon, W. D., *Biotechnol. Bioengng*, 1952, **4**, 65
[10] Constantinides, A., Spencer, J. L., & Gaden, E. L., *Biotechnol. Bioengng*, 1970, **12**, 803, 1081
[11] Yamashita, S., Hoshi, H., & Inagaki, T., 'Advances in fermentation', 1969, p. 441 (New York: Academic Press)
[12] Koga, S., Burg, C. R., & Humphrey, A. E., *Appl. Microbiol.*, 1967, **15**, 683
[13] Owen, S. P., & Johnson, M. J., *Appl. Microbiol.*, 1955, **3**, 375
[14] Kono, T., & Asai, T., *Biotechnol. Bioengng*, 1969, **11**, 393
[15] Glaxo Ltd., B.P. 868,958
[16] Rhodes, A., *Prog. ind. Microbiol.*, 1963, **4**, 167
[17] I.B.M. Ltd., Manual No. I.B.M. H20-0367-2, 1969, and revisions
[18] Report of Format Subcommittee of the SCI Simulation Software Committee, *Simulation*, 1967, Dec., p. 281
[19] Stouthamer, A. H., *Meth. Microbiol.*, 1969, **1**, 629
[20] Ringhelato, R. C., Trinci, A. P. J., Pirt, S. J., & Peat, A., *J. gen. Microbiol.*, 1968, **50**, 399
[21] Carter, B. L. A., & Bull, A. T., *Biotechnol. Bioeng.*, 1969, **11**, 785
[22] Mandelstam, J., 'Microbial growth, 19th Symposium', 1969, p. 377 (London: Soc. Gen. Microbiol.)

Part X

AUTOMATION

Editor's Comments
on Paper 21

21 NYIRI
A Philosophy of Data Acquisition, Analysis, and Computer Control of Fermentation Processes

It is proper that the last paper in this volume of Benchmark papers be one that projects an image of fermentation technology a decade or more hence. In Paper 21, Nyiri describes a successful attempt to integrate a highly instrumented fermentor with a general-purpose digital computer. This paper is valuable, however, not for the experimental results it reports but for its thorough review and thoughtful discussion of problems and its logical approach to solutions.

21

Copyright © 1972 by the Society for Industrial Microbiology
Reprinted from *Develop. Ind. Microbiol.*, **13**, 136–145 (1972)

A Philosophy of Data Acquisition, Analysis, and Computer Control of Fermentation Processes

L. K. NYIRI

Fermentation Design, Inc., Allentown, Pennsylvania 18103

Integration of a highly instrumented pilot-plant scale fermentor with a general purpose computer is described. A program makes it possible to operate the computer for on-line data acquisition, data analysis, and process control. Using such a system, the operator is able to follow the changes in conditions of the culture broth as well as some physiological and biochemical characteristics of the cells. With this flexible system the effect of environmental conditions on the physicochemical features of the culture can be investigated on a real-time basis. Results can be applied in scale-up procedures and in more sophisticated process control, where the control elements are changed to create suitable environmental conditions for growth and product biosynthesis.

INTRODUCTION

There have been successful attempts to use Direct Digital Control (DDC) in the fermentation industry on large-scale operations (Grayson, 1969; Yamashita et al., 1969). Early experiences with DDC in these processes indicate that the computer can replace analog controllers, resulting in certain advantages such as smooth start-up, less overshoot, as well as reduction in labor cost and the production of more uniform batches. During operation, the control loops were functioning in a noncooperative way.

One of the most important differences between chemical syntheses and biosyntheses is the characteristic of the microbial cell-growth and product formation. Here environment can influence metabolism in a nonpredictable way. Reports on application of DDC in the fermentation industry emphasize that the lack of coordinated operation of control variables handicaps the adjustment of environmental conditions to the status of growth and metabolism. The interrelated function of control variables requires detection of the effect of environment on cell metabolism. Although computers and computer programs have proved to be suitable tools in revealing the dynamic behavior of living processes (Heinmets, 1969; Garfinkel et al., 1970), the DDC system is considered inapplicable for this purpose because it deals with the status of control elements rather than the physiological condition of the organisms.

In this paper, a data analysis-oriented application of the computer is described. This on-line analysis of the fermentation processes provides the opportunity to determine the actual physicochemical status of the culture broth and offers insight into some of the physicochemical and biochemical conditions of the cells. By this means, the effect of control variables on the cell growth and metabolism can be studied on a real-time basis. These experiences can be converted into algorithms useful in more sophisticated process control systems.

Results and Discussion

Problems of the Flexible Control System

The final goal in the control of microbiological-biochemical processes is the development of a flexible control of the environment through which the metabolic activity of the cells is controlled and environmental conditions are adjusted according to the actual status and trends of the process. Figure 1 presents the outline of a control system which is considered to be flexible enough to serve the basis of the environmental control in microbial cultures according to the status of the metabolism. This is the adaptive control system originally described by Li (1960). It has now begun to receive acceptance in chemical plant operations. The adaptive system is a specific type of computer control where the alteration of control variables is based on the status and future trends of the process. There are two important constituents of the system: process identification; and on-line optimization on the basis of which the computer can generate the appropriate actuating signals for control. Both constituents require very complex algorithms even in the case of a simple process.

Requirements to establish adequate process identification methods are:

1) Find the sensors which can detect the status of the control and state variables.

2) Define those control and state variables which express the actual physiological and biochemical status of the process.

3) Define the patterns of metabolism with which the particular process responds to the effects of the environment.

4) Make the algorithms of the process identification based on the mathematical model of the process and verified by accumulated experimental data.

5) Try experimentally the process identification system.

To solve this complex task, a highly instrumented, pilot-scale fermentor (Harmes, 1971), integrated with a general purpose computer, we considered to be the best tool.

The strategy of the computer program construction is based on the assumption that on the pilot scale the function of the computer is primarily process analysis-oriented

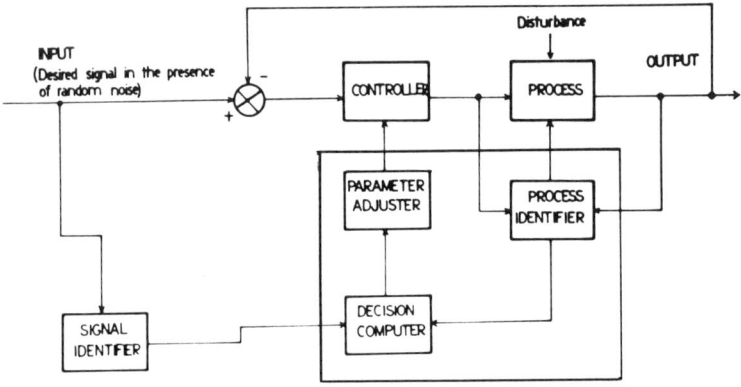

FIG. 1. Outline of adaptive process control system.

(Nyiri, 1971). Secondly, the computer performs process control, subject to the priority of the operator who can override the function of the computer any time during the process.

Data Analysis-Oriented Application of the Computer

Figure 2 shows the general scheme of the pilot-plant fermentor integrated with a computer. Here two factors are worthy of note:

1) Digital Panel Meters (DPM) serve as interfaces between the sensors and the computer. The advantages of the use of DPM are: (a) They meet the principle of modular design which is the basis of the expandability of the system. With the introduction of new analytical or control devices, new DPM units can be built into the system without difficulty. (b) The DPM converts the analog signals, arriving from the sensor-amplifier system, into binary coded digits (BCD). The BCD signals are transferable to remote places with minimal transmission problems. (c) They provide visual display of sensor values throughout the process.

2) The computer has access to the fermentor through limit switches and set points on the individual controllers. Therefore the computer does not participate directly in the process control (as is the case when DDC is performed).

Figure 3 indicates the major functions of the computer. Three on-line, real-time functions are distinguished, namely, logging of the process data (data acquisition system), analysis of process data (data reduction system), and process control (process control system). It is noted that the data acquisition and data reduction systems are operating separately from the process control system.

Table 1 presents the primarily measured control and state variables. The computer program is formulated in order to meet the requirements necessary to establish an adequate process identification method. The sequence of operation is illustrated in Figure 4. Here, the environmental conditions for the process are established by the operator and introduced into the computer prior to operation, in the form of predefined time-profiles of control variables. The computer alters the values of the control set points thereby creating new environmental conditions during the process. State and control variables are

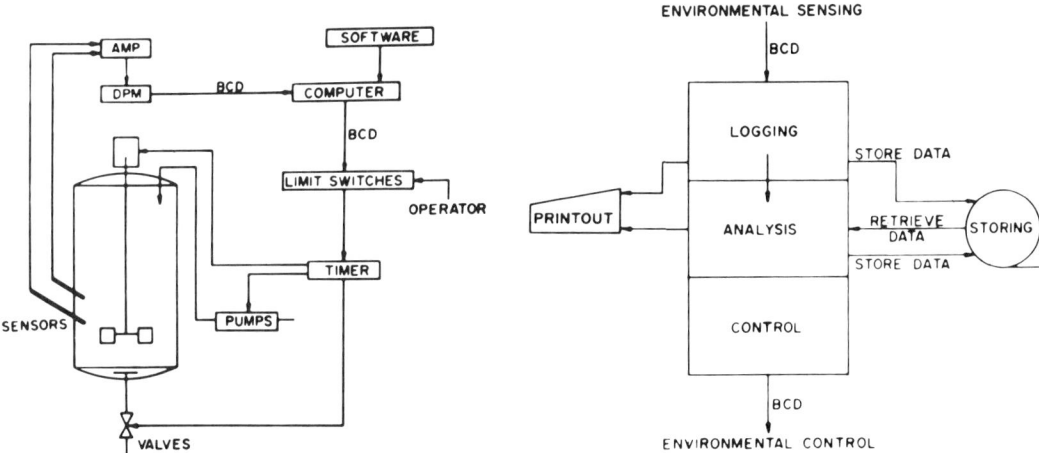

FIG. 2. Highly instrumented pilot-plant fermentor integrated with a computer.

FIG. 3. Major functions of the computer in a data analysis oriented application.

scanned and logged. A process analysis is made on the basis of the status of control and state variables by the computer. The resulting printout data enable the operator to follow the process on a real-time basis. The dynamic response of the living cells to the change of environmental conditions can thus be instantaneously followed. Should it be desirable to override the predefined control profile, the operator can modify the control profile instructing the computer via teletype or stop the computer operation and perform manual operation. In this way, computer program sequences were constructed for data acquisition, data reduction, and process control. A general outline of the program is given in Figure 5.

The data acquisition and logging routine do not differ significantly from those used in chemical pilot-plant operations (White and Hazbun, 1970). In our case, however, there are two scanning cycles. In the first cycle, the process variables are read, logged, and compared to the predefined values. If there is a deviation, the actual value of the process variable is compared to the predefined tolerance values. An alarm signal is printed if the actual value of the process variable exceeds, or is below, the tolerance value. In the second scanning cycle, the computer prints the logged values of the process variables. This arrangement has the ability to save computer time and memory for process analysis.

The use of DPM's assures that the sampled data do not contain high frequency noises so the computer core memory and execution time, usually used in a certain percentage for nonlinear filtering, can be saved. Data are stored on magnetic tape. A subroutine assures the data retrieval any time for further computations.

The data reduction routine is explicitly designed to make process analysis overcome the time-consuming and laborious manual calculations, and to extract as much information from the individual sensor data as possible. In achieving the first goal, the process is analyzed on a real-time basis which is impossible using manual calculations. With respect to the second goal, the computation permits the multivariation of the individual values of the control and state variables primarily measured and logged by the data acquisition system. In addition to this, rates, slopes, and averages of the individual data can be computed along with the validity tests. These goals can be achieved if (1) suitable algorithms and constants are available to multivariate the sensor data to obtain new information; (2) data obtained by a computation sequence can be utilized in other mathematical and logical operations; (3) data of former fermentation processes can be

TABLE 1. *Primarily measured control and state variables*

Control Variables	State Variables
Temperature – culture broth	Agitator shaft torque
	Drive motor power uptake
Temperature – liquids in addition vessels	O_2 concentration in exit air
Vessel pressure	CO_2 concentration in exit air
Agitation speed	Redoxpotential of culture liquid
Sparge air rate	Weight of the culture liquid
Flow rate of addition gases	Weight of fluids in each
DO of culture liquid	addition vessel
pH of the culture	
Rate of addition of ingredients from the addition vessels	

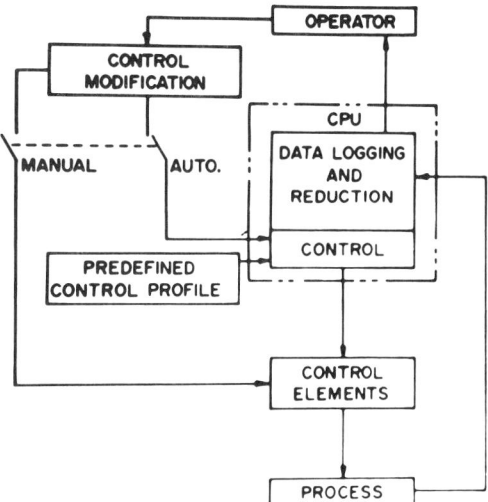

FIG. 4. Process-computer-operator relationships during a data analysis oriented application of the computer.

retrieved any time for comparison; and (4) the temporary halt of the execution of one program sequence does not influence the execution of other sequences.

In Figure 6, an example is given of the on-line, real-time data analysis program. The program consists of program sequences which are processed in a given order. These sequences make possible the analysis of the actual rheological, physiological, and biochemical characteristics of the process. The analysis is based on the logged values of

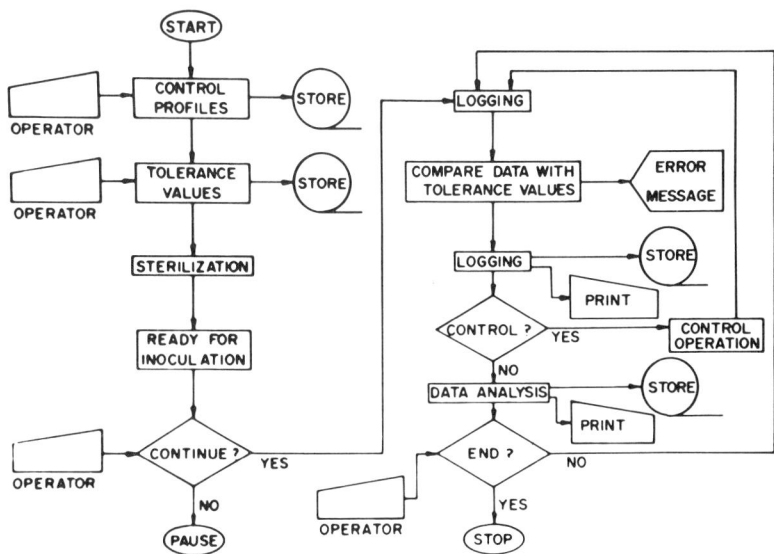

FIG. 5. General scheme of the program for pilot-plant scale data acquisition, analysis and control in fermentation processes.

the control and state variables and data obtained through off-line measurements (e.g., density of the culture). These, along with suitable constants, are introduced into the formulae of computation. In executing the program sequence, new state variables are generated (e.g., apparent viscosity, mass transfer coefficient, respiratory quotient, carbon balance, and organic energy yield). In order to use the values of these new state variables in other program sequences, a technique resembling the graph tree presentation of data (Pennington, 1970) is used. Here, the data generated by a program sequence are transferred into other sequences, e.g., shear stress:shear rate relationship results in the apparent viscosity. This value, along with the actual density, gives the Power and/or Reynolds numbers. These are introduced again into the next program sequence generating the actual power characteristics of the culture liquid.

Results of these analyses can be used in process control decisions, in scale-up calculations, or they can be compared to data obtained from off-line process simulations. It is noted that each computed analysis is the target of the data acquisition system. Data can be retrieved for mean value calculations, extremum finding, and prediction or optimization procedures. An executive routine detects the availability of all data necessary to perform the computation of the individual program sequence. In the case of lack of necessary data, the executive program steps to the following sequence. As an example, Table 2 presents a typical data printout of a simulated benzylpenicillin fermentation. In the 35th hour of the cultivation, the control variables are maintained at the preset value. The apparent viscosity indicates a viscous fermentation resulting in a slightly low $k_L a$ value. This is in accordance with the relatively large cell-mass (calculated in this particular case on the basis of carbon dioxide release). The penicillin production started in the 22-24 hr reaching 2014 IU/ml, giving a production rate of 183 IU/ml hr^{-1}. This is considered to be optimum according to the previous experiments. This rate of benzylpenicillin formation, however, requires an adenosine triphosphate (ATP) yield near

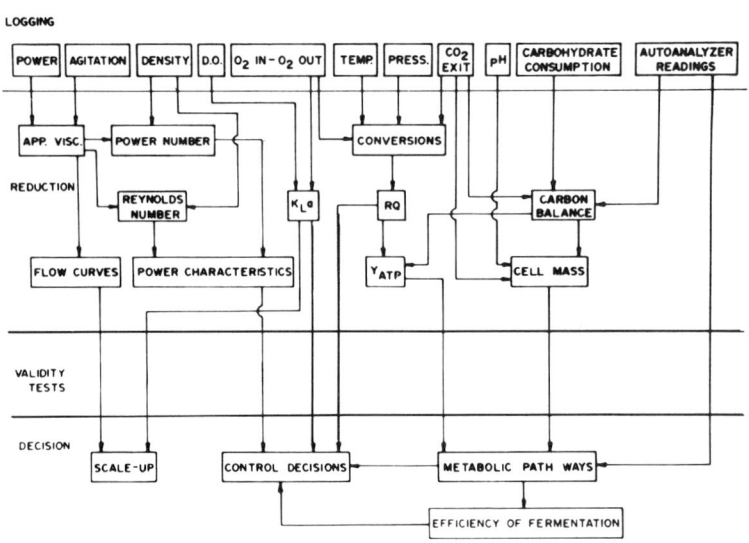

FIG. 6. Outline of data analysis program for fermentation processes.

TABLE 2. *Results from a typical data analysis made on a simulated penicillin fermentation*

Format A2	
* Values read from interface	
* Values read from data analysis system	
*	
Time	2,100 min
Temp	27.4 C
Pres	0.2 atm
Airflow	185 lpm
Agit	250 rpm
Wattmeter	0.7457 kw
Brothvol	208 l
pH	6.7
Redox	+109 mv
Powerin	0.0047 hp/l
Appvisc	103 centipoise
NP	4.3
NRE	367
NP/NRE	Transient
Flowchar	Pseudoplastic
O2 Exit	12.3 pc
O2 Uptake	124 mmole/l/hr
DO	6.7 ppm
O2 Satn Level	93 pc of satn
KLA	240 l/hr
KLA-Slope	−10 l/hr
CO2 Exit	7.8 pc
CO2 Prodn	116 mmole/l/hr
RQ	0.9
RQ av (22-34)	1.2
Lactose Concn	2.7 pc
Lactose Consumed	3.3 pc
ATP/C6 (mole)	34.4
Cellmass	7 mg/ml
Penicillin G	2014 iu/ml
Penprodrate	183 iu/ml/hr
Antifoam Added	1.70 l
Vol T1 (AF)	3.30 l
Vol T3 (Lactose)	15.00 l
Vol T4 (N NaOH	2.70 l
Vol T5 (N H_2SO_4)	8.60 l
Vol T6 (PHEAA)	6.60 l
*	
Warning: T4 runs out wthn 3 hr	

the theoretical carbohydrate/ATP conversion, providing the organic energy is obtained through the glycolytic pathway and the tricarboxylic acid cycle. This current printout information indicates that the culture is in that condition which permits studies on the effect of environmental conditions on the metabolic activity of the cells. The calculations were executed within 0.1 second and the printout took 66 seconds, using an ASR-33 teletype. This speed of calculation gives the operator an opportunity to check the effect of the change the values of control variable(s) have on the state variables.

Table 3 presents a comparison between the computers used for control-oriented and data analysis-oriented, purposes in the case of fermentation processes. The ARCH-102 was used for DDC of the temperature, airflow, pH, and foam level in 36 culture vessels producing *Penicillium chrysogenum* cell-mass as well as penicillin. Here the total number of control loops was 114. The PDP-11/20 is utilized for a data analysis-oriented usage interfaced with a highly instrumented pilot-plant fermentor. A 4K core memory for computation is a basic requirement although 8K or more is necessary for expansion of the software.

There is a possibility of application of large-scale computers to data analysis. However, the experiences in the chemical industry indicate that adequate data acquisition systems require a small-scale computer (White and Hazbun, 1970). Recently, Tuffile and Pinho (1970) used off-line computation to calculate the rheological conditions and the oxygen transfer rates in pilot-plant streptomycin fermentations. Their experiences indicate the applicability of several algorithms to define the apparent viscosity and $k_L a$. However, time-sharing of a central computer, by definition, has the limitation of priorities and queuing for data-processing time. This may handicap the real-time process evaluation and the decision making for an on-line process control.

Computer program language is dependent on the availability of assembler and compilers. The ARCH-102 was, reportedly, programmed in machine language for DDC of penicillin fermentation (Grayson, 1969); the PDP-11/20 was programmed in assembly language for data analysis-oriented purposes (Nyiri, 1971). In the case of the PDP-11/20, an additional 4K or 8K core memory and application of discs make possible the use of BASIC or FORTRAN IV languages. Here the relationships between the price of additional core memory and the manageability of the languages are the factors which finally determine the programming language. Regardless of the language, the computer/ operator communication must be as simple and as short as possible. For instance, in the

TABLE 3. *Specifications of small-scale computers involved in fermentation processes*

	ARCH-102[a]	PDP-11/20[b]
Word length	13 bits	16 bits
Mode of operation	parallel synchronous	asynchronous
Store access time	6 μsec	≈500 nanoseconds
Addition time	23 μsec	2.3 μsec
Multiply time	78.5 μsec	4.3 μsec
Instruction type	single address	double operand (memory-to-memory)
Hardware priority interrupt	3	automatic
Priority levels	4	4
Working store	ferrite store	read/write core memory
Basic capacity/words/	8K	4K
Expanded capacity/words/	16K	32K
Language	machine language	assembler, or BASIC or FORTRAN

[a] Elliott-Automation, Ltd., England
[b] Digital Equipment Co., Inc., Maynard, Mass., USA

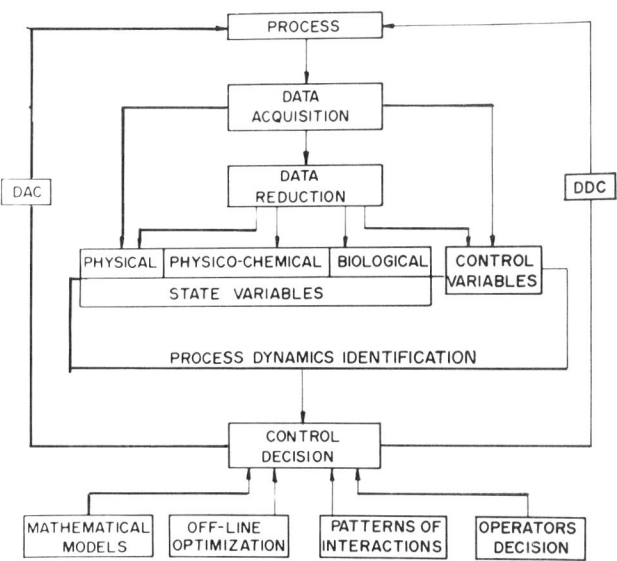

FIG. 7. Role of data analysis oriented computer operation in the development of interactive process control.

data analysis-oriented application of a PDP-11/20, the alteration of a set point takes about 30-40 sec for a skilled operator.

Role of the Data Analysis-Oriented Computer Application in the Development of Interactive Process Control

Data acquisition and data reduction systems are necessary both for direct digital control and the data analysis-oriented utilization of computers (Fig. 7). In the case of DDC, the data reduction serves the purpose of monitoring the status of the control variables only. In the case of the data analysis system, the data reduction routine provides information on the status of the state variables as well. With the multivariation of sensor data, new state variables are obtained which offer insight into the physiological and the biochemical behavior of the cells. Values describing the status of both the control and state variables serve the purpose of process identification. As it was emphasized, proper process identification is the first necessary criterion in performing adaptive-type interactive process control. The DAC variables are controlled according to their deviation from a predefined value, regarding the status of the state variables and the possible effect of the control variables and the possible effect of the control variables on the further development of the culture. As is known, control variables can interact either in synergistic or in antagonistic ways, e.g., a temperature dependency of the induction and repression of an individual enzyme synthesis (Rose, 1969). Recently, the effect of CO_2 (HCO_3^-) concentration on different biosyntheses of industrial importance was reported (Nyiri and Lengyel, 1965; Zajic and Liu, 1970). Determination of the nature of such effects will lead to refinement of control operations with a view toward the improvement of the fermentation.

SUMMARY

One of the most important advantages of the integration of a highly instrumented fermentor with a computer is the ability to observe, on a real-time basis, the effect of the environmental conditions on certain detectable parts of the cell metabolism. These on-line, real-time "effect-response" operations are tools to define the synergistic or antagonistic interactions of individual control variables. Application of mathematical models and heuristic methods offers some possibilities to resolve the numerical and logical problems of cell growth, mutation, and product formation. This information can improve the methods of process identification and furthermore will serve as a basis of algorithms for on-line optimization and prediction techniques. Elaboration of these techniques is the next necessary step on the way to interactive process control.

At this point, computer process control requires operator decision-making. The continuous development of methods of observation and the knowledge of the process dynamics will, however, result in algorithms which will step-by-step transfer the elementary decisions to the computer.

ACKNOWLEDGEMENT

The author is indebted to Arthur E. Humphrey for many valuable discussions and to John D. Wilson for his continuous help and cooperation.

LITERATURE CITED

Garfinkel, D., L. Garfinkel, M. Pring, S. B. Green, and B. Chance. 1970. Computer applications to biochemical kinetics. *Anniv. Rev. Biochem.*, **39**: 473-497.

Grayson, P. 1969. Computer control of batch fermentation. *Process Biochem.*, **4**: 43-61.

Harmes, C. S. III. 1971. Design criteria of a fully computerized fermentation system. *Develop. Ind. Microbiol.*, **13**:

Heinmets, F. (ed.). 1969. *Concepts and Models of Biomathematics*. M. Dekker, Inc., New York, Vol. 1.

Li, Y. T. 1960. *The Philosophy of Adaptive Control: Automatic and Remote Control*. Proc. 1st IFAC Congr., Moscow. Butterworths, London and Washington, D.C.

Nyiri, L. K. 1971. A philosophy of data acquisition, analysis and computer control of fermentation processes: Data analysis oriented application of the computer. Labex Symposium on Computer Control of Fermentation Processes. London.

Nyiri, L., and Z. L. Lengyel, 1965. Studies on automatically aerated biosynthetic processes. I. The effect of agitation and CO_2 on penicillin formation in automatically aerated liquid cultures. *Biotechnol. Bioeng.*, **7**: 343-354.

Pennington, R. H. 1970. *Introductory Computer Methods and Numerical Analysis*. Macmillan Co., London, 2nd Ed., 417 pp.

Rose, A. H. 1969. Temperature control of growth and metabolic activity, pp. 157-175. In: *Fermentation Advances*, D. Perlman (ed.). Academic Press, Inc., New York and London.

Tuffile, C. M., and F. Pinho. 1970. Determination of oxygen-transfer coefficients in viscous Streptomycete fermentations. Biotechnol. Bioeng. **12**: 849-871.

White, J., and E. A. Hazbun. 1970. Pilot plant data acquisition and analysis. Proc. 25th ISA Conf., Philadelphia. Paper #511.

Yamashita, S., H. Hoshi, and T. Inagaki. 1969. Automatic control and optimization of fermentation processes: glutamic acid, pp. 441-463. In: *Fermentation Advances*, D. Perlman (ed.). Academic Press, Inc., New York and London.

Zajic, J. E., and F. S. Liu. 1970. Effect of carbon dioxide upon alpha-anylase, protease and spore formation in *Bacillus subtilis*. *Develop. Ind. Microbiol.*, **11**: 350-356.

AUTHOR CITATION INDEX

Abbott, B. J., 19
Abraham, E. P., 84, 132, 135
Achorn, G. B., Jr., 214
Adams, S. L., 267
Aiba, S., 5, 185
Aida, K., 19, 247
Albrecht, F., 201
Alexander, D. F., 51, 85, 132, 135
Alexander, M. T., 47
Alikhanian, S. I., 51, 94
American Type Culture Collection, 49
Anderson, R. F., 83, 143
Anderson, W. L., 201
Arima, K., 83
Asai, T., 277
Atkinson, D. E., 246
Auerbach, C., 94

Backus, M. P., 83, 86, 143
Ball, C. O., 194
Barr, J. G., 247
Bartholomew, W. H., 205, 214, 227, 228
Baumberger, J. P., 214
Bautz, M., 267
Becze, G. de, 214
Benedict, R. G., 132, 227
Beran, K., 182
Berger, J., 132
Beukers, R., 4
Bhavaraju, S. M., 205
Bilford, H. R., 4
Birnbaum, J., 246
Blackwood, R. K., 19
Blanch, H. W., 205
Blodgett, K., 201
Boder, G., 19
Bonner, D., 104
Bottomley, R. A., 181
Bourdillon, A. G., 201
Bowden, J. P., 83
Bowers, R. H., 267
Bradley, S. G., 94

Brandle, E., 153
Brandon, B. A., 47
Brierley, M. R., 227
Briggs, R., 182
Bristol Laboratories, Inc., 201
Broda, P., 246
Brown, C. S., 86
Brown, W. E., 4, 83, 143, 214, 267
Bryson, V., 51, 94
Bufton, A. W. J., 51
Bugie, E., 139
Bull, A. T., 277
Bu'Lock, J. D., 247
Bunner, R., 247
Burg, C. R., 277
Burris, R. H., 85
Butkewitsch, W., 132
Butlin, K. R., 267

Calderbank, P. H., 205
Calam, C. T., 94, 267, 277
Callow, D. S., 5, 226, 227
Calmette, A., 135
Calvert, O. H., 83
Camici, L., 83
Campbell, T. H., 83, 143
Camposano, A., 228
Carilli, A., 228
Carrit, D. E., 181
Carter, B. L. A., 277
Časlavsky, Z., 182
Centraal Bureau Voor Schimmelcultures, 48
Chain, E. B., 83, 84, 132, 135, 226, 227, 228
Chance, B., 290
Charm, S. E., 194
Charney, J., 201
Chattaway, F. W., 247
Chen, C. Y., 201
Chen, J. W., 4, 277
Cherry, G. B., 201
Christensen, L. M., 132

Author Citation Index

Churchill, B. W., 83, 84, 86, 143
Clark, L. C., Jr., 181
Clarke, P. H., 246
Clement, M. T., 170
Clifton, C. E., 277
Clutterbuck, P. W., 132
Coghill, R. D., 84, 110, 132, 135
Commonwealth Mycological Institute, 48
Constantinides, A., 277
Cooper, C. M., 153, 205, 214, 226
Corman, J., 227
Costich, E. W., 153, 205, 228
Crawford, F. M., 19

Dale, R. F., 228
Daniel, H. S., 214
Daniels, F., 214
Datta, P., 246
Davies, C. N., 201
Davies, O. L., 94
Dawes, E. A., 248
Dawes, I. W., 248
Dawson, P. S. S., 4, 19
Day, A. A., 132
Dean, A. C. R., 277
Decker, H. M., 201
Deindorfer, F. H., 170, 185, 194, 205, 227
Demain, A. L., 19, 51, 230, 246, 247
Demerec, M., 84
Denison, F. W., Jr., 170
Dion, W. M., 228
Dodge, B. O., 104
Donovick, R., 139
Doskočil, J., 267
Doskočilová, D., 267
Driver, N., 267
Duggar, B. M., 84
Dulaney, D. D., 247
Dulaney, E. L., 214, 247, 267
Dunn, C. G., 5, 48
Dworshack, R. G., 170

Edlin, G., 246
Eisenberg, G. M., 85
Elander, R. P., 51, 247
Elberg, S., 214
Elferdink, T. H., 147, 170, 228
Ellis, J. J., 48
Elsworth, R., 4, 19, 226, 227, 267
Emerson, R. L., 85
Emmons, C. W., 84
English, A. R., 19
English, F. L., 214
Espenshade, M. A., 51
Everett, H. J., 153, 205, 228

Eyring, H., 194, 214

Farrell, L., 84
Fatt, I., 181
Fencl, Z., 248
Fennell, D. I., 48
Fernstrom, G. A., 153, 205, 214, 226
Feustel, I. C., 214
Finn, R. K., 201, 228
Fisher, W. P., 201
Fleming, A., 132
Fletcher, C. M., 132, 135
Florey, H. W., 84, 132, 135
Florey, M. E., 84
Florkin, M., 48
Floss, H. G., 247
Folkers, K., 139
Fonken, G. S., 19
Forbes, E., 51, 105
Ford, J. H., 228
Foster, J. W., 84, 110
Foust, H. C., 214
Frazier, W. C., 85
Fred, E. B., 84
Fredrikson, A. G., 277
Freedman, D., 147
Freese, E., 94
Frey, C. N., 4
Fried, J., 139
Friedland, W. C., 147
Fujikawa, K., 247
Fuld, G. J., 4

Gabriel, C. L., 19
Gaden, E. L., Jr, 153, 205, 227, 228, 267, 277
Gailey, F. B.,, 84, 86, 143, 268
Gale, E. F., 267
Gardner, A. D., 132, 135
Garfinkel, D., 290
Garfinkel, L., 290
Gastrock, E. A., 132, 267
Gee, L. L., 214
Geile, F. A., 201
Gemmel, A. R., 105
Gerhardt, P., 214
Gevers, W., 246
Gillett, W. A., 227
Gillies, R. A., 267
Glassman, H. N., 214
Glasstone, S., 214
Gledhill, W. E., 19
Glick, C. A., 201
Goebel, W., 246
Goetz, A., 201
Goldschmidt, M. C., 84

Gorman, M., 247
Gottlieb, D., 247
Goulden, S. A., 247
Graessle, O. E., 139
Grass, R. C., 194
Grayson, P., 290
Green, S. B., 290
Greene, H. C., 84
Gross, S. R., 247
Gualandi, G., 227, 228

Hagarty, C. P., 201
Hagino, H., 246
Hamill, R. L., 247
Hamlin, B. T., 248
Hamre, D., 139
Hanson, F. R., 4
Hara, M., 153
Harmes, C. S., III, 290
Harris-Smith, R., 226, 227
Hashida, W., 246
Haynes, W. C., 48
Hazbun, E. A., 290
Hearn, W. R., 247
Heatley, N. G., 84, 132, 135
Heinmets, F., 290
Hemmons, L. M., 51
Hendlin, D., 84, 110, 246
Herbert, D., 4, 267
Herold, M., 248
Herrick, H. T., 19, 267, 268
Herrman, R. G., 201
Hesseltine, C. W., 48
Higuchi, K., 84
Hinschelwood, C., 277
Hixson, A. W., 205, 228
Hockenhull, D. J. D., 277
Hodges, A. B., 267
Hogerton, J. F., 194
Hollaender, A., 84
Hopper, S. H., 201
Hoshi, H., 277, 290
Hosler, P., 110, 143, 267
Hospodka, J., 182
Hošťalék, Z., 247
Huang, H. T., 170
Hulme, M. A., 247
Humfeld, H., 214
Humphrey, A. E., 185, 194, 201, 205, 277
Hungate, R. E., 267
Hwang, S. H., 48

I. B. M. Ltd., 277
Imae, Y., 247
Inagaki, T., 277, 290

Inamine, E., 247
Ingols, R. S., 214
Ishii, K., 246
Ito, M., 247

Jackson, M., 246, 247
Jackson, R. W., 227, 268
Jacob, T. A., 246
Janglová, Z., 247
Jarvis, F. G., 84, 86, 268
Jennings, M. A., 84, 132, 135
Jensen, R. A., 246
Johnson, F. H., 194
Johnson, I. S., 19
Johnson, J. L., 228
Johnson, M. J., 19, 83, 84, 85, 86, 110, 132, 143, 181, 214, 227, 228, 247, 267, 268, 277
Johnson, R. A., 19
Johnstone, H. F., 201, 214
Jones, D., 139
Jones, G. H., 247
Jukes, T. H., 19

Kadana, T., 205
Kambe, M., 247
Kanwisher, J. M., 181
Karow, E. O., 205, 214, 227, 228
Kašparová, J., 267
Katz, E., 247
Kavanagh, F., 139
Kehoe, T. J., 170
Kemp, C. E., 227
Kempe, L. L., 267
Kendall, A. I., 132
Kern, D. Q., 194
Kihlman, B. A., 94
Kinoshita, S., 19, 246
Kinsey, D. W., 181
Kirby, G. W., 4
Kirch, E. R., 132
Kleinkauf, H., 246
Kluyver, A. J., 201
Knight, S. G., 85
Koch, R. S., 84
Koepsell, H. J., 227, 277
Koffler, H., 84, 85
Koga, S., 277
Kolachov, P. J., 4, 5
Kominek, L. A., 247
Kono, T., 277
Koplove, H. M., 5
Krieg, D. R., 94
Kuehl, F. A., Jr., 139
Kurahashi, K., 247

Author Citation Index

Lago, B. D., 247
Laidler, K. J., 214
La Mer, V. K., 201
Landhal, H. D., 201
Langlykke, A. F., 19
Langmuir, I., 201
Lawley, P. D., 94
Lee, B. K., 4
Legator, M., 247
Legg, D. A., 132
Lengyel, Z. L., 290
Lennon, R. E., 19
Lester-Smith, E., 85
Leudeking, R., 277
Levine, S., 194
Lewis, A., 19
Lewis, V. M., 214
Li, Y. T., 290
Liebmann, A. J., 214
Light, R. J., 247
Lilly, M. D., 246
Lingens, F., 246
Liu, F. S., 290
Locke, F. G., 214
Locker, A., 277
Lockwood, L. B., 132, 135
Long, R. A., 201
Losick, R., 247
Lovell, R., 132
Luedeking, R., 267
Lumb, M., 227

Mabe, J. A., 247
McCann, E. P., 201
McCormick, J. R. D., 247
McDaniel, L. E., 84, 110, 201
MacDonald, D. M., 51
Mack, D. E., 214
McKenzie, H. A., 214
MacKenzie, R. M., 277
Mackereth, F. J. H., 181
Malik, V. S., 247
Mancy, K. H., 181
Mandelstam, J., 248, 277
Marshall, R., 247
Martin, J. R., 247
Marx, A. F., 4
Masurekar, P. S., 247
Mateles, R. I., 4, 248
Matsuo, T., 246
Maxon, W. D., 4, 147, 170, 227, 228, 267, 277
May, O. E., 19, 267, 268
Meers, J. L., 248
Mellor, J. W., 214
Mercer, C. K., 227
Metzger, H. J., 139
Metzner, A. B., 205

Meyers, G. B., 227
Mikhailova, G. R., 51
Miller, A. L., 247
Miller, J., 85
Miller, S. A., 153, 214, 205, 226
Millis, N. F., 185
Mindlin, S. Z., 51
Monod, J., 277
Moorman, H. E., 201
Moo-Young, M. B., 205
Mori, K., 247
Moyer, A. J., 85, 110, 132, 135, 143, 267, 268
Muirhead, D. R., 170
Müller, O. H., 214
Murphy, H. C., 201

Nagasawa, M., 19
Nakamori, S., 246
Nakayama, K., 246
Nečinová, S., 248
Nelson, H. A., 147, 170, 228
Neville, J. R., 181
Ng, F. M. W., 248
Nimeck, M. W., 51, 94
Nimi, O., 247
Nomi, R., 247
Nyiri, L. K., 19, 290

Okabe, M., 5
Okun, D. A., 181
Oldshue, J. Y., 205
Oleson, A. P., 132
Olson, B. H., 84, 143, 214, 227
Orgel, L. E., 94
Owen, S. P., 227, 267, 277
Oyama, Y., 246

Paigen, K., 246
Paladino, S., 226
Pan, C. H., 51
Parker, A., 201
Pastan, I., 247
Pasteur Fermentation Centennial, 5
Pathak, S. G., 51
Paulus, H., 247
Peat, A., 277
Peck, R. L., 139
Pennington, R. H., 290
Peppler, H. J., 19, 48, 228
Perlman, D., 5, 19, 84, 85, 110, 267
Perlman, R., 247
Perry, J. J., 228
Petering, H. G., 214
Peterson, M. H., 147, 170
Peterson, W. H., 83, 84, 85, 132, 143, 214, 267
Phillips, D. H., 181, 227
Phillips, K. L., 19

Pinho, F., 290
Piret, E. G., 277
Pirt, S. J., 5, 227, 246, 247, 277
Pittman, R. W., 181
Pivet, E. L., 267
Pollard, E. C., 194
Pollisar, M. J., 194
Pontecorvo, G., 51, 85, 104
Porges, N., 132, 267
Pound, G. S., 83
Prescott, G. C., 228
Prescott, S. C., 5, 48
Price, K. E., 19
Pring, M., 290
Pruess, D. L., 247

Radlett, P. J., 19
Raistrick, H., 132
Rake, G., 139
Ramkrishna, D., 277
Ramskill, E. A., 201
Ranz, W. W., 201
Raper, K. B., 84, 85, 86, 132, 135
Rebello, J. L., 246
Redfield, B., 247
Reese, E., 85
Reh, M., 246
Reilley, C. N., 181
Reilly, H. C., 214
Reusser, F., 5
Rhodes, A., 277
Rhodes, R. P., 153
Řičica, J., 248
Righelato, R. C., 19, 247, 277
Robbers, J. E., 247
Roberts, M. H., 201
Robinson, H. J., 139
Rockwell, T., 194
Rodebush, W. E., 201
Roe, E. T., 132
Roegner, F. R., 83, 85, 143
Roehr, M., 247
Roper, J. A., 51, 105
Rose, A. H., 290
Rosebury, T., 201
Rosenbrook, W., 247
Rowlands, S., 85
Rowley, D., 85
Roxburgh, J. M., 227
Rushton, J. H., 153, 205, 214, 228
Ryu, D. Y., 4

Sakaguchi, K., 19
Sakamoto, Y., 247
Sallans, H. R., 227
Sanders, A. G., 84
Sanderson, K., 85

Sansome, E. R., 84, 105
Sardinas, J. L., 19
Sargeant, K., 19
Sassiver, M. L., 19
Sato, E., 247
Scalf, R. E., 4, 5
Schatz, A., 139, 214
Schlegel, H. G., 246
Schmid, A., 153
Schmidt, W. H., 85, 132, 135, 143
Schultz, A., 4
Schwab, J. L., 214
Sell, W., 201
Sermonti, G., 51, 83, 85, 94, 105, 228
Sfat, M. R., 205, 214, 227, 228
Shaeffer, P., 247
Shaffer, P. A., 143
Sharpe, E. S., 227
Shedlovsky, L., 214
Shepherd, D., 247
Sher, H. N., 170
Sherwood, T. K., 214
Shiio, I., 246
Shorenstein, R. G., 247
Shurter, R. A., Jr., 226
Sih, C. T., 19
Sikyta, B., 248, 267
Silver, R. S., 248
Sizer, I. W., 19
Slezák, J., 248
Smiley, K. L., 19
Smith, C. G., 228
Smith, D. G., 139
Smith, G., 48
Smith, R. M., 228
Solomons, G. L., 19
Somogyi, M., 143, 228
Sonenshein, A. L., 247
Spada-Sermonti, I., 51
Spencer, J. F. T., 227
Spencer, J. L., 277
Stadtman, E. R., 246
Stahley, G. L., 214
Stahmann, M. A., 83, 85, 86
Stark, W. H., 4, 5
Stauffer, J. F., 83, 85, 86, 143
Steel, R., 147, 227, 228
Stefaniak, J. J., 84, 86, 143, 268
Steinberg, R. A., 132
Steiner, H., 153
Stejskalová, E., 248
Stodola, F. H., 268
Stotz, E. H., 48
Stouthamer, A. H., 277
Stranberg, G. W., 19
Strohm, J., 228
Stros, F., 182

Stubbs, J. J., 132, 135, 267, 268
Suchý, J., 247
Sylvester, J. C., 147, 170

Tabenkin, B., 132, 135
Tabuchi, T., 19
Takahashi, H., 247
Takahashi, J., 153
Takebe, H., 153
Tanaka, H., 153, 246
Tanimoto, H., 246
Tanner, C. B., 181
Tarr Gloor, E., 105
Tashjian, A. H., 19
Taylor, J. S., 205
Telling, R. C., 4, 19, 267
Tempest, D. W., 247, 248
Terjesen, S. G., 201
Terramoto, S., 246
Terui, G., 19
Teske, G., 182
Thom, C., 85, 86
Thoma, R. W., 4, 51, 94
Thomas, D. J., 201
Tödt, F., 182
Tonolo, A., 105
Trenina, G. A., 247
Trenner, N. R., 247
Trinci, A. P. J., 277
Trutneva, E. M., 247
Tsuchiya, H. M., 227, 277
Tsunoda, T., 19
Tuffile, C. M., 290

Ueda, K., 153
Uemura, T., 247
Uesseler, H., 246
Ugolini, D., 226
Umberger, A. J., 267
Umberger, E. J., 132, 135
Unger, E. D., 5
Urbani, E., 105

Van der Sluis, J., 226
Vaněk, Z., 247
Venkatasubramanian, K., 19
Vezina, C., 19
Vieth, W., 19
Viney, M., 182
Vining, L. C., 246, 247
Visser, J., 201

Vitali, R. A., 246
Vitek, V., 214
Vladinirov, A. V., 51

Wakaki, S., 86
Waksman, S. A., 132, 139, 214
Walker, A. W., 132
Walker, J. B., 247
Walker, J. C., 83
Wallen, L. L., 268
Walti, A., 139
Wang, D. I. C., 205
Ward, G. E., 19, 132, 135
Watson, R. W., 170
Wegrich, O. G., 226
Weissbach, H., 247
Wells, P. A., 132, 135, 267, 268
West, J. C., 170
West, R. E., 267
Westlake, D. W. S., 246
White, J., 290
Whitmore, L. M., Jr., 83, 143
Whittenberg, J. V., 19
Wickerham, L. J., 48
Wilhelm, R. H., 205, 214, 228
Wilken, G. D., 227
Wilker, B. L., 84, 110, 170
Willey, C. R., 181
Williams, B., 246
Williams, P. R., 247
Williams, V., 226, 227
Wintersteiner, O., 139
Winzler, R. J., 214
Wise, W. S., 227
Witkin, E. M., 94
Wolf, F. T., 86
Wong, J. B., 201
Woodruff, H. B., 84, 110, 247
Woodward, R., 85

Yamada, K., 19, 205, 247
Yamashita, S., 277, 290
Yano, T., 205
Young, T. B., 5

Zajic, J. E., 290
Zajiček, J., 267
Zimmer, E., 84
Zinner, M., 247
Zuidweg, M. H. Z., 4

SUBJECT INDEX

Absolute filters, air, 195. See also Sterilization
Acetobacter suboxydans, sorbose from sorbitol by, 132
Acetone, 13
Acid production by Lactobacillus delbrueckii, rate processes, 254
Actinomycetes. See also Genetics; Heterokaryosis
　parasexual recombination in, 50
　transformation and transduction between, 51
Actinomycin
　enzymes in synthesis of, 244
　Streptomyces antibioticus, 243
　synthesis
　　phenoxazine synthetase and glucose repression affecting, 239
　　specific activities of key enzymes in, 244
Activated sludge, 9, 12. See also Sewage
Adaptive process control system, outline of, 282
Aeration, 9
　of culture media, measurement, sulfite oxidation vs. polarography, 227
　effect on novobiocin fermentation, 228
　effects of power input and impeller size on, 224
　requirements for the growth of microorganisms, 228
Aeration-agitation, 228
　and oxygen transfer, 203–228
　in fermentation, 205
　studies on the novobiocin fermetation, 215
　in submerged fermentation, 227
　in 30-liter fermentations, 140
Aerobacter aerogenes, 5
　use in scale-up, 227
Aerobacter cloacae, effect of oxygen supply on, 227
Aerosols, microorganisms as, 195
Aflatoxin, 42

Agricultural Research Service, culture collection, 25
Air, sterile, requirements for large volumes, 195. See also Air filtration
Air filtration. See also Sterilization
　apparatus for experimental procedures, 197, 198
　efficiency related to velocity, 198, 199, 201
　for sterilization, 195
　with various materials, 195
　velocity-efficiency relationship, 198, 199
Air flow and agitation rate, effect on fermentations with Penicillium chrosogenum and Streptomyces griseus, 205
Air sterilization, 9, 10
　by fibrous media, 9, 195–201
　methods, 195
Alcoholic fermentation, 4, 9
Algorithms, complex, in process identification and on-line optimization, 282
Alkaloid(s)
　in Claviceps paspali, 233
　production by Claviceps paspali, effect of glucose on, 239
　synthesis in Claviceps, induction by tryptophan, 240
Alkanes, normal, growth of yeast on, 12
Alkaterge C in lard oil, antifoam agent, 140
American Type Culture Collection, 25, 47
Amino acids
　in corn steep, 127
　microbial production of, 19
　produced by fermentation, 10, 16
　　from non-carbohydrate sources, 15
6α-amino-penicillanic acid derivatives, 16
Ammonia
　accumulation by microorganisms, effect of carbohydrate on, 132
　and phosphate, concentrations of, 10
Amphotericin B, 87
Amylase, 13. See also Starch

Subject Index

Anabolic agents, 17
Analog controllers, replacement by direct digital control, 281
Analog-resistant mutants, primary metabolites produced by, 236
Animal cell culture, 17
Anthranilate synthetase, resistance to feedback inhibition in *Claviceps paspali*, 233
Antibiotic-producing microorganisms
  avoidance of suicide, 230
  genetics of, 94
Antibiotic(s), 2, 13. See also Metabolites, secondary
  as animal feed supplements, 14
  in animal health, 16
  fermentation, a systems approach to design of, 5
  novel, by modification of known antibiotics, 32
  production of, 1, 5. See also Synthesis; Biosynthesis
    by fermentation, listing of, 14–15
    effects of agitation on, 228
    and spore formation, 109
    strains for, 31, 32
  screening tests, 57
  structure-activity relationships, 16
  testing against pathogens, 32
Antifoam agents
  Alkaterge C in lard oil, 140
  Dow-Corning A, 61, 62
Apparatus and equipment, 145–182
Arginine, 235
Arrhenius equation, 184, 187
Arthrobacter simplex, 87, 97
Ascorbic acid, 14
Asepsis, large scale, 10. See also Sterilization
Aspartokinase, 235
Aspergilli, 50
  culture run-down in, 41–42. See also Degeneration
  manual of, 86
  spores and sporulation, 39
Aspergillus flavus morphology, effect of mechanical agitation on, 228
Aspergillus nidulans, 95, 97, 99, 102, 104
  analysis of mitotic recombination, 105
  genetics of, 51, 104, 105
  genetic recombination without sexual reproduction in, 51
  polyploidy in, 105
Aspergillus niger, 95, 99, 102, 105
  citric acid production by, 13
  hydrolysis of soybean meal, 132
  protection of proteins by carbohydrate, 131
  study of heavy metals in nutrition of, 132
Aspergillus ochraceus, 4
Aspergillus oryzae, 9
Aspergillus terreus, induction of mutation by ultraviolet light, 84
Automation, 279–290
  of fermentation or recovery operations, 3
  of screening operations, 94
Autosynthetic processes, 231

Bacillus influenzae, effect of *Penicillium* on, 132
Bacillus megaterium, 92
Bacillus subtilis
  catabolite repression and growth rate, 244
  effect of carbon dioxide on, 290
  glucose or nitrogen limitation on growth, 244
  hypoxanthine and inosine in, 232–233
Bacteria, kinetics of inactivation of, by heat, 194. See also Sterilization
Bacterial sporulation, catabolite repression of, 239
Bacteriophage, therapeutic, in cholera, 17
Baker's yeast, 9
Basidiomycetes, isolation of basidiospores from, 32
Basidiospores, 33
Batch fermentations
  computer control of, 290
  rate processes in, 109. See also Kinetics
Biochemical genetics, computer applications to, 290
Biochemical pathways, study of, 12
Biochemical reaction systems, kinetics of, 253
Bioconversion processes, 2, 3
  metabolic regulation in, 2
Biomathematics, concepts and models, 290
Biosynthesis
  directed, 234
  of antibiotics and vitamins, 234
  improved environment for, 233
  secondary, modification of repression of, 242
Biosynthetic pathways
  catabolite regulation in, 239
  energy charge regulation of, 240
  feedback regulation in, 233, 234
  induction in, 239, 240
  to several antibiotics, 237
Biotin requirement of glutamic acid producing bacteria, 233

Blakesleea trispora, 38
  isolation from nature, 24
Branched pathways, 238. *See also* Regulation
  chemostat for study of regulation of, 243–244. *See also* Continuous culture
  yielding primary and secondary metabolites, 237
Brevibacterium flavum, threonine synthesis by, 236
Breweries and beer in South Africa, 24
Bubble size and holdup, 211. *See also* Aeration
Burk's medium, 10
2,3-butanediol, 5
n-butanol, 13

Calcium carbonate, seperate sterilization of, 134
Candida yeast, conversion of hydrocarbons to citric acid by, 15
Carbohydrate(s)
  continuous addition to penicillin fermentation, 140–143. *See also* Feeding
  conversion of, in fermentation, 14
  rapidly metabolized, feeding of, 109
Carbon and energy sources
  high molecular weight, breakdown of, 231
  other than glucose, 239
Carbon dioxide, effect of, on alpha amylase, protease, and spore formation in *Bacillus subtilis,* 290
β-carotene production by pure mixed culture, 12
Catabolite control of secondary metabolites, 239
Catabolite regulation
  in biosynthetic pathways, 239
  in continuous and batch culture, 244
  of bacterial sporulation, 239
Catabolite repression
  and growth rate, *Escherichia coli, Bacillus subtilis,* 244
  mediation by cyclic AMP in Gram-negative organisms, 239
Catalog of strains, American Type Culture Collection, 47
Cell culture, animal, 17
Cell(s), 2
  dividing action, action of chemicals on, 94
  growth and metabolism, effect of control variables on, 281
  resting, 2

  stabilized, 2
Cellulose
  anaerobic fermentation of, 11
  conversion to organic acids, 11
Cellulosic materials, organic acids from, 132
Centraalbureau list of cultures, 48
Cephalosporin C, induction of synthesis by methionine, 240
Cephalosporins, 16
Chemical kinetics, rapid development and application, 252
Chemical mutagens, 94
  classification of, 89
  methods of use, 88
Chloramphenicol
  aromatic amino acids and, 232
  biosynthesis, self-limiting, 237
  fermentation, atypical kinetic behavior, 263
  from shikimic acid in *Streptomyces* sp., 232
Chlortetracycline
  biosynthesis, methionine as precursor, 238
  production by *Streptomyces viridofaciens,* 238
  synthesis, phosphate effect in, 240
  synthesis by revertant, 238
  two-stage continuous fermentation in complex media, 245
Cholesterol and plant sterols, enzymic hydrolyses of, 14
Chromosome breakage, biochemical aspects, 94. *See also* Genetics
Citric acid
  Aspergillus niger process for, 13
  from hydrocarbon by yeast, 18
  from hydrocarbons by fermentation with *Candida,* 15
Classification of fermentations
  based on empirical examination of batch data, 255
  on basis of free energy change, 258, 259
Classification of microorganisms for production of useful products, 33–37
Claviceps, alkaloid synthesis in, induction by tryptophan, 240
Claviceps paspali
  alkaloid production by, glucose effect, 239
  resistance to anthranilate synthetase in, 233
Clostridium, isolation on corn mash, 31
Collections, special, 25. *See also* Cultures
Colloid, protective, in lyophilization, 42. *See also* Preservation

Subject Index

Commonwealth Mycological Institute, catalog of cultures, 48
Competition, fermentation and chemical processes, 14, 15
Computer
  applications to biochemical genetics, 290
  control of batch fermentations, 290
  data analysis-oriented, 281, 283
  of fermentation processes, 281-290
  methods, introductory, 290
  on-line functions of, 283
  small-scale, in fermentation, 288
  technology, 290
Conidia, *Penicillium chrysogenum*, 96
Conidial supensions, 55, 56
Continuous antibiotic fermentation units, theoretical design of, 5
Continuous culture
  of bacteria, theory of, 4
  energy charge regulation in, 245
  penicillin cell-specific productivity rate and growth rate, 244, 245
  product formation and growth in, 3, 5
  role for, in fermentation research, 243-245
  technique for determining oxygen requirement of *Aerobacter* sp., 227
Continuous environmental control, 10
Continuous fermentation
  commercial, 3, 18
  processes, 4
  in small fermentors, 157
  two-stage, 245
Continous-flow culture, 5
Continuous hydroxylation, 4
Continuous lactic acid processes, to single or multi-stage, 255
Continuous processes
  and fermentation kinetics, 263, 264
  prediction of performance of, from batch data, 255
Continuous shake flask propagator for yeast and bacteria, 227
Continuous sterilizers, principles in the design of, 194
Control mechanisms
  in enzyme action and synthesis, 231. See also Regulation
  residual, alteration of, 234
Control variables
  effect on cell growth and metabolism, 281
  need for coordinated operation of, 281
Conversion. See also Bioconversion; Transformation
  of carbohydrates in fermentation, 14
  microbial, 4

Copper-iodometric reagents for sugar determination, 143
Corn bran for spore inoculum with *Penicillium chrysogenum*, 140
Corn mash for isolation of *Clostridium*, 31
Corn steep, 11, 61, 108
  amino acids in, effect on penicillin production, 127
  ash, effect on penicillin production, 85
  complex role in penicillin production, 131
  definition, analysis of, 112
  early fermentation uses, 122
  early use in penicillin production, 122, 123
  effects on penicillin production, 123
  growth factors in, effect on penicillin production, 127
  lot-to-lot variations, 78
  in penicillin production, 108
  in penicillin test medium, 61
  replaced by cotton seed meal in penicillin production, 84
  replacement of, 109
  role in penicillin production, 83
  as source of trace elements in penicillin fermentation, 126
  substitutes for, in penicillin production, 128
  value in penicillin fermentation, 131
Corn steep-lactose medium, mycelial development in, 73
Cortisone in rheumatoid arthritis, 14
Corynebacterium glutamicum
  auxotroph, tetramethyl pyrazine excretion by, 238
  cytoplasmic membrane of, 233
  excretion of purine nucleotides by, 233
  failure to retain glutamic acid, 233
  feedback inhibition of dihydropicolinate dehydrogenase in, 232
Cotton seed meal for corn steep in penicillin production, 84, 110
Culture(s)
  acquisition and conservation, 21-48
  ARS collection, restricted category, 45
  characteristics and potency, correlation in *Penicillium chrysogenum* strains, 80
  contamination of, 27, 40, 41
  crude, 11, 12
  cultivation, principles of, 29-40
  dangerous, government regulation of, 46
  degeneration, 27, 39
    induced, 231
    and loss of strains, 40, 42

depositories, official, 25
enrichment, 11
essential part of patented process, 25
etiological agents, regulations on packing and shipping, 46, 47
feeding, practical value, 126
fees for, 26
form for deposit in ARS collection, 45
form received from depositors, 45
freezing and storing with liquid nitrogen, 38
general attributes, 23, 24
hazardous to public health and agriculture, 45, 46
heterogeneous, preservation of, 39
identification of, for patent purposes, 44
instability and heterogeneity of, 22
isolation, 32
  manual of methods, 30
  micromanipulator for, 32
  from nature, procedures for, 30, 32
  selected inhibitors used for, 33
living, shipping and packaging, 47
loss by natural selection or mutation, 41
lyophilization of, 37
  freeze-dried, 42
master stocks, 58
media for maintenance, 29–40
methods for strain development, 55
microscopic examination of, 37
mixed, 12, 39
non-pathogenic, lack of restriction on movement, 46
patented, restricted distribution of, 43–45
pathogenic, restrictions on shipping, 46
penicillin producing, Wisconsin strains, 50
Penicillium chrysogenum, non-sporulating, 115
plant pathogens, restrictions on, 46
preservation
  alternative methods, 39, 43
  by freezing, 38
  by lyophilization, 37
  soil preparation, 58
  and storage of, 22
pure, 11
pure-mixed, 12
purification of, 39
reisolation of, 87
restrictions on, in public collections, 43, 44
run-down in *Aspergilli* and *Penicillia*, 41–42
selection programs, value of, 18

serial transfer of, 43
from sewage, 12
shipment of, 42, 45
  freedom from customs duty, 46
  practical considerations, 47
sources and management, 23
stability
  maintenance of, 37–40
  viability in and over liquid nitrogen, 43
stable, 12
stock, conditions for growth, 40
storage, alternative methods, 43
stress on, 22
use of, 25
viability of, 22
Culture collections, 22, 25, 26, 29, 30. See *also specific collections*
Cyclic AMP as mediator of catabolite repression in Gram-negative organisms, 239
Cylinder plate method of assay for penicillin, 111, 112. See *also* Penicillin
Czapek-Dox medium, 10
Czapek's agar, 68

Data acquisition and analysis, 290
Deamination, prevention by carbohydrate, 131
Dehydrogenases, steroid, 4
Dextrolactic acid, rapid fermentation process for, 135
Differential equations representing fermentations, 269
Dihydropicolinate dehydrogenase in *Corynebacterium glutamicum*, 232
L-dihydroxyphenylalanine in Parkinson's disease, 14
Diploids
  heterozygous, in *Penicillium chrysogenum*, 100–102
  and mitotic recombination, 105
  from *Penicillium chrysogenum* heterokaryons, 100, 101
  from sectors, *Penicillium chrysogenum*, 101
Dissolved oxygen
  measurements in pilot and production scale novobiocin fermentations, 147
  polarographic measurement in yeast fermentations, 228
  probes, components of, 171
  probes for measurement of, 171–182. See *also* Probes
  supply in 20-liter culture vessels, 226
Distiller's yeast, continuous aerobic process for, 5

Subject Index

Diumycin, 87
Dry weight, mycelial, method for, 218

Electrode. See also Probes
  in external leg, 159
  immersion, drawbacks, 159
  oxidation-reduction, 169
  pH
    advantages of external circuit system, 168
    details of design, 160–162
    for fermentors, 158, 170
    installation in external circuit, 167
    installation and operation, 163–167
    use of, 168
  pure culture considerations, 169
  steam sterilizable, 158
  sterilization problems, 158–159
Energy charge ratio, regulation by phosphate in chlortetracycline synthesis, 240
Energy charge regulation
  in biosynthetic paths, 240
  in continuous culture, 245
Energy release in fermentations, 257
Energy sources in the penicillin fermentation, 11
Enrichment cultures, 11
Enzyme action and synthesis, control mechanisms, 231. See also Regulation
Enzyme nomenclature, 48
Enzymes, 3
  in actinomycin synthesis, 244
  integrated action, 231
  key, repressed after growth, 241
  in medicine, 16
  microbial, newer uses, 16
  as reagents, 16
  stabilized, 16
Eosin-methylene blue medium, 10
Escherichia coli
  effects of ultraviolet light on, 93, 94
  growth rate and catabolite repression, 244
Ethanol, 13
Exocellular product formation in continuous culture, 5

Feed
  exponentially increasing rate, 10
  rates, optimal, of sugars in the penicillin fermentation, 143
  systems in the penicillin fermentation, models of, 271, 272
Feedback inhibition
  elimination by altering structure of enzyme, 235, 236
  removal by sequential mutation, 236
Feedback regulaton
  in batch and continuous culture, 243–244
  in biosynthetic pathways, 233, 234
  bypassing of, 235
  of secondary metabolites, 236, 237
Feedback resistance, 236
Feeding
  of nutrients, 109
  of rapidly metabolized carbon sources, 109
Fermentation(s)
  aerobic
    development of, 226
    a three-phase system, 206
  alcoholic, 4, 9
  amino acid, 10
  apparatus, 2
  automatic control and optimization of, 254, 290
  automation of, 3
  capacity, construction, 13
  classification of, 258, 259. See also Classification of fermentations
  companies, world list of, 19. See also Industries
  of concentrated solutions of glucose to gluconic acid, 132, 135
  conditions
    effect on yield, *Penicillium chrysogenum*, 78, 79
    effects on ranking *Penicillium chrysogenum* strains, 76
  equipment, techniques, 9. See also Apparatus and equipment
  facilities, 13
  factors affecting cost per, 154
  five-liter, oxygen uptake rates in, 213
  flasks, 57. See also Flasks
  food, 9, 24, 31
  history of, 1
  industry
    cycles of activity, 13
    prospects for, 13
  kinetics, 253–258
  and continuous processes, 263, 264
  manufacturing plants, 3
  mathematical models of, 250–251, 269–277
    vs. physical models, 269
  measurements, common, 257
  media, heat sterilization time for, 194
  methods, 2
  microorganisms
    changes in permeability of, 233
    enzyme deficiency in, 233

subnormal regulation in, 232
mixed cultures, 4
mycelial, scale-up, 216
nutrient materials for, 108. See also Raw materials
organism for, selection of, 112
pattern, griseofulvin, 273
performance criteria in, 253
plant, optimization of, 5
process(es), 2
  batch, rate patterns in, 255
  competition from chemical processes, 14
  computer control of, 281–290
  development, empirical and rational basis for, 232
  for dextrolactic acid, 135
  grouping by kinetic pattern, 255
  kinetic groups, 259
  kinetics, 250, 252–268
  obsolescence of, 14
  outline of data analysis program for, 285, 286
  rate patterns in, 260, 261
  for secondary metabolites, empirical development, 232
  submerged, with molds, 133
  types of, 258–263
  variables, 258
process-based industries, 3
processing, continuous, 3
products, 9
  commercialized since 1968, 17
  derived from carbohydrates, 15
  for the food industry, 18
  from non-carbohydrate sources, 15
  non-commercialized, 17
  non-competitive, 15
  high-volume, 13
  listing of, 14–17
  uses of, 13
  various, 13
quasi-continuous, 109
rapid continuous, 4
rates, 256
  in antibiotic fermentations, 253, 261
  oxytetracycline, 262
  penicillin, 267
  streptomycin, 260
renaissance of, 10
represented by differential equations, 269
research
  centers of activity in the United States, 18
  in Japan, 18, 19
shake flasks, aerobic, 109. See also Flasks

small scale, equipment for, 147, 154–157
steroid conversion, 10
studies, equipment for, 147, 228
submerged, agitation and oxygen transfer in, 228
system, computerized, 290
technology, 4, 19
  overview of, 8
  penicillin, influence on other processes, 5
thirty-liter, 140
  pump for additions to, 141
two-stage, needed for complex processes, 264
type reactions, 258
viscous, scale-up, 216
Fermentor(s)
aerated, controls for, 18
agitated, aerated, 9
anaerobic, aerobic, 9
bench scale, 206
design, 18
  for small scale submerged fermentation, 147
  highly instrumented, pilot scale, integration with general purpose computer, 281
large, 18
pilot scale, highly instrumented, 281
small
  advantages and drawbacks of, 154
  auxiliary equipment, 155, 156
  improved features, 154, 155
  information yield per dollar, 154
  means of adding materials, 156
  sight glasses, 157
  types of, 155
stainless steel, 154
thirty-liter, aerated, stainless steel, 140, 142
Filamentous fungi
heterozygous diploids in, 105
mitotic recombination in, 104
parasexual process in, 104
Filter(s)
absolute air, 195
bed, expressions for efficiency of, 196, 197
efficiency, minimal at certain velocity, 201
fibrous. See also Sterilization
  for air sterilization, 9
  efficiency, various media, 199
penetration
  by bacteria, 201
  by spores, 200

Subject Index

Flask(s). *See also* Shake flasks
  effects of shaking speed on power consumption in, 152
  fermentation, 57
    aerobic, 109
  power consumption in, 148
  shaking machine for, 134
  shapes of, 148
  yield, run-to-run variation in, *Penicillium chryosgenum* fermentation, 78
Flavor enhancers
  nucleotides, 10
  production by microorganisms, 233
Flow patterns in agitated vessels, 205. *See also* Mixing
5-fluorouracil, inhibition of ribonucleic acid synthesis by, in *Streptomyces antibioticus*, 243
Foam control, penicillin fermentation, 140
Fungal mycelium, distribution by agitation, 148
Fungi, sexual and parasexual recombination in, 95

Gallic acid fron tannin, 135
Gas-liquid contactors, agitated, 205. *See also* Fermentors; Reactors
Gibberellin fermentation, favorable response to glucose limitation, 239
Gibberellins, 10
Genetics and selection of organisms, 18
Gluconic acid, 9, 13
  from glucose in *Penicillium notatum*, 130, 132
Glucose
  concentrated solutions of, fermented to gluconic acid, 132, 135
  or sucrose in place of lactose, penicillin fermentation, 140, 142
Glucose repression
  of mannosidase in streptomycin synthesis, 239
  of phenoxazine synthetase in actinomycin synthesis, 239
Glucose suppression of various secondary metabolites, 239
Glutamic acid, 13
  biotin requirement in bacteria producing, 233
  fermentation
    automatic control of, 290
    model of, 271, 272
  production by bacteria deficient in α-ketoglutaric dehydrogenase, 233
Glutamine synthetase, repression by intracellular ammonia, 245

Gram negative bacteria, control of, in experimental animals with streptomycin, 139
Gramicidin S biosynthesis, 237
Griseofulvin
  biochemical model, 274
  biochemistry of the fermentation, 273
  fermentation
    dynamics, model vs. experimental, 276, 277
    mechanistic model, 273
    pattern, 273
    process, biosynthetic route, 272
    simulation of, 250–251, 269–277
  mathematical model, 274
Growth, temperature control of, 290

Haemacytometer counts of spores, 56
Heat sterilization
  analytical method for calculating, 184, 186–194
  times for fermentation media, calculation of, 194
Heat transfer rate, liquid in flask, 148
Heterokaryosis, 104
  genetic proof of, *Penicillium notatum*, 105
  genetic system based on, 104
  in *Penicillium chrysogenum*, 88
  in *Penicillium notatum*, 97
Heterokaryotic vigor in *Neurospora*, 104
Homoserine dehydrogenase, 235
Honey-peptone agar medium, 55
Hyphal tip transfers, 56
Hydrocarbons, conversion to citric acid by *Candida*, 15
Hydroxylases, steroid, 4
Hydroxylation, sequential, and dehydrogenation, 4
Hypoxanthine in *Bacillus subtilis*, 232–233

Idiophase, metabolites produced during, 241
Imhoff tank in sewage disposal, 9
Impaction of particles, factors affecting, 196
Impellers
  effect of size on aeration, 224
  mixing, 205, 228
    power characteristics of, 205, 228
  size and power input, 220
Inoculation, 9
Inoculum
  for penicillin shaken flask fermentations, 134
  spore, *Penicillium chrysogenum*, 114

Subject Index

Inosine in *Bacillus subtilis*, 232–233
Insecticides, microbial, 17
Invertase, 13
Irradiation
  experiments with *Neurospora crassa*, 84
  procedure, *Penicillium chrysogenum*, 59–61
  survival after, 61
  ultraviolet, apparatus for, 88
Isolation and purification of biosynthetic or bioconversion products, 3

Ketogluconic acid from glucose, bacterial production of, 132
Kinetics
  of bioconversion processes, 2
  of biochemical reaction systems, 253
  chemical, 252
  chloramphenicol fermentation, 263
  of fermentation processes, 250, 252–268
  and modeling, 249–277
  of the penicillin process, 264–267

Lactic acid, 9
  as a carbon source for penicillin production, 119, 120
  commercial process for, 13
  fermentation
    model of, 271
    rate pattern, 255
  from starch, 14
Lactate as carbon source in penicillin fermentation, 130
Lactobacillus delbrueckii, acid production, rates and activation energy, 254
Lactose, 61, 108
  batched vs. fed, penicillin fermentation, 142, 143
  effects on penicillin production, 119, 120, and glucose feeding in penicillin production, 239
  in penicillin production, 108
  in penicillin test medium, 61
  replacement by glucose or sucrose, penicillin fermentation, 140, 142
  slow utilization in penicillin fermentation, 11
  or starch, superiority to glucose in penicillin production in surface culture, 132
Liquid culture, automatically aerated, 290
Liquid nitrogen for freezing and storage of cultures, 42, 43
Lyophile tubes, storage, tests on, records of, 38

Lyophilization
  alternatives to, 38
  apparatus, 27
  criteria of success of, 37. See also Preservation
  of cultures, 37
  of *Penicillium chrysogenum* spores, 116
  tests after carrying out, 37
Lysine, 235
  auxotroph, control of growth and penicillin synthesis by, 244
  inhibition of penicillin biosynthesis, 238
  L-, in *Corynebacterium glutamicum*, 232

Mannosidostreptomycin, conversion to streptomycin by mannosidase, 239
Mass-spore transfer, 57
Mass transfer
  from individual gas bubbles, 205. See also Aeration
  in mycelial pellets, 205
Media, various, composition of, 10
Medium
  Czapek's agar, 68
  carbon sources, for production of penicillin in surface culture, 116
  chemically defined, for penicillin, 143
  complete, addition of, 10
  corn steep-lactose
    mycelial development in, 73
    for penicillin fermentation tests, 61
  development, 107–143
  honey peptone agar, 55
  precursors and intermediates included in, 233–234
  Sabouraud, 55
Membrane probes, principle of, 171. See also Electrodes
Metabolic activity
  complex, 2
  temperature control of, 290
Metabolic changes
  during penicillin production on glucose and lactose, 121
  in the oxygetracycline fermentation, 262, 263
Metabolic regulation, 229–248
  in complex bioconversion processes, 2
Metabolism
  coordination of, 230
    effect of mutation on, 51. See also Regulation
  of penicillin-producing molds, 143
  secondary
    dependence on protein synthesis, 241
    retardation during tropophase, 241, 242

305

Subject Index

Metabolites
  cellular and environmental factors affecting excretion, 230–248
  and enzymes, microbial, overproduction of, 230
  overproduction of, 231
  primary, 234–236
    lack of catabolite regulation of, 239
    mutation to avoid further conversion, 235
    overproduction by decreasing feedback effects, 234
    produced by analog-resistant mutants, 236
    produced during idiophase, 241
    and secondary from branched pathways, 237
    synthesis of, 230
  secondary, 236–239
    catabolite control of, 239
    effect of mutation on, 51
    feedback regulation of, 236, 237
    feedback self-regulation, 237
    synthesis of, 230
  synthesis and excretion of, 231
Methane
  anaerobic production from organic materials, 12
  growth on, as sole carbon source, 12
Methionine
  induction of cephalosporin synthesis by, 240
  precursor of chlortetracycline, 238
6-methyl-salicilic acid synthesis by *Penicillium patulum*, mutation in regulatory gene, 242, 243
Microbial conversion, 4. See also Bioconversion
Microbial cultures for efficient product formation, 231
Microbial enzymes, newer uses, 16
Microbial genetics, applications in industry, 51, 94
Microbial growth, 4
  kinetics and dynamics of, 3
Microbial hydroxylation of steroids at C_{11}, 14
Microbial insecticides, 17
Microbial metabolism, coordination of, 230
Microbial metabolites, overproduction of, due to alteration of regulation, 230–232
Microbial oxidation of sorbitol to sorbose, 14
Microbial production of amino acids, 19
Microbial products, commercial, listing of, 33–37
Microbial protease in washing powder, 14
Microbial protein from hydrocarbon, 18
Microbial technology, 19
Microbiology, industrial, 4, 5
Microorganisms
  as aerosols, 195
  in air, removal by filtration, 195. See also Sterilization
  characteristics of wild type, 231
  chemical oxidations by, 19. See also Bioconversion
  commercial products produced by, listing of, 33–37
  dangerous, regulation of, 46
  fermentation, subnormal regulation in, 232
  genetics and selection of, 18
  growth, aeration requirement for, 228
  immediate sources, 24
  improvement by mutation, hybridization, and selection, 94
  industrial
    collection of, supported by government funds, 24
    physiological regulatory mechanisms of, 230
  isolation, purification, screening tests of, 24
  large-scale cultivation, 2
  physical conditions affecting, 42, 43
  screening tests on, 24
  sources and management of, 23–48
  sources, nature, collection of, 24
  strain development, selective pressures in, 94
  ultimate sources of, 24
  wild type, characteristics, 231
Minimal system, 231
Mixing of liquids, 205
Mixing, physical forces in, 204, 205
Modeling
  facilitation by availability of computers, 269
  feed systems in the penicillin fermentation, 271, 272
  of fermentations, theoretical approaches, 272
  of the glutamic acid fermentation, 271, 272
  and kinetics, 249–277
  of the lactic acid fermentation, 271
  objectives of, 271
  of the penicillin fermentation, temperature optimization, 271, 272

Mold pellet, oxygen transfer in, 205
Molds
  penicillin-producing
    metabolism of, 143
    variability in, 52, 53
  pigment-free, penicillin from, 143
  proteolytic activity, 131
  protelytic enzymes of, 132
  submerged fermentation processes, 133
  techniques for isolation of, 32
Molecular biology, kinetic basis of, 194
Mutagen-strain interaction, 93
Mutagenesis
  and DNA repair, 94
  induced, 94
    of quantitative features, 94
Mutagenic agents
  basic effects of, 93, 94
  complex nature of effects, 94
  effects on nucleic acids, 93
Mutagenic treatment
  by various agents, *Penicillium chrysogenum*, 59
  repeated, 73
  successful, with several strains, 91, 92
Mutagenization, summary of methods, 91 92
Mutagens
  chemical, listing and classification, 89
  gene specificity of, 93
  specificity of, 94
  use of, in improvement of production strains, 50, 87
Mutants
  analog-resistant, primary metabolites produced by, 236
  dwarf, of *Penicillium chrysogenum*, 95
  resistant to fluorotryptophan, feedback-resistant producers of tryptophan, 237, 238
Mutation
  biochemical, in Penicillia, 104
  chemical, methods of use, 87
  chemical basis of, 94
  effect on coordination of metabolism and production of secondary metabolites, 51
  in *Aspergillus terreus* induced by ultraviolet light, 84
  in development programs, 232
  induced by physical and chemical agents, 87
  induction methods, 87
  methods, 88, 89
  molecular mechanism of, 94
  physical agents, 81

and production of secondary metabolites, 51
in a regulatory gene of *Penicillium patulum* synthesizing 6-methylsalicilic acid, 242, 243
sequential, to remove feedback inhibition, 236
spontaneous and induced, 87
studies, 24, 25
ultraviolet, and DNA repair, 94
Mycelial dry weight, method for, 218
Mycelial pellets, mass transfer in, 205. *See also* Aeration
Mycelium
  fungal, distribution by agitation, 148
  sterile, 33
Mycotoxin, 42

Neurospora, heterokaryotic vigor in, 104
Neurospora crassa
  auxotroph, protocatechuic acid synthesis by, 238
  quantitative irradiation experiments with, 84
Nitrogen
  concentration in yeast, 10
  sources, corn steep, and penicillin production, 122
Nitrogen mustard
  induction of mutation in *Penicillium chrysogenum*, 86
  induction of variation by, 85
  line of *Penicillium chrysogenum* strains, 74, 75, 78
  treatment of *Penicillium chrysogenum*, 59
    procedure, 61
Nitrogenous compounds, degree of breakdown and availability in penicillin fermentation, 131
Nitrogenous material, organic, replacement of, 109
Non-Newtonian fluids, power consumption in agitation of, 205
Novobiocin
  assay for, 228
  fermentation, 147
    aeration-agitation studies in, 215
    behavior with different power inputs, 222
    effects of aeration on, 228
    pilot and production scale, dissolved oxygen measurements in, 147
    in twenty-liter fermentors, 215
  produced by *Streptomyces niveus*, 218

Subject Index

Nucleic acids, effects of chemical mutagens and carcinogens on, 93, 94
Nutritional control vs. fixed medium, 12
Nucleotides, food flavoring agents, 10
Nystatin, 87

On-line optimization, algorithms for, 282
Organic acids from cellulosic materials, 132
Organic nitrogen as sources of growth factors, 39
Organic nitrogenous material, replacement of, 109
Ornithine, 235
Oxidation by air, sulfite oxidation method for studying, 204
    limitations of, 216
Oxidation-reduction electrodes, 169
Oxygen, dissolved, *in situ* measurement of, 146, 147
Oxygen absorption
    mechanism of, 206
    rates in sodium sulfite solutions, 227
    theory of, 206–208
Oxygen determination, amperometric, 171
Oxygen mass transfer
    importance of control and measurement of, 204
    in pellets, 205
    in submerged fermentation of *Streptomyces griseus*, 206–214
    uptake rates, experimental, 208–210
Oxygen probes, use without potentiometric recorder, 180. See also Electrodes
Oxygen transfer
    in agitated vessels, 227
    and agitation in submerged fermentation, 227, 228
    coefficient, interfacial, 210
    coefficients in viscous *Streptomycete* fermentations, 290
    influence of suspended materials on, 216
    in penicillin fermentation, 227
    in relation to aeration and agitation, 203–228
    schematic representation of, 207
    in submerged fermentation, 205
    within a mold pellet, 205
Oxygen uptake rates during fermentation
    in five-liter fermentations, 213
    in shaken flasks, 214
Oxytetracycline fermentation
    metabolic changes during, 262, 263
    rates, 262
Oxytetracycline production by hybrid strain of *Streptomyces rimosus*, 51

Parasexual mechanisms, attempts to exploit, 50, 51
Parasexual recombination
    in actinomycetes, 50
    in fungi, 95
    in *Penicillium chrysogenum*, 50, 95–105
    successes, 51
Parkinson's disease, L-DOPA in, 14
Pasteurization, 9
Patent Office, U.S., and cultures, 25
Pathogens, antibiotics against, 32
Pediococcus in commercial sausage, 31
Pellets, oxygen mass transfer in, 205
Penicilli, green, in *Penicillium chrysogenum* heterokaryons, 100
Penicillia, 108
    biochemical mutations in, 104
    culture run-down in, 41–42
    several strains, comparative yields, 113
    X-ray induced mutation in, 83
Penicillin, 87, 132, 135
    assay, 85, 132, 135, 143
        cylinder-plate method, 62, 111, 112, 134
    biosynthesis, chemical changes and rate patterns during, 253
    from chemically defined medium, 110, 143
    discovery of, 9
    factors affecting stability, 129
    fermentation, 11
        batched vs. fed lactose, 142, 143
        energy sources in, 11
        foam control, 140
        glucose or sucrose for lactose, 140, 142
        initial pH in, 124
        kinetic processes in, 252
        lactate as carbon source, 131
        lactose in, 11
        media for 30-liter seed and fermentation batches, 140
        metabolic data obtained, 141
        modeling of, 271, 272
        nitrogenous compounds in, 131
        nutrient additions to, 111, 125, 126
        oxygen transfer in, 227
        pH rise in, factors affecting, 130
        pH rise in, prevention by glucose, 130
        potassium phenylacetate feed to, 140
        rates with pH control, 267
        simulated, data analysis on, 287
        with sucrose feed, chemical changes during, 143
        sugar feed rates in, 143
    technology, influence on other processes, 5
    temperature effects on rates, 254
    trace elements from corn steep, 126

value of corn steep in, 131
formation
  delay in batch culture, 244
  effect of agitation and carbon dioxide on, 290
precursors of, 11
process kinetics, 264–267
production
  amino acids in corn steep, 127
  in continuous culture in defined medium, 244, 245
  and corn steep, 122
  in corn steep medium with continuous carbohydrate addition, 140–143
  cotton seed meal for corn steep in, 84, 110
  from crude starch, 118
  by cultures, Wisconsin strains, 50
  during active growth, 130
  effect of corn steep ash on, 85
  effect of environment on, 86
  effect of glucose on, 239
  effect of growth factors, 127
  effect of initial pH, 124
  effect of nutrient salts, 124
  effect of phenylacetic acid on, 110
  effect of sugar feed on, 141, 142
  effect of zinc on, 127
  effects of corn steep on, 123
  effects of minerals on, 85
  effects of sodium chloride on, 122
  in laboratory-type Waldhof fermentors, 83
  on lactic acid, 119, 120
  on lactose, 119, 120
  lactose and corn steep in, 108
  media chemically defined, 110
  metabolic changes during, 121
  by molds, metabolism of, 143
  by molds, variability in, 52, 53
  mycological aspects of, 51
  in natural and synthetic medium, 84, 85
  and natural variation in *Pencillium notatum*, 135
  optimum conditions for, in surface culture, 115, 116
  by pigment-free molds, 143
  by pigmentless strains of *Penicillium chrysogenum*, 83
  in pilot plant equipment, 83
  prolongation by use of slowly utilized carbohydrate, 131
  by representative strains of *Penicillium chrysogenum*, 77
  in semi-pilot plant equipment, 143
  from several carbon sources, 117
  in shaken flasks, 134
  in shaken flasks, effect of salts on, 134
  in shaken flasks, effects of mineral elements on, 84
  shaking machine for flasks, use in, 134
  in submerged culture, 54
  in submerged culture, laboratory scale, 133
  substitutes for corn steep in, 128
  in surface culture, 53, 108–132, 135
  in surface culture, lactose or starch preferred to glucose, 132
  in surface culture by *Penicillium chrysogenum*, 115
  in surface culture by *Penicillium notatum*, 111
  in surface culture, *Penicillium notatum* Westling NRRL 1249.B21 preferred, 131
  synthetic medium for, 84
  value of nutrient addition, 132
  on whey, 121, 122
R-group of, 11
radioactive, 85
shaken flask fermentation, inoculum for, 134
submerged fermentation
  Penicillium notatum, 133, 134
  pilot plant equipment for, 86
synthesis, inhibition by lysine, 238
types, effect of phenylacetic acid derivatives, 84
yeild(s), 11
  dependence on fermentation conditions, *Penicillium chrysogenum* strains, 78, 79
  effects of surface active agents on, 84
  improvement by addition of precursors and intermediates, 234
  means of increasing, 111
  in several strains of *Penicillia*, 113
  semi-starvation effect on, 142
Penicillin G, 13
  precursor of, in *Penicillium chrysogenum* Q176, 64
  precursor of, β-phenyl-ethylamine, 62
  price, uses, world production, 113
Penicillin K, in *Penicillium chrysogenum* Q176, 64
Penicillin V, 13
Penicillins, semi-synthetic, 16
Penicillium, antibacterial action, 132
Penicillium chrysogenum, 5, 63, 68, 84, 87, 96, 99, 103, 104, 105, 132. See also Penicillin, production
  amino acid metabolism of, 86

Subject Index

balanced heterokaryons, 99, 100
breeding of strains of, 95. *See also* Parasexual recombination
colony size, relative, 72
corn bran for spore inoculum, 140
correlation between potency and culture characteristics, 80
cultural characteristics, 63
dependence of yield on fermentation conditions, 78, 79
detailed procedure for irradiation, 59–61
diploid heterozygous nuclei in, 95
diploids
  analysis of, 103
  and heterokaryons, properties, 101
  from sectors, 101
dwarf colonies, 96
dwarf mutants, 95
dwarf strains, 97
  heterokaryons between, 98, 99
early NNRL strains, Fleming strain, 112, 113
effect by environment on penicillin production, 86
effect of air flow and agitation rate on, 205
effect of zinc on, 127
experimental markers, choice of, in, 95
first and second order segregants, 103
genealogy of Wisconsin strains, 69
green penicilli in heterokaryons, 100
growth and autolysis, 83
growth factor requirements, multiple, 96, 97, 98
growth rate on agar, 97
heterokaryosis in, 85
heterozygous diploids, 100–102
hyphal tip transfers, 76
improved sporulation capacity, 77
induced variation, 84
induction of mutation in, by nitrogen mustard, 86
induction of variation in, by nitrogen mustard, 85
isolates by single-spore technique, 76
key strains, N-mustard line, 75
low growth rate, 104
marked strains, production of, 95, 96
micromanipulation of conidia, 96
mineral nutrition of, 84
mitotic segregation and recombination of color markers, 103
morphology, effect of mechanical agitation on, 228
  control of, 5

multiple growth factor requirements, 97, 98
mutagenic treatment by various agents, 59
mutant strains, 75
nitrogen mustard line, 74, 75
non-sporulating cultures, 115
parasexual genetics, use of, 51
parasexual recombination in, 50, 95–105
pigment and antibiotic production, 70
pigmentlessness, an enduring characteristic, 70
pigmentless strain BL3-D10, 68
pigmentless strains, 68, 71, 75, 83
  D-type colonies, 70, 71
population patterns by key strains, 71
practical purposes of parasexual process in, 104
recessive markers, 102
recombination in, 85, 95
recombination of recessive properties, 104
recombination without sexual reproduction, 51
repeated mutagenic treatment, 73
reproductive vigor by sectoring, 78
results of irradiation, 96
run-to-run variation in flask yield, 78
sectoring of colonies, 76
segregation and recombination of diploids, 102–104
segregation of nuclei, 99
shaken flask fermentation tests, 76
spontaneous and induced variations in, 86
spore
  maintenance of, on soil or lyophiliaed, 116
  production, inoculation, 114
  ultraviolet treatment, 64
sporulation medium, 114
strain development, 88
strain BL3-D10, 70, 73
strain NRRL 882, production of penicillin in submerged culture with, 54
strain NRRL 1249, 53
strain NRRL 1951, progenitor of the Wisconsin series, 54, 66, 68
strain NRRL 1951, X1612, NRRL 1951·B25, 66, 68
strain Q176, 64, 65, 68, 70, 237
  and pigmentless descendants, 75
  cultural characteristics, 66
  eighty-gallon fermentation, 64
  penicillin G and precursor of penicillin K, 64
  population pattern, spontaneous variation, 66
  population patterns, 67

Subject Index

variation and mutation in, 85
    from X1612, 63
strain X1612, a "super-strain," derived by X-ray treatment, 54, 55
strains, 71–75
    by breeding, 95
    commercial properties, 80
    derived from Q176, 78
    efficiency of utilization of phenylacetic acid, 74, 75
    inverse correlation of productivity and reproductive vigor, 80
    production and selection, 50
    production of a family of, 52
    prototrophic, wild, 96
    ranking, 76
    selection without mutagenic treatment, 73
    Squibb family, 88, 90
    standard conditions for evaluation, 79
    testing of, 62, 63
    by a variety of methods, 80
    48-701, 48-479, D-type colonies, 71
    in submerged culture, 105
synthesis of balanced heterokaryons, 96–100
syntrophic growth, 97
three-way selection program, 72
ultraviolet-derived strains, yield and fermentation conditions, 78, 79
ultraviolet treatment, survival after, 81
valine synthesis in, 237
variants, maintenance of, 115
Wisconsin family, 53
Wisconsin strain 47-1564, 71
Wisconsin strain 49-133, greatest improvement, 74
Wisconsin strain 51-20, key strain, 75, 76
    side-branch of N-mustard line, 78
Wisconsin strains
    relative colony size, 72
    genealogy, 69
    49-133, and Q176, 140
yellow pigment in, 114
Penicillium luteum-purpurogenum, 131
    hydrolysis of soybean meal, 132
Penicillium notatum, 63, 68, 126
    genetic proof of heterokaryosis in, 105
    gluconic acid production from glucose, 130
    heterokaryosis, 97
    natural variation and penicillin production, 135
    spontaneous variation in, 105
    spore production on whole wheat flour, 131

strain NRRL 832, 72, 122
strain NRRL 1249-1321, 53
    for surface culture, 115
    submerged fermentation for penicillin, media and conditions, 133, 134
    Westling NRRL 1249-1321 perferred for penicillin production in surface culture, 111, 131
Penicillium patulum, mutation in regulatory gene affecting synthesis of 6-methyl-salicilic acid, 242, 243
Petrochemicals as sources of solvents, 14
pH control
    continuous, 11
    penicillin fermentation, 267
pH electrodes. See Electrode, pH
pH, and dissolved oxygen, *in situ* measurement of, 146, 147. See also Electrode; Probe
pH gland installation
    in a ten-gallon fermentor, 164
    in a 1300-gallon fermentor, 165
Phenoxazine synthetase
    in actinomycin synthesis, glucose repression by, 239
    in *Streptomyces antibioticus*, 243
Phenylacetic acid, 11
    addition of, continuous, repeated, 11
    continuous addition, toxicity of, 11
    derivatives, effects on types of penicillins, 84
    derivatives of, 11
    effect on penicillin production, 110
    efficiency of utilization by *Penicillium chrysogenum* strains, 74, 75
Phenylalanine synthesis by tyrosine-less mutants, 235
Phoma lingam, induced variability in, 83
Phosphatase, feedback inhibition and repression of, by phosphate, 238
Phosphate
    ammonia, concentrations of, 10
    in chlortetracycline synthesis, effect on energy charge, 240
    fermentations sensitive to, 238
    inhibition and repression of phosphatase, 238
Phosphorus concentration in yeast, 10
Physiological regulatory mechanisms, 230
Potassium phenylacetate solutions, feed to penicillin fermentation, 140
Power consumption
    estimation of, in shaken flasks, 150
    in flasks
        effect of shaking speed on, 152
        factors affecting, 146, 148

Subject Index

flasks vs. tanks, 152, 153
  measurement in shaken flasks, 149
  prediction of, in agitation of non-Newtonian fluids, 205
Power input
  effect on aeration, 224
  and impeller size, 220
  novobiocin fermentation, 215
  per unit volume, twenty-liter fermentors, 215
Power measurements with a torsion dynamometer, 219
Prasinomycin, 87
  biosynthesis by *Streptomyces prasinus*, 92, 93
Probes. See also Dissolved oxygen; Electrodes
  membrane
    construction of, 172–176
    linearity of response in, 178, 179
    performance of, 176–178
    principle of, 171
    use of, 178–181
  steam-sterilizable, 147
  for dissolved oxygen, 171–182
Process-computer-operator relationship, schematic, 285
Process control, by computer, over-ride by operator, 283
Process design considerations, chemical reactions, physical processes, 252
  approaches to, 2
  facets of, 232
Process identification, methods for, 282
Process variables in fermentations, 258
Processes
  autosynthetic, 231
  bioconversion, 3
  biosynthetic, 2
  fermentation, 2
  large-scale, highly-aerated, 2
Product formation
  cultures for efficient production, 231
  direction of, by precursors and intermediates, 234
  and growth in continuous culture, 3, 5
  instantaneous rates of, 253
  rates of, lactic acid fermentation, 255
Productivity
  definition of, 253
  and rates, 256, 257
Progesterone, 4
Prostaglandins by combined chemical and microbiological synthesis, 14

Protease, 13
  formation, effect of carbon dioxide on, 290
  microbial, in washing powder, 14
Protein, microbial, from hydrocarbon, 18
Protein-sparing action of carbohydrate in *Aspergillus niger*, *Citromyces*, 131
Protein synthesis, inhibition or repression of, inimical to secondary metabolism, 241
Proteolytic activity of molds, 131, 132
Protocatechuic acid, synthesis by *Neurospora crassa* auxotroph, 238
Purine nucleotides, excretion by *Corynebacterium glutamicum*, 233
Pyrrolnitrin, optimum production by tryptophan addition, 237

Rate expressions, rate processes in fermentations, 255–258
Rate patterns
  basic types in fermentation, 252
  in batch fermentation processes, 255
  complex, in streptomycin fermentation, 257, 260
  in fermentations, 260, 261
Rate processes
  concepts in batch fermentation, 109
  in fermentations, 257
  in the penicillin fermentation, pace-setting enzymes, 254
Rates and productivity, 256, 257
Raw material
  basis of choice, 109
  cyclic costs, 18
Reaction systems, biochemical, kinetics of, 253
Reactor, homogeneous overflow, 263. See also Continuous culture
Regulation, metabolic, 229–248
  in complex bioconversion processes, 2
Regulatory controls, modification of, 232
Regulatory genes, modification of, to eliminate repression, 235, 236
Regulatory mechanisms, 231
  important, in microorganisms, 232
  physiological, 230
Reineckates, crystalline, of streptomycin and streptothricin, 139
Repressor protein, lack of production in synthetic media, several antibiotics, 243
Respiratory activity, 147
Revertant(s)

312

Subject Index

enzyme content of, 237
   superior chlortetracycline synthesis by, 238
Rheological measurements, 219
Rheumatoid arthritis, cortisone in, 14
Riboflavin, 10
   fermentative production of, 9
   from starch, 14
Ribonucleic acid synthesis, inhibition by 5-fluoruracil in *Streptomyces antibioticus*, 243

Sabouraud medium, 55
Scale-up from flasks to tank, 154
Scale-up studies, 146, 147
Scale-up of viscous, mycelial fermentations, 216
Screening programs, efficiency of, 94
Segregants, first and second order, *Penicillium chrysogenum*, 103
Serratia marcescens, areobic propagation of, 227
Sewage digestion, anaerobic, aerobic, 9
Sewage disposal
   in Imhoff tank, 9
   trickling filter for, 9
Shake flasks, 10
Shaken flask
   effect of volume and shape on power consumption, 151
   effects of shape and size, 149
   fermentation, streptomycin, importance of volume, 137, 139
   fermentation tests, *Penicillium chrysogenum*, 76
   graphical method of estimationg power consumption, 150
   invention of, 9
   method of measuring power consumption in, 149
   methods in streptomycin fermentation, 137
   oxygen uptake rates in, 214
   performance of, power consumption, 146, 148–143
   power study on, 149
   sulfite oxidation measurements, 149
Shikimic acid and aromatic acid biosynthesis, 232
Simulation languages and computers, 269, 274, 275
Single-spore isolation, 56
Single-spore technnique for isolation of *Penicillium chrysogenum* strains, 76

Siomycin production by resting cells of *Streptomyces sioyaensis*, glucose effect on, 239
Sodium chloride, effects on growth and penicillin production, 122
Sodium ion, need for, in streptomycin fermentation, 138, 139
Sodium sulfite oxidation and oxygen absorption rates, 227
Solid disperse phase, effect on oxygen absorption in a fermentor, 227
Solvent production by *Clostridium*, 31
Solvents from petrochemicals, 14
Sorbose, 9
   by microbial oxidation, 14
   from sorbitol, semi-plant-scale production by *Acetobacter suboxydans*, 132
Soybean meal, 108
   cost of, 15
   hydrolytic cleavage by *Penicillium luteumpurpurogenum* and *Aspergillus niger*, 132
   principal nitrogen source in streptomycin fermentation medium, 136
   in streptomycin production, 108
Spore(s)
   formation
      and antibiotic production, 109
      effect of carbon dioxide on, 290
   penetration of filters by, 200
   production on whole wheat flour, *Penicillium notatum*, 131
   survival after irradiation, 61
   suspension, plating of, 56
   various types, 32
   viable, after ultraviolet treatment, 64
Sporulation
   in *Aspergilli* and *Penicillia*, 39
   of *Bacillus subtilis* effected by growth limitation, 77
   bacterial, repression of, 239
   capacity, improved in *Penicillium chrysogenum*, 77
   medium, *Penicillium chrysogenum*, 114
   and streptomycin production, 138
Starch
   crude sources for penicillin production, 118
   liquifaction by malt or acid, 117
   as source of riboflavin and lactic acid, 14
Starter cultures, 40
   in food fermentations, 31
Steady-state low sugar concentration, 10

Subject Index

Steady-state operating conditions, prediction from batch data, 263, 264
Sterility, practical standards of, 195
Sterilization, 183–201
  batch and continuous, 186–188
  heat
    analytical method for calculating, 184, 186–194
    applications of design equation, 191–193
    time-temperature profiles, 188–190
  time, heat, calculation of, 186–194
Sterilizers, continuous, design of, 194
Steroid(s)
  bioconversion, simulation of, 4
  conversions, 10
    models of, 271
    sequential, 4
  dehydrogenation of, 87
  hydroxylases and dehydrogenases, 4
  microbial hydroxylation of, 14, 87
  synthesis, 14
  transformations, 2
Steroid-1-dehydrogenase, 87
Steroid-16α-hydroxylase, 87
  in *Streptomyces roseochromogenes*, 92, 93
Stoichiometry, or lack of, between reactants and products, 257
Strain(s)
  catalogs, lists, 47–49
  degeneration and loss, 40–42
  development, 49–106
    culture methods for, 55
    in fermentation-based pharmaceutical companies, 50
    Penicillium chrysogenum, 88
    selective pressures in, 94
  mutagen interaction, 93
  of *Penicillium chrysogenum*
    effects of fermentation conditions on ranking, 76
    production and selection of, 52
  production, weak control mechanisms, 231
  for production of antibiotics, 31, 32
  wasteful, overproductive, 231
Streptomyces
  antibiotics, crystalline salts of streptomycin and streptothricin, 139
  nutrition of, 109
Streptomyces antibioticus
  actinomycin synthesis and phenoxazine synthetase, 243
  effect of inhibition of synthesis of RNA by 5-fluorouracil, 243

Streptomyces aureofaciens, chlortetracycline synthesis by, energy charge and ATP effects, 240
Streptomyces coelicolor, gene recombination in, 53
Streptomyces griseus, 87–89
  effect of air flow and agitation rate on, 205
  gamma irradiation, 89
  nutritional requirements for streptomycin production, 136–139
  oxygen transfer in submerged fermentation, 206–214
  strains, development of, 88
Streptomyces niveus, novobiocin producer, 218
Streptomyces nodosus, 87
  strain development, 89
Streptomyces noursei, 87, 89, 92
  recombinant strain, 89, 90
  strain development, 89
Streptomyces prasinus, 87
  prasinomycin biosynthesis by, 92, 93
Streptomyces rimosus, oxytetracycline production by a hybrid strain, 51
Streptomyces roseochromogenes, 87
  16α-hydroxylase, 92, 93
Streptomyces sioyaensis, siomycin production by resting cells, glucose effect on, 239
Streptomyces umbrinus, 87, 92
Streptomyces viridofaciens, chlortetracycline production by, 238
Streptomycete fermentations, viscous, oxygen transfer coefficients in, 290
Streptomycetes, genetic recombination in, 51
Streptomycin, 87
  assay for, in supernatant liquor, 137
  biosynthesis, role of phosphate in, 238
  chemical studies on, 136
  chemotherapeutic properties, 139
  in control of gram-negative bacteria, 139
  fermentation, 109
    need for sodium ion in, 138, 139
    phases of rate patterns in, 253, 254
    rate patterns in, 257, 260
    rates, 260
    shaken flask, importance of volume, 137, 139
    soybean meal principal nitrogen source in, 136
  in vivo and *in vitro* antibiotic activity, 136
  phosphate, effect of phosphate on, 239

-producing strains, development of, 88
production
by resistant mutants, 237
shaken flasks methods, 137
soybean meal in, 108
sporulation during, 138
by *Streptomyces griseus*, nutritional requirements for, 136–139
in viscous media, 137
recoverability of, 136
and streptothricin
broth dilution method for assaying, 139
crystalline reineckates of, 139
synthesis in complex medium with added phosphate, 238, 239
Sugar determination, reagents for, 143
Sugars, determination of, by a new reagent, 228
Sulfite oxidation. See also Oxidation by air
oxygen transfer method in shaken flasks, 149
rate
effects of power input and impeller size on, 221
inadequacy for measuring oxygen transfer, 225, 226
modified method, 219
values, twenty-liter fermentors, 215
Suspensions, conidial, 55, 56
Syntheses, combined chemical and microbiological, 14

Tank turnover, operational factors, 256
Tannin, transformation to gallic acid, 135
Tetracycline, 13, 16, 237
Tetramethyl pyrazine, excretion by *Corynebacterium glutamicum* auxotroph, 238
Thermal death rate of bacteria, determination of, 194
Thermal processing time for canned food, 194
Threonine fermentation by *Brevibacterium flavum*, 236
Torsion dynamometer for power measurements, 219
Trophophase
retardation of secondary metabolism during, 241, 242
reversion of enzymes in, 241
Trophophase-idiophase relationship, 241
Tryptophan
as precursor and inducer of alkaloids in *Claviceps*, 240

synthesis by fluorotryptophan-resistant mutants, 237, 238

Ultraviolet irradiation
apparatus for, 88
effect on *Eschericia coli*, 93, 94
of *Penicillium chrysogenum*, 96
stimulation by, 70
Ultraviolet mutation and DNA repair, 94
Ultraviolet treatment
of *Penicillium chrysogenum*, 59, 73, 81
of spores of *Penicillium chrysogenum*, 64
useful wavelengths, 81
United Nations Industrial Development Organization, 26
Ustilago zeae, production of ustilagic acid by, 227

Valine synthesis
and feedback inhibition in penicillin-producing *Penicillium chrysogenum*, 237
by isoleucine-less mutants, 235
Variables, control and state, measured, 283, 284
Vessels, aerated and agitated, 10. See also Reactor; Fermentor(s)
Vinegar
generator, 9
manufacture, 12
Viruses, physics of, 194
Vitamin B_{12}, fermentative production of, 9
Vitamins and other products produced by fermentation, list of, 17
Volumetric rates, 256

Waldhof fermentor, penicillin production in, 83
War Production Board program, 52, 53
Washing powder, protease in, 14
Whey, salt content and inhibitory effects in penicillin production, 121, 122
Wisconsin series of *Penicillium chrysogenum* strains, 54. See also *Penicillium chrysogenum*, Wisconsin strains
World Directory of Culture Collections, 26

X-ray induced mutation in *Penicillium*, 83
X-ray treatment of *Penicillium* strains, 55

Subject Index

Yeast, 13
   aerobic production, continuous sugar addition, 9
   baker's, 9
      aeration studies on propagation, 227
   environment for maximum yield or other characteristic, 10
   fermentations, polarographic measurement of dissolved oxygen in, 228
   grown in high sugar concentrations, 10
   growth on normal alkanes, 12
   high yields in synthetic medium, 227
   industry, history of, 4
   nitrogen and phosphorus concentrations in, 10
   polypeptidase, isolation and properties, 143
   production from sulfite waste liquor, 12
   strains, for beer in South Africa, 24
   yield, maximum, 10

About the Editor

RICHARD W. THOMA has been a practicing industrial microbiologist at The Squibb Institute for Medical Research, New Brunswick, N.J., since 1951. He began his higher education at the University of Chicago in 1939, saw military service from 1942 to 1946, and continued his education at the University of Wisconsin, Madison, from 1946 to 1951, where he earned the B.S. in Chemistry and the M.S. and Ph.D. in Biochemistry. The training that oriented him toward a career in industrial microbiology was obtained in the fermentation laboratory of the well-known and highly respected biochemist, William H. Peterson. While he was an undergraduate and graduate student under Peterson he came under the tutelage of Marvin J. Johnson as well.

In 1951 Thoma joined the Microbial Biochemistry Section (then known as the Division of Microbiology) in The Squibb Institute for Medical Research, which was and still is synonymous with the Research and Development arm of E. R. Squibb and Sons. As a result of several years of collaborative effort with other Squibb scientists he is named as co-inventor on more than 30 patents on microbiological transformation of steroids.

Since 1958 Thoma has been involved, as an investigator and director, with the design and development of new processes, and improvement of existing fermentation processes for production of steroids, antibiotics, and enzymes. During his tenure as a scientist in process development he has avoided technological obsolescence by maintaining active membership in a number of scientific and professional groups, by being an avid reader of the literature, and by close collaboration with a number of consulting scientists and engineers of considerable stature.

QR
53
T48

APR 14 1978